KU-194-824

CERTIFICATE MATHEMATICS

VOLUME II

II

CERTIFICATE MATHEMATICS

By

CLEMENT V. DURELL, M.A.

Author of *General Arithmetic, School Certificate Algebra, A New Geometry*, etc.

VOLUME II

Second Edition
decimalised and metricated

LONDON
G. BELL & SONS LTD
1972

First published February 1958
Reprinted 1958, 1959, 1962, 1965, 1966, 1967, 1968
Second Edition, decimalised and metricated
Published 1971
Reprinted 1972

India
Orient Longmans Ltd
Calcutta, Bombay, Madras and New Delhi

Canada
Clarke, Irwin & Co. Ltd, Toronto

Australia
John Cochrane Pty. Ltd, 373 Bay Street, Port Melbourne

New Zealand
Book Reps (New Zealand) Ltd, 46 Lake Road, Northcote, Auckland

East Africa
J. E. Budds, P.O. Box 4536, Nairobi

West Africa
Thos. Nelson (Nigeria) Ltd, P.O. Box 336, Apapa, Lagos

South and Central Africa
Book Promotions (Pty), Ltd, 311 Sanlam Centre, Main Road, Wynberg, Cape Province

Vol I with Answers ISBN 0 7135 1602 x
Vol I without Answers ISBN 0 7135 1970 3

Printed in Great Britain by
J. W. Arrowsmith Ltd, Bristol 3

PREFACE

Certificate Mathematics is planned to meet the requirements of the 'General Certificate of Education' *at ordinary level*, as specified by the syllabuses of the various examining bodies. It is arranged in four volumes, each of which forms a year's work for the four years from 11 plus to 15 plus.

Certificate Mathematics is derived from the author's original *General Mathematics*. There are, however, especially in the later volumes, substantial changes of order, treatment, and subject-matter, designed to meet the special purpose of this course; advantage has also been taken of suggestions made by teachers. Fresh material has been introduced in order to meet all the requirements of the relevant examination syllabuses, while some topics and processes not required in the examinations have been excluded.

Attention may be called to the following features:

(i) There is frequent revision to secure familiarity with basic ideas, useful facts and standard processes. Such work is given at appropriate places in the text and is emphasised by the sets of 'Quick Revision' papers at the end of each volume. These papers should form a valuable preparation for examinations, particularly if access to previous volumes is impracticable.

(ii) The primary object of a homogeneous course is to encourage the pupil to select on every occasion the most appropriate method, whether algebraic or geometrical or graphical or trigonometrical. The text and exercises are planned with this end in view. At the same time, great care has been taken to secure that the explanation of a new process is followed by *straightforward* examples which enforce the method without the distraction of miscellaneous applications. In cases where the drill element is of the first importance, practice is given in examples of a single type before another type is introduced, miscellaneous types being included later in the exercise; this applies to all subjects alike.

(iii) A number of exercises are headed 'oral' or 'class discussion'; these are constructed so that much of the working can be done mentally and are designed to accustom the pupil to the *routine of the argument*. It is suggested that these exercises should be taken *viva voce*, but that all pupils should be required to write down the answers.

The examples in the other exercises are classified under two main heads:

(a) A first course: plain numbers.

(b) A parallel course: numbers enclosed in brackets.

The parallel course provides extra practice, if needed; it does not extend the ground covered in the first course. Some examples marked with an asterisk are intended for those pupils who run ahead of the class.

(iv) The geometry in Volume I is restricted to the work of Stage A because experience has shown that progress in the work of Stage B is far more rapid if the pupil is thoroughly familiar with the basic ideas and the fundamental facts.

This is secured by extensive training in the use of instruments. The standard of accuracy in drawing demanded by examining bodies, although properly far below that of the professional draftsman, can only be attained by frequent practice; such work is used to investigate and illustrate the group of properties on which Stage B geometry is based.

(v) The customary practice of numbering theorems has been discontinued. This removes the risk of a pupil referring to a theorem by a number and forces him to give reasons in verbal form; an 'abbreviation for reference' is suggested for each theorem to help him to do so.

(vi) Comprehensive *Tests in Computation* and sets of miscellaneous *Revision Papers* are given at the end of each volume.

The author acknowledges with thanks permission to include questions set in the examinations for the *General Certificate of Education* by the following bodies: Cambridge Local Examinations Syndicate, Senate of the University of London, Joint Matriculation Board of the Northern Universities, Delegates of the Oxford Local Examinations, Oxford and Cambridge Schools Examinations Board, Welsh Joint Education Committee. Such questions are marked C, L, N, O, OC, W in the text.

NOTE ON NEW EDITION

In this new edition the currency has been decimalised, and SI metric units of weight and measure, with their proper abbreviations, have been used throughout.

The essential characteristics of the order and treatment of *Certificate Mathematics* have been preserved, but a few extra topics have been added which are now generally accepted as belonging to the traditional course. Volume I contains a chapter on Number Systems; Volume II includes a fuller treatment of the linear function, half planes and linear inequalities; Volume III includes the trigonometry of angles of any magnitude, an introduction to three-dimensional coordinates, and quadratic inequalities; Volume IV contains a section on arithmetical and geometrical progressions.

Certificate Mathematics was originally available in two forms. Volumes I and II were common to both courses, but Volumes IIIA, IVA were intended for the traditional syllabus, and Volumes IIIB, IVB for the alternative syllabus. Since the first of these has largely disappeared, and has been replaced by what was at first the alternative syllabus, there is no longer any need for two courses, and *Certificate Mathematics* is now published in four volumes.

CONTENTS

TABLES

LENGTH

10 millimetres (mm) = 1 centimetre (cm)
100 centimetres = 1 metre (m)
1000 metres = 1 kilometre (km)

CAPACITY

1000 millilitres (ml) = 1 litre (l) = 1000 cm^3

WEIGHT

1000 milligrammes (mg) = 1 gramme (g)
1000 grammes = 1 kilogramme (kg)
1000 kilogrammes = 1 tonne (t)

MONEY

British

100 new pence (p) = 1 pound (£)

American

100 cents = 1 dollar ($)

French

100 centimes (c) = 1 franc (fr)

TIME

60 seconds (s) = 1 minute (min)
60 minutes = 1 hour (h)
24 hours = 1 day
7 days = 1 week
365 days = 1 ordinary year
366 days = 1 leap year
100 years = 1 century

ANGLE

60 seconds (60″) = 1 minute (1′)
60 minutes = 1 degree (1°)
90 degrees = 1 right angle

CHAPTER 1
DIRECTED NUMBERS

Signless Numbers In the previous volume, the signs + and − have been used solely as orders: 'add', 'subtract'. The expression, 5 − 3, means 'From 5 subtract 3'; the numbers, 5 and 3, are signless. For many kinds of measurement, signless numbers supply all that is needed, *e.g.* the number of days in a week, the number of kilometres from London to Land's End, etc. In quantities of this kind there is no idea of 'up and down' or 'backwards and forwards' or 'clockwise and counter-clockwise'. But when quantities which involve the idea of direction occur, it saves time to give further meanings to the symbols + and −.

Positive and Negative Numbers Suppose I buy a number of things and sell them again, the results of the transactions may be recorded as follows:

	House	Car	Picture	Horse	Field	Carpet
Gain	£80			0	£60	
Loss		£40	£15	0		£12

But it is *shorter* to write:

	House	Car	Picture	Horse	Field	Carpet
Gain	£(+80)	£(−40)	£(−15)	0	£(+60)	£(−12)

The symbols + and − in this table are not instructions to add or subtract; they are called the signs of the numbers.

The number (+80) is called a **positive number,** the number (−40) is called a **negative number.** My gain £(+80) is a short way of saying that I am £80 better off, or that my capital *goes up* £80; my gain £(−40) is a short way of saying that I am £40 worse off, or that my capital *comes down* £40. Thus this new notation represents in a short form up-and-down movements, backwards-and-forwards movements, etc., in fact, quantities with which the idea of direction is associated; for this reason positive and negative numbers are called **directed numbers.** The symbol 0 means that there is no change either way, (+0) is the same as (−0) and so each is simply denoted by 0.

EXERCISE 1 (Oral)

1 On a Celsius thermometer (see Fig. 1) the freezing-point of water is indicated by 0°.

(i) Express in shorthand form the following temperatures: 4° below freezing-point; 20° above freezing-point; 15° of frost.

(ii) What is the meaning of (−5) degrees, of (+100) degrees?

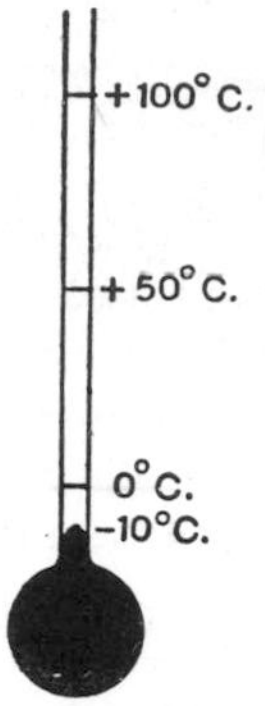

+100° C.
+50° C.
0° C.
−10° C.

FIG. 1

2 Explain the meaning of the following:

	Winchester	The Dead Sea
Height in metres above sea-level	+39	−396

3 A gun is being ranged by an aeroplane on a target; the direction is correct; the distance of the fall of the shell from the target is signalled as follows:

Round	I	II	III	IV	V	VI
Distance in metres	+260	−180	+140	−70	+35	O.K.

What do these signals mean?

4 A stone is thrown vertically upwards with velocity 12 metres per second. Explain the meaning of the following table, which shows its velocity at half-second intervals:

Time in seconds	0	$\frac{1}{2}$	1	$1\frac{1}{2}$	2	$2\frac{1}{2}$
Velocity in m per second	+12	+7	+2	−3	−8	−13

5 Taking 12 noon as zero hour, and 1 hour as the unit, express by directed numbers: 3 p.m.; 11 a.m., 8 a.m.; 4.30 p.m.; 10.30 a.m.

6 Rewrite using signless numbers: (i) A is (−5) km east of B; (ii) C is (−8) km north of B; (iii) my watch is (−3) min fast; (iv) my bank balance is £(−42).

7 With Greenwich time as the standard, the following variations occur in local time: Brussels (+1) hour; New York (−5) hours; Hong Kong (+8) hours. What does this mean?

Addition and Subtraction

On the Celsius scale, the temperature at which water freezes is marked 0 (zero); temperatures above zero are represented by positive numbers, temperatures below zero by negative numbers. This is an example of what is called the **number-scale,** which can be used to perform addition and subtraction, moving up and down it, as on a ladder. Thus, to add (+2) to (−5), start at (−5) and move 2 steps up the ladder: (−5)+(+2) = (−3).

We therefore also say that $(+2)+(-5) = (-3)$, and this means that to add (-5) to $(+2)$, we start at $(+2)$ and move 5 steps down the ladder.

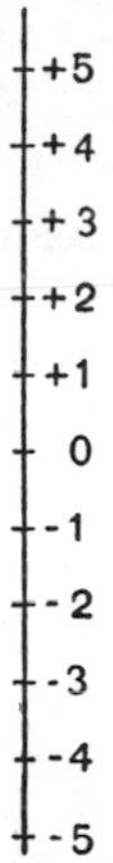

In general, $+(+N)$ means 'add $(+N)$'; to do this, move N steps up the scale;

and $+(-N)$ means 'add $(-N)$'; to do this, move N steps down the scale.

For *subtraction*,

since $(-3) = (-5)+(+2)$,

we say that $(-3)-(+2) = (-5)$.

But if we start at (-3), we arrive at (-5) by moving 2 steps down the scale; thus to subtract $(+2)$, we move 2 steps down the scale.

Since $(-3) = (+2)+(-5)$, we say that $(-3)-(-5) = (+2)$. But if we start at (-3), we arrive at $(+2)$ by moving up 5 steps; thus to subtract (-5), we move 5 steps up the scale.

+5 +4 +3 +2 +1 0 -1 -2 -3 -4 -5

FIG. 2

In general, $-(+N)$ means 'subtract $(+N)$'; to do this, move N steps down the scale;

and $-(-N)$ means 'subtract $(-N)$'; to do this, move N steps up the scale.

Moving N steps up the scale may be represented more shortly by writing $+N$, and moving N steps down the scale by $-N$. We therefore make the following rule of signs:

$$+(+N) = +N; \quad -(-N) = +N; \quad \text{move up } N \text{ steps.}$$

$$+(-N) = -N; \quad -(+N) = -N; \quad \text{move down } N \text{ steps.}$$

Example 1 State how the following operations can be performed on the number-scale, and give the results:

(i) $(+3)-(+7)$; (ii) $(+2)+(-3)$; (iii) $(-4)-(-6)$.

(i) Start at $(+3)$ and move 7 steps down the scale, we arrive at (-4); $\therefore (+3)-(+7) = (-4)$.

(ii) Start at $(+2)$ and move 3 steps down the scale, we arrive at (-1); $\therefore (+2)+(-3) = (-1)$.

(iii) Start at (-4) and move 6 steps up the scale, we arrive at $(+2)$; $\therefore (-4)-(-6) = (+2)$.

EXERCISE 2 (Oral)

Give in full expressions for the final temperatures on the Celsius scale, and then simplify:

1 First temperature $(-3°)$; a rise of $(+5°)$.

2 First temperature $(+5°)$; a fall of $(+8°)$.

3 First temperature $(-4°)$; a rise of $(-3°)$.
4 First temperature $(-7°)$; a fall of $(-2°)$.

Give in full expressions for the *rise* in temperature on the Celsius scale, and then simplify:

5 First temperature $(-4°)$; second temperature $(+3°)$.
6 First temperature $(+5°)$; second temperature $(-3°)$.
7 First temperature $(-6°)$; second temperature $(-2°)$.

8 A is 100 metres above sea-level, B is 60 metres below sea-level. What is the height of A above B? What is the value of $(+100)-(-60)$?

9 A man starts the year £100 in debt and ends the year £30 in debt. How much has he gained during the year? Simplify $(-30)-(-100)$.

10 The temperature of a freezing mixture is $(-12°)$ C. It is heated up to $(+4°)$ C. Find the rise in temperature.

11 Which is the greater, (-3) or (-7), and by how much?

12 What must be added to
(i) $(+5)$ to give $(+2)$; (ii) (-4) to give (-1);
(iii) (-5) to give (-9); (iv) $(+3)$ to give 0?

Write down the values of:

13 $(+8)+(-4)$ **14** $(-7)+(-3)$ **15** $(+2)-(+4)$
16 $(-2)-(-3)$ **17** $(-6)-(-5)$ **18** $(-9)+(+9)$
19 $(-2)-(+3)$ **20** $0-(+2)$ **21** $(-2)-(+2)$

Simplify the following:

22 $(+5t)+(-t)$ **23** $(-3c)+(+c)$ **24** $(-2e)+(-4e)$
25 $(+r)+(-r)$ **26** $(+5s)-(-2s)$ **27** $(+t)-(+2t)$
28 $(-4n)-(-7n)$ **29** $0-(-3p)$ **30** $(-3x)-(-3x)$
31 $d-5d+3d$ **32** $2e-5e+3e$ **33** $3f-5f-4f$

State shortly the results of the following operations:

34 Add $(-6a)$, then add $(+2a)$.
35 Subtract $(-4b)$, then add $(+3b)$.
36 Add $(-8c)$, then subtract $(-2c)$.
37 Subtract $(-2d)$ and $(-6d)$.

Multiplication

The temperature of the water in a boiler is being raised at a steady rate of 5°C. per hour throughout the day. If we regard mid-day as zero hour, the temperature at n o'clock is $(+5°)\times n$ above that at mid-day (zero hour).

Here n is a directed number: thus at 2 p.m. $n=(+2)$, and at 9 a.m. $n=(-3)$.

At 2 p.m. the temperature is evidently 10° *above* that at mid-day;

$$\therefore (+\mathbf{5})\times(+\mathbf{2}) = (+\mathbf{10}).$$

And at 9 a.m. the temperature is evidently 15° *below* that at mid-day;

$$\therefore (+\mathbf{5})\times(-\mathbf{3}) = (-\mathbf{15}).$$

Next, suppose that the temperature of the water in the boiler is *falling* at a steady rate of 5°C. per hour throughout the day. Then the *rise* per hour is $(-5°)$; $\therefore$ the temperature at n o'clock is $(-5°)\times n$ above that at mid-day (zero hour).

At 2 p.m. the temperature is evidently 10° *below* that at mid-day;

$$\therefore (-\mathbf{5})\times(+\mathbf{2}) = (-\mathbf{10}).$$

And at 9 a.m. the temperature is evidently 15° *above* that at mid-day;

$$\therefore (-\mathbf{5})\times(-\mathbf{3}) = (+\mathbf{15}).$$

This argument can be applied to any directed numbers; we therefore make the following **rule of signs:**

$$(+\boldsymbol{a})\times(+\boldsymbol{b}) = (+\boldsymbol{ab}) = \boldsymbol{ab}; \qquad (-\boldsymbol{a})\times(-\boldsymbol{b}) = (+\boldsymbol{ab}) = \boldsymbol{ab};$$

$$(+\boldsymbol{a})\times(-\boldsymbol{b}) = (-\boldsymbol{ab}) = -\boldsymbol{ab}; \qquad (-\boldsymbol{a})\times(+\boldsymbol{b}) = (-\boldsymbol{ab}) = -\boldsymbol{ab},$$

where, for simplicity, we write ab for $+ab$.

Division

Since $(+5)\times(+2) = (+10)$, $\therefore (+10)\div(+2) = (+5)$.
Since $(+5)\times(-2) = (-10)$, $\therefore (-10)\div(-2) = (+5)$.
Since $(-5)\times(+2) = (-10)$, $\therefore (-10)\div(+2) = (-5)$.
Since $(-5)\times(-2) = (+10)$, $\therefore (+10)\div(-2) = (-5)$.
We therefore make the following **rule of signs:**

$$(+\boldsymbol{a})\div(+\boldsymbol{b}) = \left(+\frac{\boldsymbol{a}}{\boldsymbol{b}}\right) = \frac{\boldsymbol{a}}{\boldsymbol{b}}; \qquad (-\boldsymbol{a})\div(-\boldsymbol{b}) = \left(+\frac{\boldsymbol{a}}{\boldsymbol{b}}\right) = \frac{\boldsymbol{a}}{\boldsymbol{b}};$$

$$(+\boldsymbol{a})\div(-\boldsymbol{b}) = \left(-\frac{\boldsymbol{a}}{\boldsymbol{b}}\right) = -\frac{\boldsymbol{a}}{\boldsymbol{b}}; \qquad (-\boldsymbol{a})\div(+\boldsymbol{b}) = \left(-\frac{\boldsymbol{a}}{\boldsymbol{b}}\right) = -\frac{\boldsymbol{a}}{\boldsymbol{b}}.$$

provided that b is not zero.

The rules of sign given above may be stated as follows:

In multiplication and division of one directed number by another, like signs give a positive sign, and unlike signs give a negative sign.

Square Roots From the rule of signs we see that

$$(+a)\times(+a) = (+a^2) = a^2 \quad \text{and} \quad (-a)\times(-a) = (+a^2) = a^2.$$

This shows that a positive number, a^2, has two square roots, $(+a)$ and $(-a)$; these are usually written in the form $\pm a$. The symbol $\sqrt{(+a^2)}$ or $\sqrt{a^2}$ is used to represent the *positive* square root.

For example, if $x^2 = 9$, then $x = \pm 3$; but $\sqrt{9} = 3$.

There is no square root of a negative number.

EXERCISE 3

[*Nos. 1–40 are intended for oral work*]

1 A train travelling due east at v metres per minute (see Fig. 3) passes a signal-box B at mid-day, zero hour, and therefore at t minutes after mid-day it is $(v \times t)$ metres *east* of B. Evaluate $v \times t$ and state the meanings of the data and the results if

(i) $v = +\ 800, t = +10$; (ii) $v = +\ 900, t = -4$;
(iii) $v = -1200, t = +5$; (iv) $v = -1000, t = -7$.

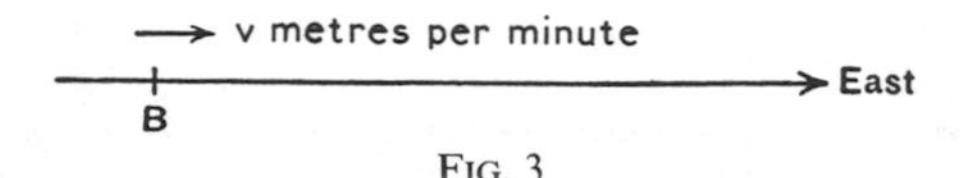

FIG. 3

2 What are the square roots of (i) 25; (ii) $9a^2$; (iii) $\frac{1}{4}b^2$?

Find the values of the following:

3 $(-2)\times(+3)$ **4** $(-3)\times(-4)$ **5** $(+5)\times(-2)$
[**6**] $(+5)\times(+1)$ [**7**] $(-3)\times(+5)$ [**8**] $(-2)\times(-1)$
9 $(-3)^2$ **10** $0\times(-5)$ **11** $(-2)^3$
12 $(+6)\div(-2)$ **13** $(-12)\div(-1)$ **14** $(-8)\div(+8)$
[**15**] $\dfrac{(-6)}{(+3)}$ [**16**] $\dfrac{(-10)}{(-1)}$ **17** $\dfrac{(+12)}{(-3)}$ **18** $\dfrac{(-2)}{(-2)}$

If $a = (+2)$, $b = (-4)$, $p = (+1)$, $q = (-1)$, $r = 0$, evaluate:

19 ab **20** br **21** q^2 **22** bpq **23** b^2q
[**24**] pq [**25**] b^2 [**26**] $3ab$ [**27**] pqr [**28**] bq^2
29 $\dfrac{1}{b}$ **30** $\dfrac{b}{a}$ **31** $\dfrac{p}{q}$ **32** $\dfrac{r}{b}$ **33** $\dfrac{bp}{q^2}$ **34** $\dfrac{2pq}{b}$
[**35**] $\dfrac{q}{a}$ [**36**] $\dfrac{r}{q}$ [**37**] $\dfrac{bq}{a}$ [**38**] $\dfrac{b^2}{q}$ [**39**] $\dfrac{a^2q}{b}$ [**40**] $\dfrac{b^3}{q^3}$

Find the values of the following, for the stated values:

41 (i) a^2+b^2, (ii) $(a+b)^2$, when $a = (+4)$, $b = (-5)$.
42 (i) $2c^3+d^3$, (ii) $(2c+d)^3$, when $c = (-1)$, $d = (+2)$.
43 (i) $(x+y)(2x-y)$, (ii) $x^2+y^2-2x-3y$ when $x = (+2)$, $y = (-1)$.
[**44**] $(a-b)^2-(a^2-b^2)$, when $a = (+3)$, $b = (-2)$.

45 (i) $\dfrac{c+2d}{c-2d}$, (ii) $\dfrac{c^2-d^2}{c^2-2cd+d^2}$, when $c = (-4)$, $d = (+6)$.

46 $x\left(\dfrac{x-4}{x-1}-\dfrac{x-1}{x+2}\right)$ when $x = (-5)$.

Brackets The object of using brackets in such cases as $(+3)$, (-4) is to distinguish between the positive number 'plus three' and the order 'add three', or between the negative number 'minus four' and the order 'subtract four'. Usually this distinction need not be made because the rules of sign have been chosen so that the rules for removing brackets containing directed numbers are precisely the same as the rules already given for signless numbers, and therefore for simplicity we write $(+3)$, (-4) as 3, -4. Also

$+(-2a)$ is written $-2a$; $-(+3b)$ is written $-3b$;
$-(-4c)$ is written $+4c$; $+(+5d)$ is written $+5d$.

Example 2 Simplify $(+3a)-4\{(-2b)-(+3c)\}$.

$$\text{The expression} = (+3a)-4(-2b)+4(+3c) = \mathbf{3a+8b+12c}.$$

Example 3 Simplify (i) $(-2xy)\times(+3y)$; (ii) $\dfrac{(-6a^2b)}{(-4ab^2)}$.

(i) $(-2xy)\times(+3y) = -2\times 3\times xy\times y = \mathbf{-6xy^2}$.

(ii) $\dfrac{(-6a^2b)}{(-4ab^2)} = \dfrac{(-6a^2b)\times(-1)}{(-4ab^2)\times(-1)} = \dfrac{6a^2b}{4ab^2} = \dfrac{\mathbf{3a}}{\mathbf{2b}}$.

Example 4 Fill in the blanks in:

(i) $a-b = -(\cdots)$; (ii) $\dfrac{b-a}{c-d} = \dfrac{a-b}{\cdots}$.

(i) $a-b = -(-a)-(+b) = \mathbf{-(-a+b)}$.

(ii) $\dfrac{b-a}{c-d} = \dfrac{(b-a)\times(-1)}{(c-d)\times(-1)} = \dfrac{-b+a}{-c+d} = \dfrac{\mathbf{a-b}}{\mathbf{d-c}}$.

It is customary to arrange an expression so that when possible it starts with a positive sign, unless there is some reason for not doing so; thus, instead of $-a+b$, we write $+b-a$, that is $b-a$; but it is better to write $3x-2x^2-5$ either as $-2x^2+3x-5$ or as $-5+3x-2x^2$; this is called arranging the expressions in *descending powers* of x or in *ascending powers* of x.

EXERCISE 4

Write more shortly:

1 $+(-2a)$ **[2]** $-(-b)$ **3** $(+2)(-5c)$
4 $(-5)(-2d)$ **[5]** $(+2e)(-e)$ **6** $(-f)(+3f)$
7 $(-3g)(-3h)$ **[8]** $(-km)(-1)$ **9** $(-2z^2)(0)$
10 $(-6a^2)\div(+2a)$ **[11]** $(-12b^2)\div(-3b)$
12 $(+10c^2)\div(-5c)$ **13** $(-4d^6)\div(-2d^3)$

[14] $\dfrac{(-12mn)}{(-4)}$ **15** $\dfrac{(-6rs^2)}{(-2rs)}$ [16] $\dfrac{(+5t^2)}{(-1)}$ **17** $\dfrac{(-xy)}{(+xy)}$

18 $a+(-3a)$ **19** $2b-(-b)$ **20** $c-(+c)+(-c)$

[21] $(-d)+(-d)$ [22] $e-(-e)+(+e)$ [23] $f+(-f)-(-f)$

24 $(g)+(-2g)-(-3g)-(+2g)$ **25** $-3(h-2k)+k$

26 $2(m-n)-3(m+n)$ **27** $0-2(r-s)-(r+s)$

[28] $q+(+4p)+(-3q)-(-2r)$ [29] $2s-t-3(s+t)$

30 $3y^2-5y-2(1-y+y^2)$ [31] $2(1-z)-3z(1-z)$

32 $(a-b)(-1)-3(b-a)(-1)$ [33] $(c^2-3c+2)(-c)-(-2c)$

34 $(-3a)^2$ **35** $(-2b)^3$ **36** $-(-cd)^2$ **37** $(-3ef^2)^3$

[38] $g^2-(-g)^2$ **39** $h^3+(-h)^3$ [40] $(4k)^2\div(-2k)$

[41] $\dfrac{3mn}{-m}$ [42] $\dfrac{-p^2}{-pq}$ **43** $\dfrac{8r^3}{-4rs}$ **44** $\dfrac{(-2t)^3}{2t}$

45 $(-e)\div(ef)$ **46** $(-2h)^3\times h^2$ **47** $2x^3\div(-2x)^2$

48 $\dfrac{r}{s}\times(-s)$ [49] $\left(-\dfrac{s}{t}\right)\div(-s)$ **50** $y\div\left(-\dfrac{1}{y}\right)$

Copy and complete the following:

51 $2a-6b = 2(\cdots) = -2(\cdots)$ **52** $3d-c = -(\cdots)$

53 $\dfrac{-p}{-q} = \dfrac{p}{\cdots}$ **54** $\dfrac{(-p)(-q)}{r(-s)} = \cdots$ **55** $\dfrac{a(-b)}{(-c)d} = \cdots$

56 Find x if (i) $x^2 = 1$; (ii) $x^2 = \frac{1}{4}$.

57 Find y if (i) $2y^2 = 50$; (ii) $y = \dfrac{1}{y}$.

58 Simplify (i) $\dfrac{a-b}{b-a}$; (ii) $\dfrac{c+d}{-c-d}$; (iii) $\dfrac{(e-f)^2}{(f-e)^2}$.

Simplification In previous examples, brackets have been used for performing addition, subtraction, multiplication and division. The working may, however, be arranged as in Arithmetic.

Example 5 Add $3y-x-5z$ to $3x-3y+z$.

Arrange the terms of the two expressions in similar orders.

$$\begin{array}{r} -x+3y-5z \\ 3x-3y+\ z \\ \hline \mathbf{2x}\qquad \mathbf{-4z} \\ \hline \end{array}$$

Example 6 Subtract $3y-x-5z$ from $3x-3y+z$.
Arrange the terms of the two expressions in similar orders.

Subtracting $(-x)$ from $3x$ is the same as adding $(+x)$ to $(+3x)$, giving $(+4x)$ or $4x$, etc.

$$\begin{array}{r} 3x-3y-z \\ -x+3y-5z \\ \hline \mathbf{4x-6y+6z} \\ \hline \end{array}$$

Alternatively, if we use brackets, we write

$$(3x-3y+z)-(3y-x-5z) = 3x-3y+z-3y+x+5z,$$

and this shows that if the expressions are set down as in Arithmetic the result is obtained by *changing the sign of each term in the lower line (mentally) and then adding.*

Example 7 Divide $12a^3b-6a^2b^2+18ab^3$ by $-6ab$

$$\begin{array}{r|l} -6ab & 12a^3b-6a^2b^2+18ab^3 \\ \hline & \mathbf{-2a^2 \quad +ab \quad -3b^2} \\ \hline \end{array}$$

EXERCISE 5

Add:

1 $3a-b$ and $3b-a$ **2** $2c+3d-e$ and $2d-e-c$

[3] $2f^2-1$ and $-f^2$ **[4]** $g+h-k$ and $h+k-g$

5 $1-(n+p)$ and $p-(n-1)$ **[6]** $t-3r+2s$ and $s-r-3t$

7 $2m^2-5m-4$ and $1+3m-3m^2$

8 $2(3a-1-a^2)$ and $-3(2a-a^2)$

[9] $3c-2d-e$, $3d-2e-c$, $3e-2d-c$

10 $1-m-3m^2$, $2m-5-m^2$, $2m^2-m+2$, $1-m^2$

Subtract:

11 $2c-d$ from $2d-c$ **[12]** b^2-1 from $2b^2$

13 e^2+f^2 from e^2-f^2 **[14]** $2g$ from $1+g$

15 $r-s+t$ from $s+t-r$ **16** $3x+y+1$ from $4x-y-1$

[17] $2a-3b-c$ from $b-a+c$ **[18]** $2c-d-e$ from $d-2e$

19 $3f+g+1$ from $4f-g-1$ **20** $3h^2-5h-6$ from h^3+h^2

[21] $1+m-m^2$ from 0 **[22]** x^3-2x^2+5x from x^2+3x-2

23 Multiply (i) $x-b$ by $-z$; (ii) $-r+2s$ by $-3s$.

[24] Multiply (i) $1-t$ by $-2t$; (ii) $1-3h+h^2$ by -1.

25 Divide (i) a^2-ab by $-a$; (ii) $6r^3-4r^2$ by $-2r$.

[26] Divide (i) p^2+pq by $-p$; (ii) $12pq-9q^2$ by $-3q$.

27 Divide (i) s^6-s^2 by $-s^2$; (ii) $6x^3-8x^2y-4xy^2$ by $-2x$.

Subtract:

	(i)	(ii)	(iii)	(iv)
28	$4a+b$ $a+3b$	$5c-3d$ $3c+2d$	$7e+4f$ $3e-5f$	$6g-4h$ $4g-5h$
29	$3b+2c$ $5b+2c$	$4d+3e$ $6d-3e$	$2f$ $3f-g$	$m-n$ n
[**30**]	$2q-5r$ $2q-3r$	$s-3t$ $4s-3t$	$3x$ $k+5x$	$5y+2z$ $6y$

[**31**] $2a-b+1$ from $a+2b-1$

32 $3c-4d-3$ from $c+4d+1$

[**33**] $x+3y+2$ from $x-y-1$

[**34**] $5-2y-z$ from $3-y-2z$

Equations

Example 8 Solve the equation:

$$1-\frac{1}{6}(t+5) = \frac{2t+7}{3}-\frac{t-10}{4}.$$

Multiply each side by 12 (the L.C.M. of 6, 3, 4).

$$\therefore\ 12-\frac{12}{6}(t+5) = \frac{12(2t+7)}{3}-\frac{12(t-10)}{4};$$

$$\therefore\ 12-2(t+5) = 4(2t+7)-3(t-10);$$

$$\therefore\ 12-2t-10 = 8t+28-3t+30;$$

$$\therefore\ 2-2t = 5t+58;$$

$$\therefore\ -2t-5t = 58-2;$$

$$\therefore\ -7t = 56;$$

$$\text{multiply each side by } -1,\ \therefore\ 7t = -56;$$

$$\therefore\ t = \mathbf{-8.}$$

Check: If $t = -8$,

$$\text{left side} = 1-\tfrac{1}{6}(-8+5) = 1-\tfrac{1}{6}(-3) = 1+\tfrac{3}{6} = 1\tfrac{1}{2};$$

$$\text{right side} = \frac{-16+7}{3}-\frac{-8-10}{4} = \frac{-9}{3}-\frac{-18}{4} = -3+4\tfrac{1}{2} = 1\tfrac{1}{2}.$$

$$\therefore \text{ if } t = -8, \text{ left side} = \text{right side}.$$

Note If the root of an equation is an awkward fraction, it is better to look carefully through the working or to do it all over again than to check by substitution.

EXERCISE 6

Solve the following equations:

1 $1\frac{1}{2}p = -12$

2 $35 = -1\frac{3}{4}q$

3 $\frac{1}{r} = -2\frac{1}{2}$

4 $5(x-2)-3(x-1) = -1$

5 $y+1 = 2(y-3)-3(y-1)$

[6] $3(6-m)-4(m+8) = 0$

[7] $2(n-1)-3(3+n) = 5+n$

8 $4-4(r-5) = 2(2-r)-6$

[9] $4(s-1)-3(s-2) = 4-5s$

10 $x(2+x)-3(1+x) = 1+x^2$

11 $0{\cdot}6x+0{\cdot}72 = 0$

[12] $z(2z-1)-3(5+z) = 2z^2+2z+3$

[13] $1{\cdot}8t+0{\cdot}63 = 0$

14 $0{\cdot}3(6-n) = 0{\cdot}4(n+8)$

[15] $1{\cdot}4t+0{\cdot}6t = 0{\cdot}7$

16 $\frac{1}{2}(p-1)-\frac{1}{4}(3-p) = 2$

[17] $\frac{1}{2}(1+x) = \frac{1}{5}(2x-1)-1$

18 $\frac{3}{2m}-4 = 3-\frac{9}{m}$

[19] $\frac{7}{3z}-5 = \frac{1}{2}-\frac{5}{z}$

20 $\frac{p+1}{3}-\frac{p-1}{2} = 1+\frac{2p}{3}$

[21] $\frac{2x-1}{5}-\frac{3x+1}{2} = \frac{2}{5}$

22 $\frac{2z+7}{4}-\frac{z+1}{3} = \frac{3}{4}$

23 $\frac{t}{2}-\frac{5t+4}{3} = \frac{4t-9}{3}$

24 $\frac{3x+2}{5}-\frac{2x+5}{3} = x+3$

[25] $\frac{1-y}{2} = \frac{3+y}{3}-\frac{9+y}{4}$

26 What number must be added to both numerator and denominator of $\frac{11}{17}$ so that the new fraction may equal $\frac{3}{5}$?

[27] Find six consecutive odd numbers whose sum is 12.

28 The present ages of A and B are 21 and 35 years. In n years' time, B will be just twice as old as A will be then. Find the value of n and interpret the answer.

29 If $C°$ Celsius is the same temperature as $F°$ Fahrenheit, $C = \frac{5}{9}(F-32)$. What temperature is measured by the same number on both scales?

[30] A starts with £100 in the bank and draws out £6 a week; B starts with £120 at the same time and draws out £7 a week. How do their bank accounts stand when they are the same?

31 A stone is thrown upwards from the top of a tower and it is $(20t-5t^2)$ metres above the starting point after t seconds. Find the distance between its positions after 3 seconds and 4 seconds.

CHAPTER 2

FIRST STEPS IN STAGE B GEOMETRY

Some fundamental geometrical facts have been discussed and illustrated practically in the previous volume; these deal with

(i) angles at a point,
(ii) angles made by a transversal with parallel lines,
(iii) angle-tests for lines to be parallel,
(iv) tests for triangles to be congruent.

Familiarity with these facts is essential. By making use of them, it is possible to prove other important geometrical properties, called **theorems.**

EXERCISE 7 (Class Discussion)

Draw and letter diagrams to illustrate the statements in Nos. 1–5, denoting angles by small letters; then state the facts in terms of the letters of your diagrams.

1 Adjacent angles on a straight line are supplementary.

Reference: adj. ∠s on st. line.

2 If the adjacent angles COA, COB are supplementary, then AOB is a straight line.

Reference: adj. ∠s supp.

3 If two lines intersect, the vertically opposite angles are equal.

Reference: vert. opp. ∠s.

4 If a transversal meets two *parallel* lines, then

(i) corresponding angles are equal, (ii) alternate angles are equal, (iii) allied angles are supplementary.

Reference: (i) *corr. ∠s,* (ii) *alt. ∠s,* (iii) *allied ∠s.*

5 Two lines are parallel *if* a transversal makes *either* two corresponding angles equal *or* two alternate angles equal *or* two allied angles supplementary.

Reference: (i) *corr. ∠s equal,* (ii) *alt. ∠s equal,* (iii) *allied ∠s supp.*

6 Draw diagrams to show, by suitable marking, the meanings of the tests for congruence of two triangles, expressed by the references,

(i) **SAS;** (ii) **ASA** or **AAS;** (iii) **SSS**; (iv) **RHS.**

Angles at a Point

Example 1 (Oral Discussion) If a straight line EC meets another straight line ACB at C and if CH, CK are bisectors of ∠ACE, ∠BCE, prove that ∠HCK is a right angle.

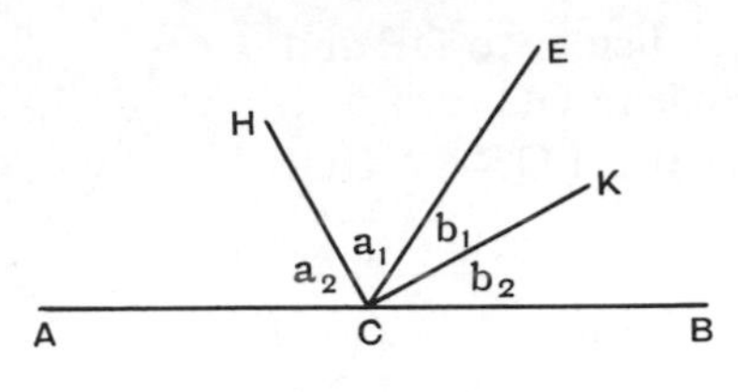

FIG. 4

Since ∠s HCE, HCA are given equal, we denote them by a_1, a_2; similarly for ∠s KCE, KCB.

(i) Express with small letters the fact that has to be proved.

(ii) What do you know about ∠ECA and ∠ECB? Give the reason. Express this fact with small letters.

(iii) What do you get by putting $a_2 = a_1$ and $b_2 = b_1$? Complete the proof.

Always state all the necessary reasons

The proof may be set out as follows:

$(a_1+a_2)+(b_1+b_2) = 2$ rt. ∠s *adj. ∠s on st. line,*

but $a_1 = a_2$ and $b_1 = b_2$ *given*

$\therefore 2a_1+2b_1 = 2$ rt. ∠s,

$\therefore \quad a_1+b_1 = 1$ rt. ∠,

$\therefore$ ∠HCK = 1 rt. ∠.

EXERCISE 8

Nos. 1–8 refer to Fig. 5.

1 Write in small letters, ∠QOS + ∠ROT, ∠POS − ∠POR.

2 Write in capital letters, $c+d$, $a+b+c$.

3 Express the following statements with small letters: (i) ∠QOS = ∠ROT, (ii) OS is perpendicular to OQ.

4 If ∠POQ = ∠ROS, prove that ∠POR = ∠QOS.

[5] Prove that ∠POR + QOS = ∠POS + QOR.

6 If OS bisects ∠ROT, prove ∠QOS − ∠TOS = ∠QOR.

7 If OR bisects ∠QOT, prove ∠ROS = ½(∠QOS − ∠TOS).

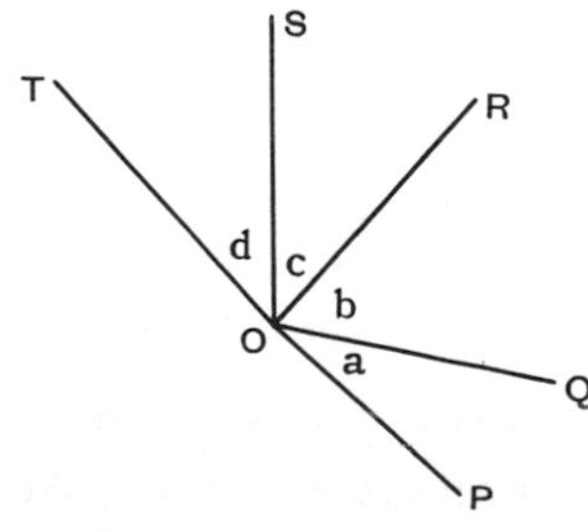

FIG. 5

8 If OR is perpendicular to OP and if OS is perpendicular to OQ, prove (i) ∠POQ = ∠ROS, (ii) ∠POS + ∠QOR = 2 rt. ∠s.

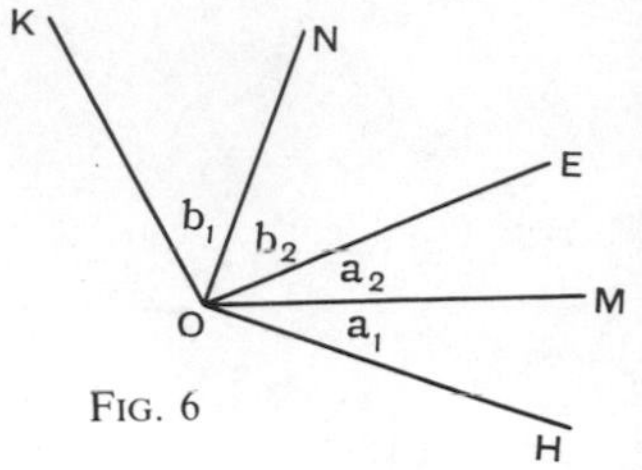

Fig. 6

Nos. 9–11 refer to Fig. 6, in which OM, ON are the *bisectors* of ∠EOH, ∠EOK.

9 Prove that ∠MON = $\frac{1}{2}$∠HOK.

[**10**] Prove that ∠MOE + ∠MOK = 2∠MON.

[**11**] Prove that ∠HON + ∠KOM = 3∠MON.

12 Fig. 7 represents three straight lines. (i) If $p = q$, prove $a = b$. (ii) If $a = c$, prove $q + t = 2$ rt. ∠s. *Give reasons.*

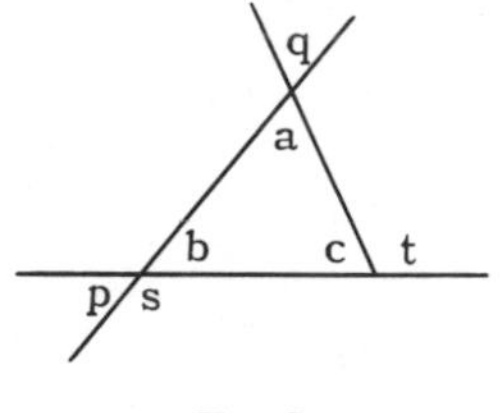

Fig. 7

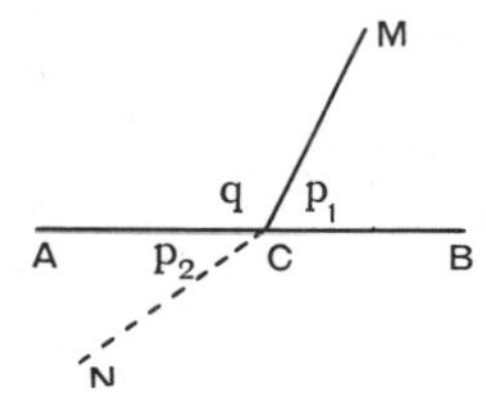

Fig. 8

13 In Fig. 8, not drawn accurately, ACB is a straight line and $p_1 = p_2$. Prove that MCN is a straight line.

14 Draw a triangle ABC in which ∠B = ∠C. Produce BA to E. Draw AP parallel to BC. Prove that AP bisects ∠CAE.

15 Draw a quad. ABCD in which AB is parallel to DC and AD is parallel to BC. Prove ∠BAD = ∠BCD. [Produce AB to E.]

[**16**] Draw a quad. ABCD in which AB is parallel to DC and take a point K inside ABCD. Prove ∠AKD = ∠KAB + ∠KDC.

[Draw KN parallel to BA.]

THEOREM

(1) If one side of a triangle is produced, the exterior angle so formed is equal to the sum of the interior opposite angles.

(2) The sum of the angles of a triangle is equal to two right angles.

Given a triangle ABC with BC produced to D. (See Fig. 9.)

To prove that (i) ∠ACD = ∠A + ∠B;

(ii) ∠A + ∠B + ∠ACB = 2 right angles.

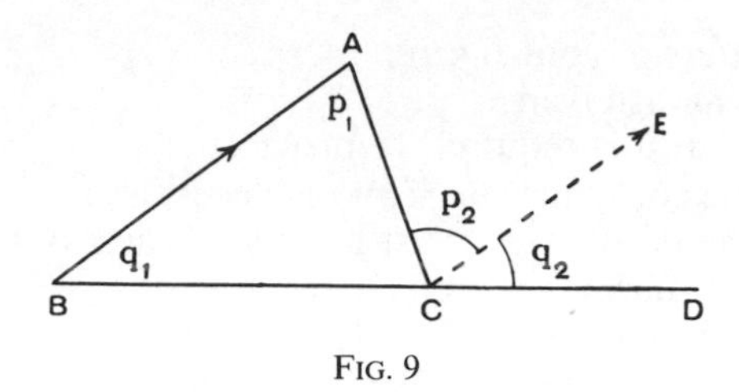

FIG. 9

Construction Through C draw CE∥BA.

Proof (i) With the notation in the figure,

$$p_2 = p_1 \qquad \textit{alt. } \angle s, \text{ CE} \parallel \text{BA},$$
$$q_2 = q_1 \qquad \textit{corr. } \angle s, \text{ CE} \parallel \text{BA},$$

$\therefore$ by addition, $p_2 + q_2 = p_1 + q_1$,

$$\therefore \angle ACD = \angle A + \angle B.$$

(ii) Add $\angle ACB$ to each side,

$\therefore \angle ACD + \angle ACB = \angle A + \angle B + \angle ACB.$

But $\angle ACD + \angle ACB = 2$ rt. $\angle$s. *adj. $\angle$s on st. line*

$\therefore \angle A + \angle B + \angle ACB = 2$ rt. $\angle$s.

Abbreviations for reference: (1) ext. $\angle$ of $\triangle$, (2) $\angle$ sum of $\triangle$.

Corollary If the triangles ABC, PQR are such that

$$\angle A = \angle P \quad \text{and} \quad \angle B = \angle Q,$$

then

$$\angle C = \angle R.$$

$\angle A + \angle B + \angle C = 2$ rt. $\angle$s and $\angle P + \angle Q + \angle R = 2$ rt. $\angle$s,

$\angle$ sum of $\triangle$,

$\therefore \angle A + \angle B + \angle C = \angle P + \angle Q + \angle R$ and so $\angle C = \angle R$.

A simple deduction from a theorem is called a **corollary.**

It is convenient to use a *dotted* line for the construction.

A general statement of the fact which it is required to prove is called the **general enunciation** of the theorem. If this fact is stated in terms of the letters of a particular diagram, it is called the **particular enunciation**.

In the theorem just proved, the statement in **heavy type** at the beginning is the general enunciation, and the statements of what is given and what it is required to prove form a particular enunciation. If a particular enunciation is given in a question, the letters used in the figures must be those in the given enunciation.

Writing out the Proof of a Theorem

If what it is required to prove is expressed in the form of a *general enunciation*, set out the work in the following order:

Using the letters of your figure,

(1) state what is given;
(2) state what it is required to prove;
(3) state the construction, if any is necessary;
(4) state the proof; this must include all necessary reasons, using suitable abbreviated references.

Thus there are four stages:

(1) Given (2) To Prove (3) Construction (4) Proof.

It is necessary to distinguish carefully between what is *given in the figure* and what is *added to the figure* for purposes of proof.

In the property of the exterior angle of a triangle on p. 14, the fact that BC is produced to D is part of what is given and must be included in the particular enunciation because it is needed for stating what has to be proved; it is *not* part of the 'construction', *i.e.* it is not an addition to the figure which is made to help in the proof.

The theorems in this book are not numbered because reference to a theorem in the course of a proof should always be given in some such abbreviated form as is suggested.

Examination Requirements In the theorem on p. 14, the enunciation contains two statements, and the proof of the second depends on that of the first. If in an examination a candidate is asked to prove only the second statement, the wording of the question will usually make it clear whether or not the first part can be assumed. Throughout this book, the theorems are so arranged that this remark applies to all enunciations which contain more than one part.

Angles of a Triangle It is important to bear in mind that it is often simpler to use the property of the exterior angle instead of the angle-sum of the triangle.

EXERCISE 9

Find the value of x and the angle a in Figs. 10, 11:

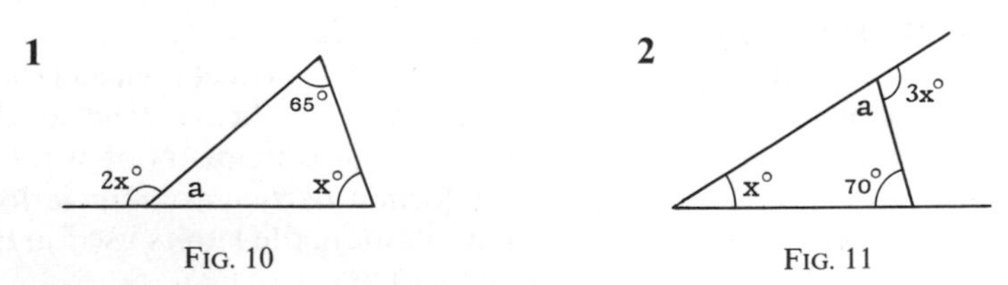

FIG. 10 FIG. 11

3 In a right-angled triangle, one angle is 43°; find the other acute angle.

Nos. 4–11 refer to the angles A, B, C of a triangle ABC.

4 Find C if A $= 40°$, B $= 105°$; also if A $= 90°$, B $= x°$.

[**5**] Find A if (i) B $= 15°$, C $= 18°$, (ii) B $=$ C $= 52°$.

6 Find A if A $=$ B and C $= 48°$.

7 Find C if A + B $=$ 3C.

[**8**] Find A if A − B $= 15°$ and B − C $= 30°$.

9 Express $\frac{1}{2}$A + $\frac{1}{2}$B in terms of C.

[**10**] Simplify $\frac{1}{2}$(B + C − A).

11 If A + B $=$ C, prove that C $= 90°$.

12 Can a triangle be drawn having its angles equal to
(i) 45°, 65°, 80°; (ii) 43°, 64°, 73°; (iii) 100°, 110°, x°?

[**13**] The angles of a triangle are $2x°$, $3x°$, $4x°$. Find x.

14 Two exterior angles of a triangle are 120°, 130°. Find the third unequal exterior angle.

15 In △ABC, AX bisects ∠BAC and AD is perpendicular to BC. If ∠A – 60° and ∠B = 70°, find ∠DAX.

[**16**] In △ABC, ∠B = 110°, ∠C = 50°; AD is the perpendicular from A to CB produced. Prove that ∠DAC = 2∠DAB.

17 If BP, CP bisects the angles ABC, ACB of a triangle ABC, and if ∠BAC = 80°, find ∠BPC.

EXERCISE 10

1 If, in Fig. 12, $p - 2c$, prove that $b = c$.

[**2**] If, in Fig. 12, $b = 2c$, prove that $c = \frac{1}{3}p$.

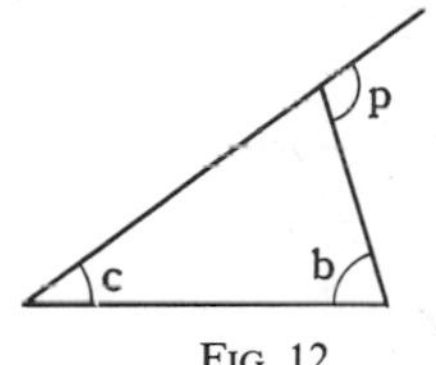

FIG. 12

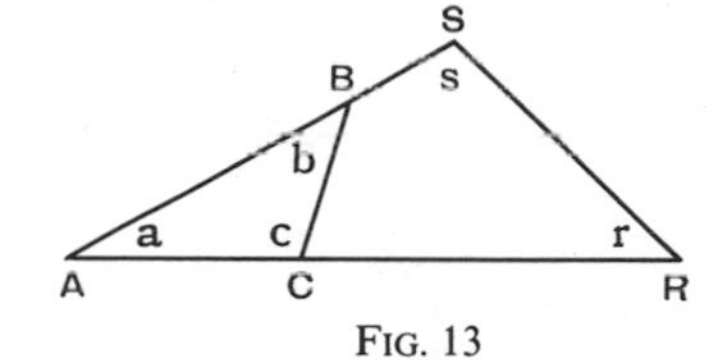

FIG. 13

3 In Fig. 13, if $b = r$, prove that $c = s$.

4 In Fig. 13, find r in terms of b, c, s.

[**5**] ABCD is a quadrilateral such that the diagonal AC bisects ∠DAB and bisects ∠DCB. Prove that ∠ABC = ∠ADC.

6 ABCD is a quadrilateral such that AB is parallel to DC; AC cuts BD at K. Prove
∠AKD = ∠ABD + ∠ACD.

7 In Fig. 14, BE and CF are perpendicular to AC and AB. Prove
(i) ∠FBE = ∠FCE,
(ii) ∠FHB = ∠BAC,
(iii) ∠BHC + ∠BAC = 2 rt. ∠s.

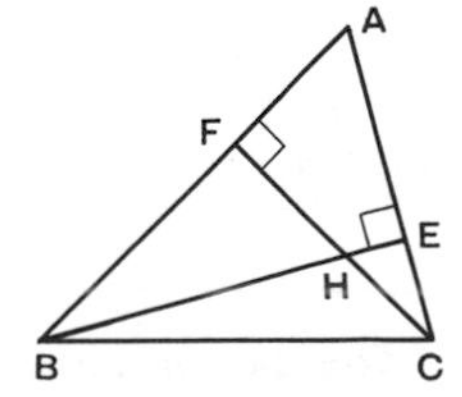

FIG. 14

8 ABCD is a parallelogram; the bisectors of ∠BAD, ∠ABC meet at P; prove ∠APB is a right angle.

9 Draw a triangle ABC right-angled at A and draw the perpendicular AD from A to BC. What angle in your figure is equal to ∠DAC? Give reasons.

10 If, in the triangle ABC, the bisectors of the angles ABC, ACB meet at I, prove that $\angle BIC = 90° + \frac{1}{2}\angle BAC$.

***11** The side BC of △ABC is produced to D; the bisector of ∠BAC cuts BC at K; prove that $\angle ABD + \angle ACD = 2\angle AKD$.

***12** The side BC of △ABC is produced to D; the bisectors of the angles ABC, ACD meet at Q; prove that $\angle BQC = \frac{1}{2}\angle BAC$.

Congruent Triangles The statement,

$$\triangle s \begin{matrix} ABC \\ QXK \end{matrix} \text{ are congruent,}$$

means that △ABC can be placed on the top of △QXK so that A is at Q, and B is at X, and C is at K. If two congruent triangles are written in this way, it is unnecessary to look at the figure to pick out a pair of equal sides or a pair of equal angles.

Example 2 (Oral Discussion) ABCD is a quadrilateral such that AB is parallel to DC and AB = DC. Prove

(i) AD is parallel to BC, (ii) AD = BC.

Join BD and insert small letters for angles, as shown in Fig. 15.

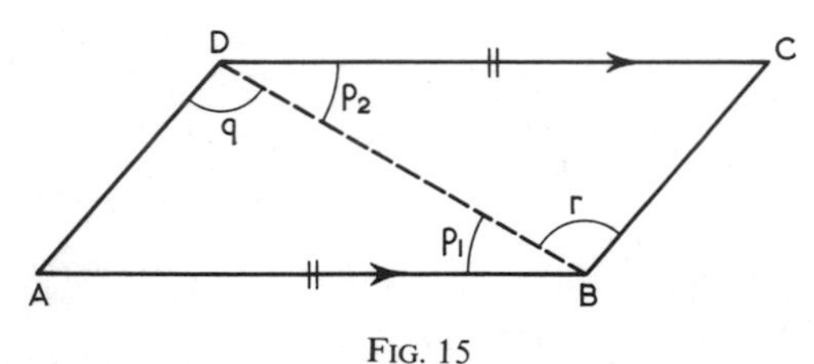

FIG. 15

(1) What do you know about p_1, p_2?

(2) Give the reasons why $\triangle s \begin{matrix} ABD \\ CDB \end{matrix}$ are congruent.

(3) Complete the proof.

Compare your proof with the way in which the reasons are given and the proof is set out on p. 269.

EXERCISE 11

The data in Nos. 1–9 refer to two triangles ABC, DEF. In each case show the data *on your own figure* by suitable marking. Do the data show that the triangles *must* be congruent? If so, state the test used and express the fact in the proper way.

1 AB = EF, AC = DF, ∠A = ∠F
2 CA = FD, CB = FE, ∠B = ∠E
3 AC = EF, ∠A = ∠E, ∠C = ∠F
4 AC = DF, ∠B = ∠D, ∠C = ∠F
5 BC = DF, ∠B = ∠F, ∠A = ∠E
6 ∠A = ∠D, ∠B = ∠E, ∠C = ∠F
7 AB = AC, DE = DF, ∠A = ∠D
8 BC = DF, AC = DE, AB = EF
9 AB = AC, DE = DF, BC = EF

10 If △s $\frac{\text{ABC}}{\text{YKX}}$ are congruent, which side is equal to XY and which angle is equal to ∠ACB?

11 In Fig. 16, the lines ANB, CND bisect each other. Prove (i) AD = CB, (ii) AD is parallel to CB.

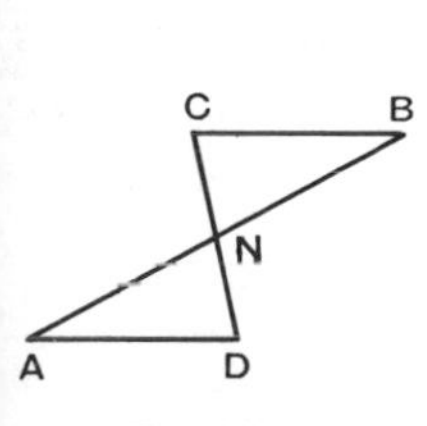

FIG. 16

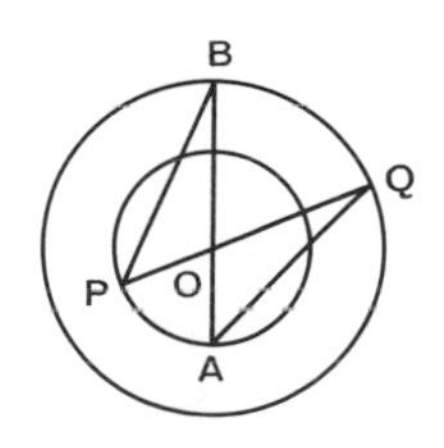

FIG. 17

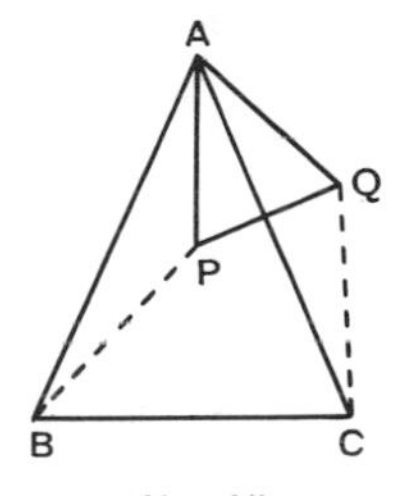

FIG. 18

12 O is the centre of each of the circles in Fig. 17; AOB, POQ are straight lines. Prove that AQ = PB.

13 In Fig. 18, AB = AC, AP = AQ and ∠BAC = ∠PAQ. Which two triangles in the figure are congruent? Give reasons.

14 ABCD is a quadrilateral such that AB = DC and AD = BC. Prove ABCD is a parallelogram (*i.e.* its opposite sides are parallel).

[**15**] K is a point inside the triangle ABC. If AB = AC and KB = KC, prove that ∠ABK = ∠ACK.

16 AB is a chord of a circle, centre O; N is the mid-point of AB. Prove that ON is perpendicular to AB.

[**17**] D is the mid-point of the base BC of the triangle ABC; BX, CY are the perpendiculars from B, C to AD, produced if necessary. Prove that BX = CY.

[**18**] In △ABC, AB = AC and ∠BAC = 1 rt. ∠; XAY is any line through A; BH, CK are the perpendiculars from B, C to XAY. Prove AH = CK.

19 X is the centre of each semicircle AQB, CPD in Fig. 19; XPQ, ACXDB are straight lines.

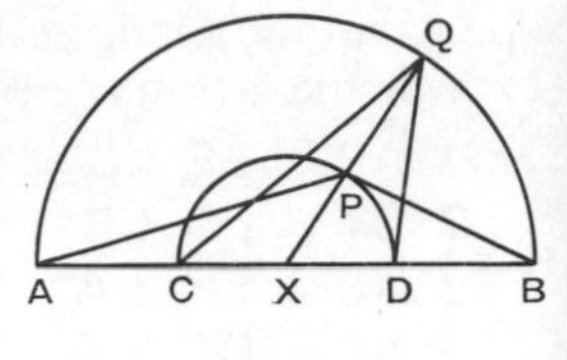

Fig. 19

(i) Which triangle in the figure is congruent to △AXP? Give reasons.

(ii) Name two other congruent triangles. Which angle is equal to ∠XPB? Give reasons.

(iii) Prove that ∠APB + CQD = 2 rt. ∠s.

20 In Fig. 20, AB = BH, BC = BK, AK = CH and ABC, BHK are straight lines. Prove that (i) ∠ABK = 1 rt. ∠, (ii) CH is perpendicular to AK.

21 In Fig. 21, AB = AD, AC = AE and BC = DE. Prove that BE = CD.

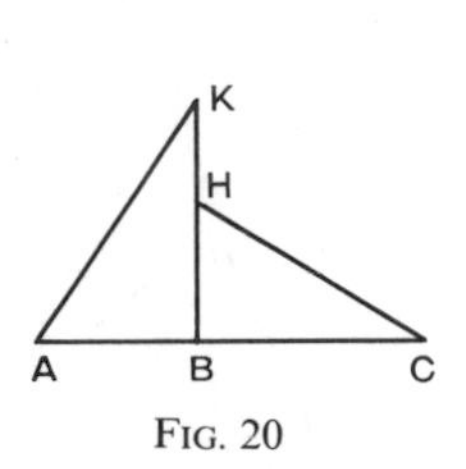

Fig. 20

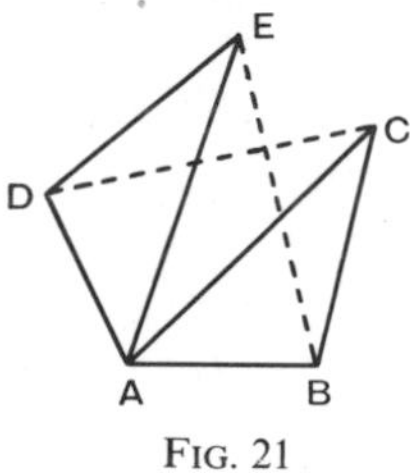

Fig. 21

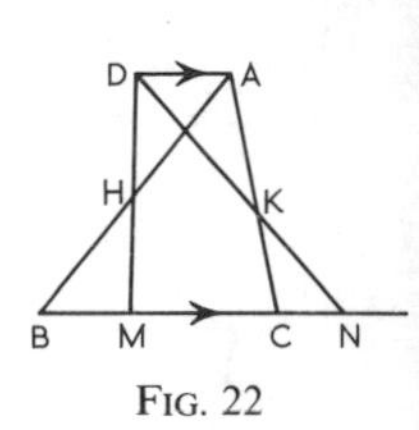

Fig. 22

22 In Fig. 22, H, K are the mid-points of AB, AC; AD is parallel to BC; DH, DK meet BC, BC produced in M, N. Prove that (i) AD = BM, (ii) MN = BC.

***23** In Fig. 23, AB = AC and APQR is a straight line. BP and CQ are drawn so that ∠BPR = ∠BAC = ∠CQR. Prove that AP = CQ.

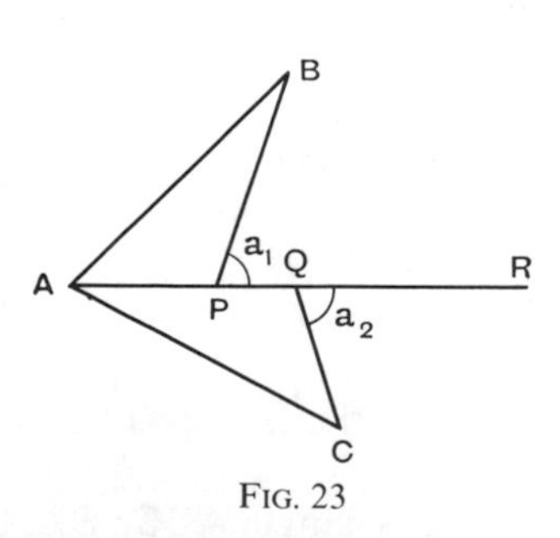

Fig. 23

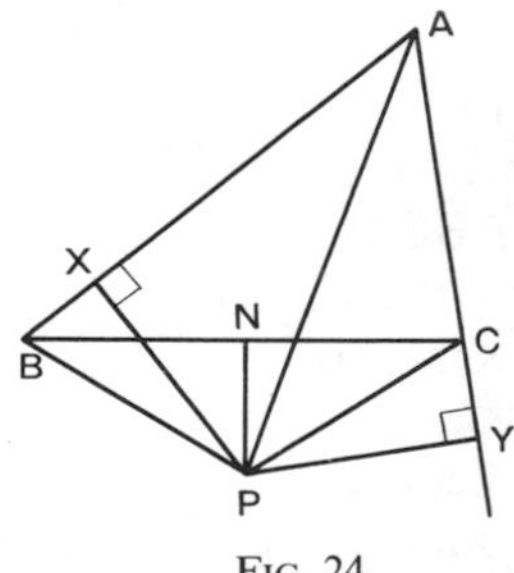

Fig. 24

***24** In Fig. 24, the bisector of ∠BAC meets at P the line NP which bisects BC at right angles; PX, PY are the perpendiculars from P to AB, AC. Find in the figure three pairs of congruent triangles and prove the congruence.

Prove also that AX = AY = $\frac{1}{2}$(AB + AC).

CHAPTER 3
RATE AND RATIO

Rate If a man walks 10 km in 2 hours, we say that he walks at the *rate* of 5 km per hour. The word **rate** is here used to state how the distance is altering with the *time*; it is, however, also used in other ways: If I buy 3 kg of tea for 216p, I pay at the *rate* of 72p per kg; if a laundry charges 36p for washing 12 towels, the *rate* of charge is 3p per towel.

The sizes of two quantities of the *same kind* may be compared in two ways. The ages of two children might be compared by saying one is 9 months older than the other; in this case the comparison is made in the form of a *difference*. On the other hand, the scale of the model of a ship would be described as $\frac{1}{100}$, if 1 metre length of the model represents 100 m length of the ship. Here the comparison is made in the form of a **ratio;** that is, the fraction which the first quantity is of the second.

Suppose a school contains 150 boys and 200 girls, then the number of boys is $\frac{3}{4}$ of the number of girls, and we say that the *ratio* of the number of boys to the number of girls is 3 to 4, written 3:4; and this ratio can be represented by the fraction $\frac{3}{4}$.

Ratios should be expressed as simply as possible; just as the fraction $\frac{8}{36}$ is reduced to $\frac{2}{9}$, so the ratio 8:36 is equivalent to 2:9. Thus a ratio is unaltered if the two numbers (or quantities) of the ratio are both multiplied, or both divided, by the same number. For example,

the ratio $\frac{5}{6}:\frac{3}{4}$ equals the ratio $\frac{5}{6}\times 12:\frac{3}{4}\times 12$, that is 10:9.

If the prices of two cars A, B are £720, £960 respectively,

$$\frac{\text{price of A}}{\text{price of B}} = \frac{£720}{£960} = \frac{3}{4} \quad \text{and} \quad \frac{\text{price of B}}{\text{price of A}} = \frac{£960}{£720} = \frac{4}{3};$$

and we write price of A: price of B = 3:4 and price of B: price of A = 4:3. Conversely, the statement that the ratio of the price of A to the price of B is 3:4 means that the price of A is $\frac{3}{4}$ of the price of B, and that the price of B is $\frac{4}{3}$ of the price of A.

EXERCISE 12 (Oral)

Write the following rates in a simple form:

1 60 km in 5 hours
2 90p for 3 hours' work
3 4 kg for 48p
4 180p for 12 tickets
5 12p for washing 6 cloths
6 £30 rent for 4 months
7 5 m^3 of wood weigh 4000 kg
8 Income £1000, tax £400
9 House rent £60, tax £18
10 10 cm^3 weigh 136 g

Express the following ratios as simply as possible in the form $a:b$.

11 A length of 8 cm to a length of 12 cm
12 A cost of 9p per kg to a cost of 12p per kg
13 12 cm to 4 cm **14** 15 m to 10 m **15** 14 kg to 18 kg
16 16 cm^2 to 36 cm^2 **17** 20 kg to 5 kg **18** 75p to £2
19 18 to 36 **20** 32 to 24 **21** 0·4:1
22 60p:£1·60 **23** 250 cm:3 m **24** 4 cm:25 mm
25 $\frac{1}{6}:\frac{1}{2}$ **26** $\frac{1}{4}:\frac{3}{4}$ **27** $1\frac{1}{2}:2$
28 Wages at 42p per hour to wages at 432p for 8 hours
29 A price of £90 per kg to a price of 12p per g

EXERCISE 13

Express the ratio of the first quantity to the second as simply as possible (i) as a fraction; (ii) in the form $a:b$ (Nos. 1–5).

1 960 km; 1320 km [**2**] £1200; £640 **3** 28p; 210p
[**4**] 4·2 kg; 56 kg **5** 35 m per second; 1·5 km per min

[**6**] Find the ratio of the length of the Ganges (2400 km) to that of the Amazon (6400 km).

[**7**] The tax on an income of £840 is £280. Find the ratio of the tax to the income.

8 A school contains 360 girls and 240 boys. Find the ratio of (i) the number of girls to the number of boys; (ii) the number of boys to the number of pupils.

9 A man earns £750 a year and spends £630 a year. Find the ratio of (i) his income to his expenditure; (ii) his savings to his income.

[**10**] Two cars cost £720, £840 respectively when new. After 2 years the market value of each has fallen £120. Find the ratio of their values (i) when new; (ii) after 2 years.

11 The sides of two squares are 6 cm, 8 cm long. Find the ratio of (i) their perimeters; (ii) their areas.

12 Pressed pork is quoted at 22p per $\frac{1}{2}$ kg and roast pork at 66p per kg. Find the ratio of their prices.

[**13**] An agent's commission is £25·20 if he sells goods of value £189. Find the ratio of the commission to the sales.

14 A hotel charges £1·20 a day in winter and £12·60 a week in summer. Find the ratio of summer to winter charges.

[**15**] An alloy consists of 275 g of copper and 27·5 g of tin. Find the ratio by weight of tin to copper in the alloy.

16 A car travels 224 km in $3\frac{1}{2}$ hours; a train travels 48 km in 50 min. Find the ratio of the car's speed to the train's speed.

17 A bottle weighs 16 g if empty, 61 g if full of water, 52 g if full of alcohol. Find the ratio of the weight of 1 cm^3 of alcohol to that of 1 cm^3 of water.

Scales Scales of maps are often given as ratios in such forms as 1:20 000, 1:100 000 etc. The scale 1:20 000 means that 1 cm on the map represents 20 000 cm on the ground, or more generally that 1:20 000 is the ratio of the distance between any two points on the map to the actual distance between the two places they represent; and the fraction $\frac{1}{20\,000}$ is called the *representative fraction*, or more shortly the R.F. of the map.

Example 1 The scale of a map is 5 cm to 1 km. Express this ratio in the form 1:*n*.

$$\frac{5\text{ cm}}{1\text{ km}} = \frac{5\text{ cm}}{100\,000\text{ cm}} = \frac{1}{20\,000}.$$

$\therefore$ the scale of the map is **1:20 000.**

Example 2 The scale of a plan is 1:200. Find the dimensions of a room which measures 4 cm by 2·5 cm on the plan.

1 cm on the plan represents 200 cm = 2 m on the ground.

$\therefore$ 4 cm represent 8 m and 2·5 cm represent 5 m.

$\therefore$ the room is **8 m by 5 m.**

Example 3 The scale of a map is 2 cm to the km. Find the area of an estate represented by an area 5·4 cm^2 on the map.

A square, side 2 cm, represents 1 km^2,

i.e. 4 cm^2 represent 1 km^2.

$$\therefore\ 5{\cdot}4\text{ cm}^2 \text{ represent } \frac{5{\cdot}4}{4}\text{ km}^2$$

$$= \mathbf{1{\cdot}35\ km^2}.$$

EXERCISE 14

1 The scale of a map is 2 cm to 1 km. Find its R.F.

[**2**] The scale of a plan is 1 cm to 200 m. Find its R.F.

3 The scale of a map is 10 cm to 1 km. Find its R.F.

4 On a map of scale 1:2 500 000, the distance between Calais and Marseilles is 35·3 cm. Find the actual distance in km.

5 The scale of a map is 1 cm to 4 km. How far apart on the map are two churches distant $9\frac{1}{2}$ km from one another?

[**6**] The scale of a map is 1:100 000. Find in km the length of a road, represented by a line 4·7 cm long on the map.

7 The scale of a map is 1:40 000. Find in cm the length on the map of a road 700 m long.

[**8**] The scale of a map is 1 cm to 2 km. Find the length of a road 3·6 cm long on the map.

9 A ground plan of a house is made on the scale 1 cm to 2 m. Find the R.F. of the plan. Find the length and breadth on the plan of a room 6 m by 4 m. What area on the ground is represented by 1 cm^2 on the plan?

10 On a map, scale 1 cm to the km, an island has an area of 3·5 cm^2. Find its actual area in hectares, if 1 hectare = 10 000 m^2.

[**11**] On a map, scale 1 cm to 2 km, a wood has an area of 0·43 cm^2. Find its actual area in hectares.

12 What area is occupied by a farm of area 4·1 km^2 on a map of scale 5 cm to the km?

13 The gradient of a railway up a slope 1 km long from A to B is 1 in 240 (*i.e.* the ground rises 1 m vertically for each 240 m measured along the slope). Find the height in metres of B above A.

[**14**] A train runs at 60 km an hour up a gradient of 1 in 120. How many metres does it rise in 1 minute?

Increase and Decrease in a given Ratio If the annual rent of a house is raised from £60 to £80, the ratio of the new rent to the old rent = 80:60 = 4:3, and we say that the rent has been *increased in the ratio* 4:3. In other words, the new rent is $\frac{4}{3}$ times the old rent. If the annual rent of a house is lowered from £60 to £48, the ratio of the new rent to the old rent = 48:60 = 4:5, and we say that the rent has been *decreased in the ratio* 4:5. In other words, the new rent is $\frac{4}{5}$ times the old rent.

The fraction $\frac{4}{5}$ by which the old rent £60 must be multiplied to give the new rent £48 is called a *multiplying factor*.

$$\frac{\textbf{New Quantity}}{\textbf{Old Quantity}} = \textbf{Multiplying Factor.}$$

The multiplying factor is *less* than 1 if the new quantity is *less* than the old quantity; it is *greater* than 1 if the new quantity is *greater* than the old quantity.

Example 4 Increase 56p in the ratio 10:7.

$$\text{Increased value} = 56\text{p} \times \tfrac{10}{7} = \tfrac{56 \times 10}{7}\text{p}$$
$$= \mathbf{80p.}$$

Example 5 Decrease 2 hours in the ratio 5:6.

$$\text{Decreased time} = 2\text{ h} \times \tfrac{5}{6} = \tfrac{2 \times 5}{6}\text{ h}$$
$$= \tfrac{5}{3}\text{ h} = \mathbf{1\ h\ 40\ min.}$$

Example 6 In what ratio must £75 be increased to become £100? The ratio, £100:£75 = 100:75 = 4:3;

∴ if £75 is increased in the ratio **4:3**, it becomes £100.

Check: £(75 × $\frac{4}{3}$) = £(25 × 4) = £100.

Example 7 Find the multiplying factor which alters 90 kg into 63 kg. The ratio, 63 kg:90 kg = 63:90 = 7:10;

$$\therefore 90 \text{ kg} \times \tfrac{7}{10} = 63 \text{ kg};$$

$$\therefore \text{the required multiplying factor is } \tfrac{7}{10}.$$

EXERCISE 15 (Class Discussion)

1 Increase 144 in the ratio 7:4.

[**2**] Decrease 105 in the ratio 3:5.

[**3**] Increase 75p in the ratio 6:5.

4 Decrease 150p in the ratio 7:10.

[**5**] In a sale, prices are reduced in the ratio 3:5. Find the sale prices of articles whose ordinary prices are (i) £1; (ii) $7\frac{1}{2}$p.

6 A photograph 3·6 cm by 2·4 cm is enlarged in the ratio 8:3. Find the dimensions of the enlargement.

7 In what ratio must 24 be increased to become 32?

8 In what ratio must 100 be decreased to become 80?

9 In what ratio must £1 be increased to become £1·50?

[**10**] In what ratio must 2·5 kg be decreased to become 2 kg?

11 What multiplying factor alters 72 into 96?

[**12**] What multiplying factor alters 60 into 48?

[**13**] What multiplying factor increases £100 to £120?

14 What multiplying factor reduces 30p to 18p?

[**15**] What multiplying factor increases 80p by 20p?

16 If the price of petrol rises from $6\frac{1}{2}$p to $7\frac{1}{2}$p per litre, find the ratio in which the price increases.

17 Two sums of money are in the ratio 4:5; the smaller is 72p, what is the larger?

[**18**] Two distances are in the ratio 12:7; the larger is 21 km, what is the smaller?

19 What multiplying factor makes a number half as large again?

[**20**] What multiplying factor diminishes a number by $\frac{3}{10}$ of itself?

21 A man works 8 hours a day; in what ratio have his earnings changed if his pay is altered from 42p per hour to £4·20 a day?

22 In what ratio has the average speed changed if the time for a certain journey is reduced from 2 h to 1 h 40 min?

23 When the price of electricity is reduced from 3p to $2\frac{1}{2}$p per unit, I increase my annual consumption from 300 units to 400 units. In what ratio does my bill for electricity alter?

24 A photograph measuring 7·5 cm by 5 cm is enlarged so that the longer side becomes 18 cm; what does the shorter side become? In what ratio is the area increased?

Direct and Inverse Variation It is shorter to work with *multiplying factors* than to use the unitary method.

Example 8 If 4 kg of syrup cost 36p, find the cost of 7 kg.

If the number of kg is increased in the ratio 7:4, the cost is also increased in the ratio 7:4; but 4 kg cost 36p;

$$\therefore \text{ 7 kg cost 36p} \times \tfrac{7}{4}, \text{ that is } \mathbf{63p}.$$

Example 9 A man takes 18 min for a journey if he travels at 20 km per h; how long will he take if he travels at 24 km per h?

If the speed is *increased* in the ratio 24:20, the time is *decreased* in the ratio 20:24; but at 20 km per h the time is 18 min.

$$\therefore \text{ at 24 km per h, the time is 18 min} \times \tfrac{20}{24},$$

that is $\frac{18 \times 5}{6}$ min or **15 min.**

It is essential to consider whether the required multiplying factor is greater than 1 or less than 1, but the argument which has been written down in Examples 8, 9 can be abbreviated.

EXERCISE 16

[*Assume that the rates in this exercise are uniform*]

Use **multiplying factors** *for the following examples:*

1 14 kg of salt cost 48p; find the cost of 21 kg.

[**2**] 20 tablets of soap cost 44p; find the cost of 25 tablets.

3 50 kg of ground rice cost £7·20; find the cost of 35 kg.

4 12 men weed a field in 10 days; how long will 15 men take?

[**5**] A journey takes 6 h if I travel at 30 km an hour; how long will it take at 45 km an hour?

6 Five equal pipes fill a swimming-bath in 40 min; how long does it take if only 4 of the pipes are used?

7 $1\frac{1}{2}$ tonnes of coal cost £12·60; find the cost of $2\frac{1}{2}$ tonnes.

8 Railway fares were increased from $1\frac{1}{2}$p per km to $2\frac{1}{2}$p per km. What is the new fare for a journey which used to cost 42p?

[**9**] A train takes 50 min for a journey if it runs at 72 km per h. At what rate must it run to do the journey in 40 min?

10 A train takes t minutes for a journey if it runs at v km per h. How long will it take if it runs at x km per h?

11 A garrison has enough food for 24 days. How long will it last if each person's daily ration is reduced in the ratio 2:3?

[**12**] A man earns £16·50 for a working week of 44 hours. If he is absent for 8 hours, how much should he receive?

13 Explain why the travel graphs in Fig. 25 correspond to *constant* speeds. If the distance travelled in x min is y km, find y in terms of x for the graphs OP, OQ, OR, OS, OA and find the corresponding speeds in km per hour.

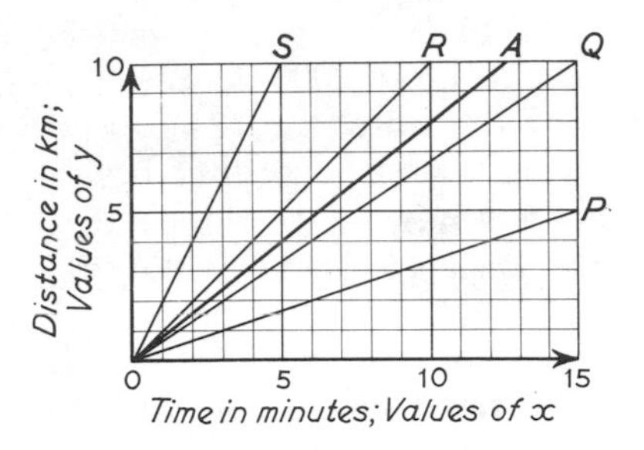

FIG. 25

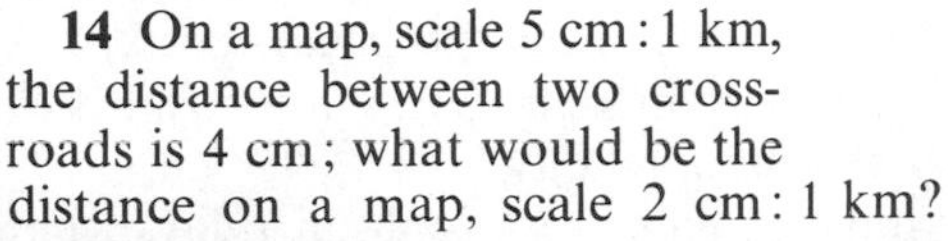

14 On a map, scale 5 cm : 1 km, the distance between two cross-roads is 4 cm; what would be the distance on a map, scale 2 cm : 1 km?

[**15**] At 39p per hour, the weekly wage of a number of workmen is £1027. What will be the increase in the wages bill if the rate of pay is raised to 42p per hour?

16 The lengths of the rims of 2 wheels of a carriage are 324 cm, 252 cm. In a certain journey, the larger makes 315 revolutions; how many revolutions does the smaller make?

17 A cog-wheel which has 24 cogs fits into another which has 45 cogs. If the former turns 5 times in 4 seconds, how many times does the latter turn in 12 seconds?

Compound units It is often convenient to invent special units for particular problems. For example, the amount of work required to regulate the traffic in a town each day might be described as 120 policeman-hours; this would mean that if 20 policemen were used, each would be on traffic duty for 6 hours a day; if 24 were used, each would be on traffic duty for 5 hours a day, and so on.

Example 10 If 12 men earn £132 in 10 days, how much will 14 men earn in 8 days, if the daily wage is the same for each man?

Call the wages of 1 man for 1 day the cost of 1 man-day.

(12×10) man-days cost £132,

$\therefore$ (14×8) man-days cost $£(132 \times \frac{14 \times 8}{12 \times 10}) = £\frac{1232}{10}$;

$\therefore$ 14 men in 8 days earn **£123·20.**

EXERCISE 17

[*Assume that the rates given in this exercise are uniform*]

1 If 12 boys earn £9 in 6 days, how much will 8 boys earn in 9 days?

[**2**] If 6 men dig a trench 60 m long in 6 days, what length of trench would 10 men dig in 9 days?

3 The gas for 10 gas-fires for 18 days cost £6. What is the cost of gas for 12 gas-fires for 5 days?

[**4**] 16 men can make 800 boxes in 9 days; how long will 15 men take to make 1000 boxes?

5 Two fires use 12 kg of coal in 4 hours. How much coal is used by three similar fires in 5 hours?

6 A mowing machine if driven at 8 km per h will cut a certain field in 6 hours; how long will it take to cut a field half as big again at 6 km per h?

7 The cost of carriage of 36 tonnes for 28 km is £15. Find the cost of carriage of 42 tonnes for 54 km at a fixed rate per tonne per km.

[**8**] 25 men working 8 hours a day remake a road in 63 days; how long would 45 men take, working 7 hours a day?

9 How many hours a day must 45 men work to do in 7 days what 19 men working $7\frac{1}{2}$ hours a day can do in 21 days?

[**10**] A man pays £5 for keeping 10 horses in a field for 8 weeks; how much will he have to pay for keeping 12 horses in the field for 6 weeks?

11 150 men make 2750 articles in a 44-hour week. How many men are needed to make 3000 in a 40-hour week?

[**12**] A moneylender charges £3 for lending £36 for 8 months; what will he charge for lending £40 for 9 months?

13 If b men earn £c in d days, how much will x men earn in y days, at the same rate?

[**14**] If b men can mow c square metres in d days, how long will x men take to mow y square metres, at the same rate?

15 For how many days can 17 men be maintained when the index price-figure is 156 by the money needed to maintain 12 men for 39 days when the index price-figure is 136? [Assume cost of living is proportional to the index figure.]

Taxation The cost of the various national services, such as defence, the upkeep of the Army, Navy and Air Force, general administration carried out by the Civil Service, national debt interest, etc., is met by taxation, levied in accordance with the regulations contained in the Budget which is laid before the House of Commons each April by the Chancellor of the Exchequer.

Revenue is collected in many different forms, such as duties on commodities (customs and excise), special taxes (motor tax, etc.), stamps on legal documents, death duties, etc., but the largest source is income tax.

Income Tax Detailed rules for calculating the amount of the tax levied on a person's income, called income tax, are complicated and change from year to year.

From a person's total income, various deductions are made which depend on individual circumstances, such as for a married man with children. After all permitted deductions have been made, the remainder is called the *taxable income.* This is not taxed at a uniform rate. The statement that the first £260 of taxable income is taxed at 30p in the £ means that the tax on this £260 is £$\frac{30 \times 260}{100}$, or £78. Tax is charged at the full rate on the rest of the taxable income. On small incomes, there is no tax or only a small tax; but the larger the income, the greater is the fraction of the income paid in tax.

Example 11 A man with 2 children has an earned income of £1710 and an unearned income of £95. His taxable income is calculated by making the following deductions from his total income:

(i) a personal allowance of £375,
(ii) $\frac{2}{9}$ of his earned income,
(iii) £140 for the first child and £115 for the second.

On the taxable income, he pays tax at 30p in the £ on the first £260 and 41p in the £ on the rest.

How much does he pay in income tax?

Allowances:	Personal allowance	£375
	$\frac{2}{9}$ of earned income ($\frac{2}{9}$ of £1710)	£380
	children's allowance (£140+£115)	£255
	Total	£1010

$\therefore$ taxable income = £(1710+95−1010) = £795.

Tax on £260 at 30p in the £	£78
Tax on £535 at 41p in the £	£219·35
Tax	**£297·35**

Local Government Services Those who live in towns enjoy certain advantages such as street-lighting, efficient drainage and sanitation, public parks, public libraries, etc. The cost of these municipal services and other charges is met by the **Rates,** that is a tax levied on the householders in the town; there is a similar tax in rural districts.

The principle on which this tax is levied is that a man who lives in a house of 'annual value' £60 a year ought to pay twice as much as a man whose house has an annual value of only £30 a year. The annual value is not necessarily the same as the annual rent, but is fixed by periodical assessments under Act of Parliament in which allowance is made for cost of maintenance and other charges and is called the *rateable value* or *assessed value.* The tax varies with the locality, from as little as 50p in the £ up to as much as 150p in the £, or even more, for every £1 in the rateable value. For example, if a house, whose rent may be £60 a year, is assessed at £50 a year and if a rate of 70p in the £ is levied, the tax for that year would be (70 × 50)p, or £35.

Example 12 A council levies a rate of 95p in the £ to meet the year's expenditure. A householder pays half-yearly £20·90 in rates. At what amount is his house assessed?

The householder pays for the year £20·90 $\times$ 2 = £41·80.

The ratio, assessed value : rates paid = 100 : 95 = 20 : 19.

$\therefore$ assessed value = £41·80 $\times \frac{20}{19}$ = £$\frac{836}{19}$

= **£44.**

EXERCISE 18

1 Find the tax on an income of £1430 if no tax is paid on the first £693, and tax is paid at 30p in the £ on the rest.

2 Find the tax on an income of £900 if no tax is paid on the first £575 and if tax is paid at 30p in the £ on the next £260 and at 41p in the £ on the rest.

3 Find the tax on an income of £480 if no tax is paid on the first £250 and if tax is paid at 15p in the £ on the next £150 and at 30p in the £ on the rest.

[**4**] How much is obtained from a rate of 8p in the £ on a rateable value of £156 000?

5 The rateable value of a town is £265 800. Find the money obtained from a rate of 81p in the £.

6 The rates on a house whose rateable value is £42 are £44·94. At what rate in the £ are the rates levied?

[**7**] The rateable value of a town is £384 000. What rate to the nearest penny in the £ must be levied to meet expenditure estimated at £352 000?

8 When a man's house is assessed for rates at £56, he has to pay rates at 120p in the £. When the assessment of his house is raised to £64, he pays rates at 115p in the £. Find the change in the amount paid in rates.

[**9**] A man's earned income is £2754. The following allowances are free of tax, (i) $\frac{2}{9}$ of his earned income, (ii) personal allowance £375, (iii) £115 for each of his 3 children. The remainder is taxed at 30p in the £ on the first £260 and at 40p in the £ on the rest. Calculate the total tax on his income.

10 The earned incomes of a man and his wife were £560, £380. The following allowances were free of tax, (i) $\frac{1}{5}$ of total earned income, (ii) £180 of the man's income, (iii) £110 of his wife's income. The remaining joint income was taxed at 12p in the £ on the first £50, at 25p in the £ on the next £200, at 45p in the £ on the rest. Calculate the tax paid and the net income after payment of tax.

11 The rent of a house is £65 per annum and the rateable value is £32. The annual rates are 86p in the £. The water rate is £1·42 half-yearly and the ground rent is £3·20 per annum. Find the total annual cost of these rates and rents. How much is this per week? [1 year = 52 weeks.]

[**12**] In a certain borough a rate of 1p in the £ brings in £6350. If 32p in the £ is the rate allocated to Health services, calculate (i) the cost of these services, (ii) the amount a householder pays towards them if the rateable value of his house is £44. It is estimated that an extra £53 500 is required to meet the next year's expenditure. Calculate the extra charge in pence per £, correct to 2 figures, which must be added to the rates.

13 A rate of 91p in the £ was levied to meet a year's expenditure of £154 000. Of this amount, £7200 was spent on the public library. What is the rate in pence per £, to the nearest tenth, levied for the library? How much does a householder whose house is rated at £35 contribute to the cost of the library? How much more than this would it cost to join a private library at 3p a week for 52 weeks?

14 The total rateable value of a town is £30 000. How much is yielded in a year from a rate of 80p in the £? If next year an extra £3750 is required, by how much must the rate in the £ be increased if the total rateable value is unchanged?

If it is decided to leave the rate at 80p in the £ but to get the extra £3750 by increasing the rateable value of all property in the same ratio, find the new rateable value of a house, formerly rated at £32.

CHAPTER 4

PROBLEMS AND FORMULAS

Problems

Example 1 I walk at $4\frac{1}{2}$ km per hour from my house to a town by a path through the fields. After waiting 40 minutes, I return in a bus travelling at $13\frac{1}{2}$ km per hour. If the road adds another kilometre to the journey and if the total time taken is 4 hours, find the distance by road from my house to the town.

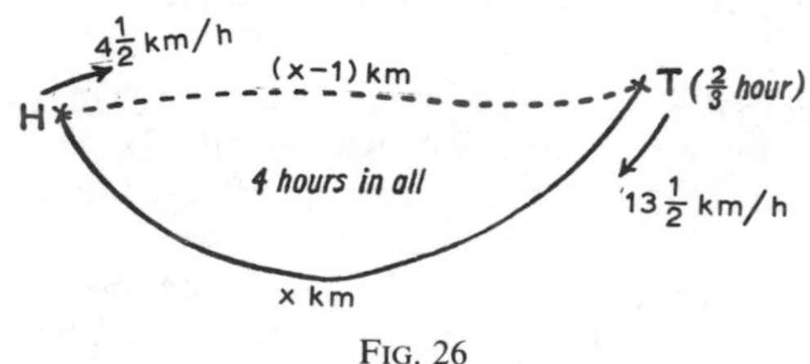

FIG. 26

Let the distance *by road* from my house to the town be x km; then the distance by the path is $(x-1)$ km.

I walk $(x-1)$ km at $4\frac{1}{2}$ km per hour, wait $\frac{2}{3}$ hour, and ride back x km at $13\frac{1}{2}$ km per hour; the total time is 4 hours.

To walk $(x-1)$ km at $4\frac{1}{2}$ km/h takes $\dfrac{x-1}{4\frac{1}{2}}$ hours;

to ride x km at $13\frac{1}{2}$ km/h takes $\dfrac{x}{13\frac{1}{2}}$ hours;

$\therefore$ the total time taken is $\dfrac{x-1}{4\frac{1}{2}}$ hours $+\dfrac{2}{3}$ hour $+\dfrac{x}{13\frac{1}{2}}$ hours;

but this is 4 hours, $\quad \therefore \dfrac{x-1}{4\frac{1}{2}}$ hours $+\dfrac{2}{3}$ hour $+\dfrac{x}{13\frac{1}{2}}$ hours $=4$ hours;

but $$\frac{x-1}{4\frac{1}{2}} = \frac{2(x-1)}{2\times 4\frac{1}{2}} = \frac{2(x-1)}{9} \quad \text{and} \quad \frac{x}{13\frac{1}{2}} = \frac{2x}{2\times 13\frac{1}{2}} = \frac{2x}{27}.$$

$$\therefore \frac{2(x-1)}{9}+\frac{2}{3}+\frac{2x}{27} = 4;$$

multiply each side by 27, $\quad \therefore 6(x-1)+18+2x = 108;$

$$\therefore 6x-6+18+2x = 108, \quad \therefore 8x = 96; \quad \therefore x = \frac{96}{8} = 12.$$

$\therefore$ the distance *by road* from the house to the town is **12 km.**

When *solving* a problem,

(i) If possible, make a rough diagram and show the data on it.

(ii) Choose a letter for some unknown number the problem involves, and state precisely what this letter represents.

(iii) Rewrite the question, using the letter you have chosen for the unknown to make the statement of the problem more detailed.

When *checking* the answer to a problem,

Use the actual data of the problem. It is not sufficient to check by substituting in the equation, because your equation may be wrong.

EXERCISE 19

1 Find a number such that if you add 7 and divide the sum by 5 you will get the same answer as if you had subtracted 1 and then divided by 3.

2 Fig. 27 gives the lengths of three sides of a rectangle in centimetres. Find the value of x and the perimeter of the rectangle.

FIG. 27 FIG. 28

[3] Fig. 28 gives the lengths of the sides of a triangle in centimetres. If the triangle is isosceles, find the value of x and the perimeter of the triangle.

4 At a fair, a boy receives 8p for a hit and pays 3p for a miss. After 24 shots he has to pay 6p; how many hits did he score?

[5] A boy counts 2 marks for each sum he gets right and loses 1 mark for each he gets wrong. He does 18 sums and obtains 15 marks. How many sums did he get right?

6 In Fig. 28, if the perimeter of the triangle is 24·5 cm, find x.

7 If x 10p coins and $(2x-1)$ 5p coins together make up £1·95, find the value of x.

[8] If $(y-1)$ coins of 50p each and $(1\frac{1}{2}y+3)$ coins of 10p each together make £5, find the value of y.

9 Find a value of n for which the fraction $\dfrac{5n+2}{7n+1}$ reduces to $\dfrac{3}{4}$.

[10] What is the number which when added both to the numerator and denominator of the fraction $\frac{6}{13}$ makes a new fraction whose value is $\frac{3}{5}$?

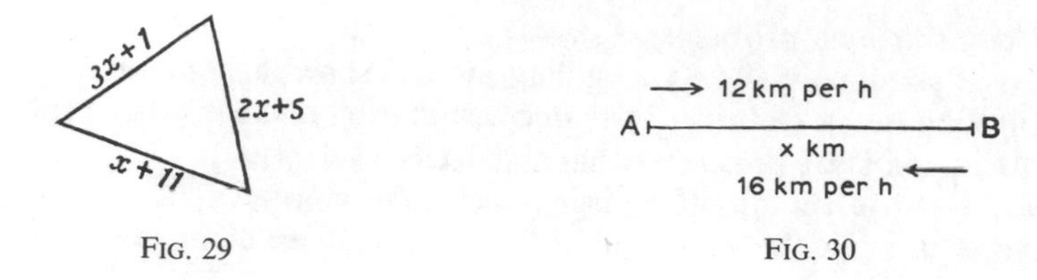

Fig. 29 Fig. 30

[**11**] A man rows upstream at 5 km per h and back to the same place at 6 km per h; he takes 33 minutes altogether. How far upstream did he go?

12 Fig. 29 gives the lengths of the sides of a triangle in centimetres. Find the value of x if the triangle is isosceles, and find the perimeter of the triangle. *There is more than one answer.*

13 Fig. 30 represents a cyclist riding from A to B and back again; the double journey takes $3\frac{1}{2}$ hours. Find x.

14 In a factory, the men each get £3 a day and the women each £2·50 a day; 200 people are employed, and their wages amount to £552 a day. How many men are there?

[**15**] A man buys one lot of eggs at 20p a dozen, and a second lot, which is 3 dozen more than the first lot, at 30p a dozen. He sells then all at 35p a dozen and makes £1·95 profit. How many eggs of each quality did he buy?

16 Fig. 31 represents a cyclist leaving A at the same moment as a pedestrian leaves B. If they meet x km from A, find x.

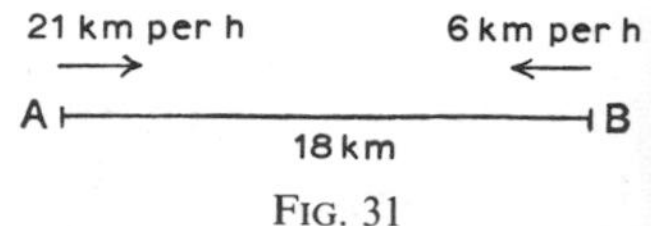

Fig. 31

17 The total time for riding $2\frac{1}{2}n$ km at 20 km per h and $5n$ km at 50 km per h is $4\frac{1}{2}$ hours. Find n.

[**18**] The total time for walking n km at $7\frac{1}{2}$ km per h and riding $(n+4)$ km at $18\frac{3}{4}$ km per h is 1 h 20 min. Find n.

[**19**] A man buys 800 bulbs for £5·04, some of them at 48p for a hundred and the rest at 72p a hundred. How many of the first kind did he buy?

[**20**] A river flows at 6 km an hour. What is the speed through the water of a steamer that can go downstream twice as fast as upstream?

21 A first-class ticket costs *two-thirds as much again* as a second-class ticket for a certain journey. If the total cost of 8 first-class and 20 second-class tickets is £15, find the cost of a second-class ticket.

22 A goods train travelling at 40 km per h passes a level-crossing at 2 p.m., and an express train, travelling at 96 km per h in the same direction, passes the crossing at 2.14 p.m. At what time will the first train be overtaken by the second?

[**23**] A messenger goes on an errand at 6 km per h; 20 minutes later a boy bicycles after him at 18 km per h. How far must the boy go to overtake the messenger?

24 My train starts in 12 minutes and the station is 2 km away. I walk at 6 km per h and run at 12 km per h. How far must I run to reach the station in 12 minutes?

[**25**] A rectangle is l metres long and its perimeter is 5 metres; find its breadth in terms of l. Another rectangle is twice as long and half as broad as the first and its perimeter is 7 metres. Find l.

26 If $C°$ Celsius is the same temperature as $F°$ Fahrenheit, $C = \frac{5}{9}(F-32)$. Find x if $x°$ C is the same temperature as $2x°$ F.

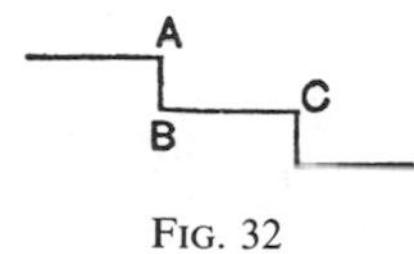

FIG. 32

[**27**] The 'rise' AB, R cm, and the 'tread' BC, T cm, of a staircase (see Fig. 32) are connected by the formula, $R = \frac{1}{2}(60-T)$. Find the rise if it equals $\frac{3}{4}$ of the tread.

28 A kettle of water is placed on a stove. After t minutes the temperature of the water is $(16+5t)$ degrees Celsius. Use this formula to *write down* the temperature after 10 minutes and to find how long the water takes to boil. [Water boils at 100° Celsius.]

***29** 5 kg of tea at a certain price is mixed with 4 kg of tea costing 27p per kg *more*. The value of the mixture is 93p per kg. Find the price per kg of the cheaper kind.

***30** How much tea at 90p per kg must be mixed with 12 kg of tea at 66p per kg so that the mixture may be worth 81p per kg?

***31** If I walk to the station at 6 km per h to catch a train, I shall have 3 minutes to spare; but if I walk at 4 km per h, I shall miss the train by 5 minutes. How far off is the station?

***32** Find v if v km per hour is the same speed as $(v-8\frac{2}{3})$ metres per second.

Use of Formulas

Example 2 If a man is photographed when standing u centimetres from the lens of a camera, focal length f centimetres, the plate should be v cm from the lens, where $\frac{1}{u}+\frac{1}{v}=\frac{1}{f}$. Find the distance of the plate from a lens of focal length 4·5 cm, if the man is 72 cm from the lens.

Here $f = 4{\cdot}5$ and $u = 72$.

$$\therefore \frac{1}{72}+\frac{1}{v}=\frac{1}{4\frac{1}{2}}=\frac{2}{9}.$$

Multiply each side by $72v$,

$$\therefore\ v+72 = 16v;\quad \therefore\ 15v = 72;$$

$$\therefore\ v = \frac{72}{15} = 4{\cdot}8.$$

$\therefore$ the distance of the plate from the lens is **4·8 cm.**

EXERCISE 20

1 The area of a triangle is given by the formula, $A = \frac{1}{2}hb$; find b if $A = 6\frac{3}{4}$ and $h = 1\frac{4}{5}$.

[**2**] The area of a trapezium is given by the formula, $A = \frac{1}{2}h(a+b)$; find b if $A = 5{\cdot}4$, $a = 1{\cdot}7$, $h = 3{\cdot}6$.

3 If $F°$ Fahrenheit is the same temperature as $C°$ Celsius, then $F = 32+\frac{9C}{5}$; find C if $F = 68$.

4 If the velocity of a body sliding downhill increases steadily from u metres per second to v metres per second in t seconds, the distance, s metres, it moves in that time is given by $s = \frac{1}{2}(u+v)t$. Find v if $s = 84$, $u = 9$, $t = 3\frac{1}{2}$.

[**5**] From the formula, $\frac{pv}{273+t} = c$, find t if $p = 16$, $v = 180$, $c = 9{\cdot}6$.

6 From the formula, $d = t\cdot\frac{\mu-1}{\mu}$, find μ if $d = 0{\cdot}25$, $t = 0{\cdot}8$.

7 A machine-test showed that a load of W tonnes could be raised by an effort of P tonnes, where $P = 0{\cdot}075\,W+0{\cdot}05$. What load could be raised by an effort of 2 tonnes?

8 From the formulas, $u = (n+1)f$ and $v = \left(1+\frac{1}{n}\right)f$, find u if $n = 3$, $v = 8$.

Transformation of Formulas

Most formulas are given or remembered in some standard form; but for the purposes of a particular problem it may be convenient to express the formula differently.

For example, the volume, V cm^3, of a rectangular block l cm long, b cm broad, h cm high, is given by the formula

$$\boldsymbol{V = lbh.}$$

Here V is called the **subject** of the formula.

Suppose, however, we are given the volume of the block and also its length and breadth and are asked to find the height. With the same notation as before, $lbh = V$.

Divide each side by lb, $\quad\therefore\ \boldsymbol{h = \frac{V}{lb}}.$

This is a formula whose subject is ***h***. It has been obtained from the first formula *by using precisely the same methods as are employed in solving equations.* The process is called **changing the subject of the formula.**

If V cm^3 is the volume of a rectangular block, h cm high, whose base is a square, side l cm, then $\boldsymbol{V = hl^2}$.

To make l the subject of this formula, proceed as follows:

$$hl^2 = V, \quad \therefore l^2 = \frac{V}{h};$$

take the square root of each side,

$$\therefore \boldsymbol{l = \sqrt{\left(\frac{V}{h}\right)}}.$$

A number has *two* square roots, but here we must take the positive square root because l cm is a length.

It will be shown in Chapter 20 that the length of the circumference, C cm, of a circle is *approximately* 3·14 times the length of the diameter, d cm. The *exact* number of which 3·14 is an approximation is always denoted by the Greek letter π; to 4 places of decimals, it is 3·1416. Hence the formula for C in terms of d is $\boldsymbol{C = \pi d}$.

Example 3 The volume V cm^3 of a sphere, radius r cm, is given by the formula, $V = \frac{4}{3}\pi r^3$. Make r the subject.

$$\frac{4}{3}\pi r^3 = V, \qquad \therefore 4\pi r^3 = 3V, \qquad \therefore r^3 = \frac{3V}{4\pi};$$

take the cube root of each side,

$$\therefore \boldsymbol{r = \sqrt[3]{\left(\frac{3V}{4\pi}\right)}}.$$

EXERCISE 21

1 If $x°$, $y°$, $z°$ are the angles of a triangle, $x+y+z = 180$. Make z the subject of this formula.

2 If V cm^3 of gas in a cylinder is under a pressure of P grammes weight, then $PV = 500$. Make P the subject of this formula.

[**3**] If the sum of the angles of an n-sided polygon is r right angles, then $r = 2n-4$. Make n the subject.

4 The rise R cm and the tread T cm of a staircase are connected by the formula $R = \frac{1}{2}(60-T)$. Make T the subject.

5 With the data of Fig. 33, $A = lb$. Make b the subject.

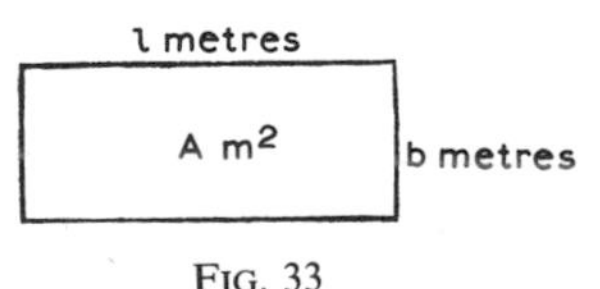

FIG. 33

6 If the perimeter of the rectangle in Fig. 33 is p metres, express p in terms of l and b, and then make l the subject of the formula.

[**7**] What relation connects x and y in Fig. 34? Make x the subject of the formula.

[**8**] If the perimeter of the triangle in Fig. 34 is p cm, express p in terms of a and b, and then make a the subject of the formula.

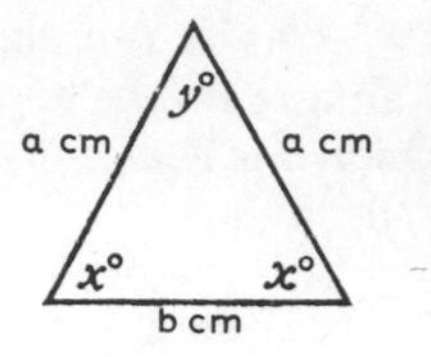

FIG. 34

9 What relation connects a and b in Fig. 35? (i) Make a the subject; (ii) make b the subject.

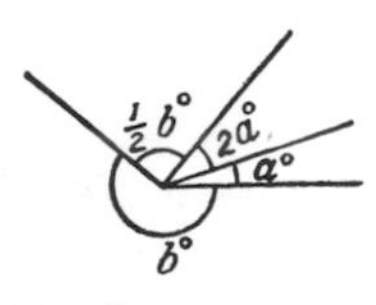

FIG. 35

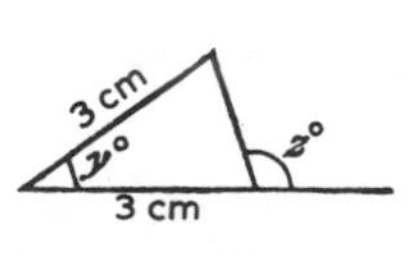

FIG. 36

[**10**] What relation connects y and z in Fig. 36? (i) Make y the subject; (ii) make z the subject.

[**11**] If $C°$ Celsius is the same temperature as $F°$ Fahrenheit, $C = \frac{5}{9}(F-32)$. Make F the subject.

12 If the circumference of a circle of radius r cm is C cm, then $C = 2\pi r$. Make r the subject.

[**13**] The area A cm^2 of a triangle whose base is b cm and height is h cm is given by the formula $A = \frac{1}{2}bh$. Make h the subject.

14 Make t the subject of the formula $s = (u+v)t$.

[**15**] Make P the subject of the formula $A = P(1+R)$.

16 (i) Find a value of x if $x^2 = 9$.

(ii) Make l the subject of the formulas (a) $A = l^2$; (b) $V = l^3$.

17 If the area of a circle of radius r cm is A cm^2, then $A = \pi r^2$. Make r the subject.

18 Make r the subject of the formula $V = \pi r^2 h$.

19 Make d the subject of the horse-power formula $H = \frac{2}{5}nd^2$.

20 Make d the subject of the formula $V = \frac{1}{6}\pi d^3$.

21 (i) Find N if $\sqrt{N} = 5$; (ii) find A if $\sqrt{A} = d$.

***22** Interpret the formulas $A = lb$, $p = 2(l+b)$; and express p in terms of A, b only, and A in terms of l, p only.

CHAPTER 5

PERCENTAGE AND RATIO

Percentage Suppose that in a consignment of eggs, 1 in every 20 is bad, then it follows that 5 in every 100 are bad, and we say that 5 *per cent* of the eggs in the consignment are bad. The ratio of the number of bad eggs to the total number is 1:20 or 5:100; a percentage is simply a ratio in which the second number is 100, and may be represented by a fraction whose denominator is 100.

For example, 7 per cent, *written* 7%, may denote the ratio 7:100 or may be represented by the fraction $\frac{7}{100}$.

It should be noted that in some cases it is customary to take the second number of the ratio as 1000; birth-rates and death-rates are usually quoted in this form. For example, in the United Kingdom in 1953 the birth-rate was 15·9 per *thousand* of the population and the death-rate was 11·5 per *thousand* of the population. These rates are equivalent to the ratios 15·9:1000 and 11·5:1000 and could be expressed as 1·59% and 1·15% respectively.

Example 1 Express in percentage form the statement:
At a certain school, 3 boys in every 4 own bicycles.

$\frac{3}{4}$ of the total number of boys own bicycles;
$\therefore$ in every 100 boys, ($\frac{3}{4}$ of 100) boys own bicycles;
$\therefore$ in every 100 boys, 75 boys own bicycles.

The statement may therefore be written:

At a certain school, **75**% of the boys own bicycles.

Example 2 If 30% of the pupils in a school are boys, what percentage of the pupils are girls?

In every 100 pupils, there are 30 boys and therefore there are (100 – 30) girls, that is 70 girls;

$\therefore$ **70**% of the pupils are girls.

Or we may say,

the ratio of the number of boys to the number of pupils is 30:100,
$\therefore$ the ratio of the number of girls to the number of pupils is 70:100,

$\therefore$ **70**% of the pupils are girls.

EXERCISE 22 (Oral)

Express in percentage form the statements in Nos. 1–10:

1 1 day in every 5 is wet.

2 A boy obtains 7 marks out of 10.

3 3 men in every 10 men possess motor vehicles.

4 3 pupils in every 50 pupils are absent from school.

5 7 people in every 25 die before the age of 50.

6 A tax is at the rate of 5p in every £.

7 A boy gets 1 sum in every 3 sums wrong.

8 1 orange in every 8 oranges in a box is bad.

9 The interest is £1 on every £40.

10 In 1953, the birth-rate in Scotland was 17·8 per thousand of the population.

Write down the answers to the following:

11 If 5% of the pupils in a school are absent, what percentage are present?

12 If a man spends 70% of his income, what percentage does he save?

13 In a railway accident 85% of the passengers were unhurt; what percentage were injured?

14 If 16% of those who enter for a competition get prizes, what percentage get nothing? How many per thousand get nothing?

15 If 28% of the population are men and 30% are women, what percentage are children? How many per thousand are children?

16 A man spends 10% of his income on rent and 55% on household expenses; what percentage remains?

17 A boy eats 35% of a cake and gives away 25% of it; what percentage of the cake is left?

18 On a boat 24% of the passengers travel 1st class, 32% travel cabin class, and the rest tourist class. What percentage travel tourist class?

Percentages and Fractions Percentages can be represented by fractions. For example, 40% denotes the ratio 40 : 100, and is therefore represented by $\frac{40}{100}$, that is $\frac{2}{5}$ or 0·4.

Conversely, any fraction or decimal can be expressed as a percentage by transforming it so that its denominator is 100. For example, $\frac{3}{4} = \frac{75}{100}$; $\therefore$ $\frac{3}{4}$ is equivalent to 75%.

The following table illustrates different ways used for representing fractions in everyday life:

Fraction	Decimal	Per cent	Per thousand
$\frac{3}{5}$	0·6	60	600

Example 3 Express (i) $\frac{2}{3}$, (ii) 0·225 as percentages.

$$\text{(i)}\quad \frac{2}{3} = \frac{\frac{2}{3} \times 100}{100}$$

$$= \frac{66\frac{2}{3}}{100}$$

$$= \mathbf{66\tfrac{2}{3}\%}$$

(ii) $0{\cdot}225 = 0{\cdot}225 \times 100$ per cent $= \mathbf{22{\cdot}5\%}$.

Example 4 Find the value of $62\frac{1}{2}\%$ of £4.

$$62\tfrac{1}{2}\% \text{ of } £4 = \frac{62\frac{1}{2}}{100} \text{ of } £4 = £(4 \times \tfrac{125}{200})$$

$$= £\tfrac{5}{2} = \mathbf{£2{\cdot}50.}$$

Notice that, since £1 = 100p, it follows that 1% of £1 = 1p, 2% of £1 = 2p, and so on.

$\therefore$ 1% of £A is Ap, 2% of £A is $2A$p, etc.

$62\frac{1}{2}\%$ of £4 $= (62\frac{1}{2} \times 4)$p $= 250$p $= £2{\cdot}50$,

as before.

Example 5 A boy obtains 52 marks out of 80 for a paper. What percentage is this?

The boy obtains $\frac{52}{80}$ of the total, that is $(\frac{52}{80} \times 100)$ per cent;

$\therefore$ he obtains **65%** of the total.

Example 6 Express £2·55 as a percentage of £30.

$$\frac{£2{\cdot}55}{£30} = \frac{2{\cdot}55}{30} = \frac{255}{3000} = \frac{51}{600} = \frac{17}{200};$$

but $\frac{17}{200}$ is equivalent to $\frac{17}{200} \times 100$ per cent $= 8\frac{1}{2}\%$.

$\therefore$ £2·55 is $\mathbf{8\frac{1}{2}\%}$ of £30.

Example 7 Find correct to the nearest penny the value of $6\frac{3}{4}\%$ of £7·84.

$$\text{The required value} = \frac{6\frac{3}{4}}{100} \text{ of } £7{\cdot}84 = £(0{\cdot}0784 \times 6\tfrac{3}{4}).$$

The value can now be found by ordinary multiplication, or by a practice method as follows:

Since the answer is required to the nearest penny, we keep 3 places of decimals in the working.

	£
	0·078
6 times,	0·468
$\frac{1}{2}$ times,	0·039
$\frac{1}{4}$ times,	0·019
$6\frac{3}{4}$ times,	0·526

$\therefore$ value = £0·53, to 2 places,

= **53p**, to nearest penny.

Insurance The chance of a man's house being burnt down is small, but if it should happen, the consequences are very serious. The object of insurance is to enable a man to protect himself against small risks of heavy losses, by making comparatively small annual payments. It is the pooling or averaging out of a common risk.

For example, the owner of a house, by paying an insurance company 7p per each £100 at which his house is valued, can insure it against loss by fire. Thus, if the house is valued at £2000, the owner pays each year 7p × 20, that is £1·40, and in the event of a fire will be paid the sum of money at which the damage caused by the fire is assessed.

The sum paid each year to cover the risk is called the *premium*, the document issued by the company binding them to cover the risk is called the *policy*, and in the example given above we say that the insurance is at the rate of '7p per cent', an abbreviation for '7p for each £100 of the insured value'.

Some forms of insurance are now compulsory: the owner of a car, when applying for a licence, is required to prove that he is insured against third-party risks so that anyone who is injured by his negligent driving is sure of receiving compensation. *National Insurance* is compulsory for everyone, subject to certain statutory exceptions.

The other principal form of insurance is life insurance; the commonest practice is to make an annual payment, the *premium*, which secures either a lump sum at death or at a stated age or an annual payment, called a *life-annuity*, which starts at a selected age and continues as long as the insured person lives. A man who buys a house through a Building Society is often required to take out a life-insurance as a security for the loan.

Example 8 A merchant insures goods in his warehouse valued at £12 500 at the rate of 20p per cent. What annual premium does he pay?

In *insurance*, the phrase, 20p per cent, means 20p per £100.

The premium for a policy of £100 is £$\frac{1}{5}$;

∴ the premium for a policy of £12 500 is £($\frac{1}{5}$ × 125) = **£25.**

EXERCISE 23

[*Nos. 1–28 are suitable for class discussion*]

Express the following percentages as fractions and as decimals:

1 25%; 50%; 75% **[2]** 5%; 15%; 85%

[3] 2%; 4%; 300% **4** $2\frac{1}{2}$%; $112\frac{1}{2}$%; $37\frac{1}{2}$%

Express the following as percentages:

5 $\frac{1}{2}$, $\frac{1}{3}$, $\frac{1}{4}$, $\frac{1}{5}$, $\frac{7}{5}$ **[6]** $\frac{1}{20}$, $\frac{1}{25}$, $\frac{1}{30}$, $\frac{1}{40}$, $\frac{36}{25}$

7 0·35, 0·08, 1·4, 0·065 **[8]** $1\frac{1}{2}$, 1·75, 4, $1\frac{1}{3}$, 2·3

9 $\frac{3}{4}$, $\frac{4}{5}$, $\frac{5}{6}$, $\frac{7}{8}$ **[10]** $\frac{8}{15}$, $\frac{3}{16}$, $\frac{17}{20}$, $\frac{11}{25}$

Find the values of the following:

11 60% of 25p	**12** $33\frac{1}{3}$% of 24p	**13** 180% of £1·25
[**14**] 75% of 12p	[**15**] $12\frac{1}{2}$% of £40	[**16**] $166\frac{2}{3}$% of 3 m
17 34% of 2 m	[**18**] 55% of 1 kg	**19** 135% of 1 km

Express the first quantity in each pair as a percentage of the second:

20 2p; 8p	**21** 20p; 60p	**22** 20 g; 16 g
[**23**] 9p; 12p	[**24**] 4 m; 3 m	[**25**] 3 cm; 20 cm
26 48 cm; 1 m	[**27**] £1·50; £60	**28** £6; £4·50

29 A boy obtains 72% in an examination, full marks being 450. How many marks did he get?

[**30**] A salesman receives $7\frac{1}{2}$ per cent commission on his sales. What is his commission for selling a wireless set for £70?

31 A boy scores 45 marks out of 60. What percentage is this? How much per thousand is this?

[**32**] A man whose income is £1728 a year spends £108 a year on rent. What percentage of his income is spent on rent?

33 The components of gunpowder are as follows: nitre 75%, charcoal 15%, sulphur 10%. How many kg of each are there in 1 tonne of gunpowder?

34 x and y are two numbers. (i) What fraction is x of y? (ii) What percentage is x of y?

Find correct to the nearest penny the value of:

35 4% of £3·42	[**36**] 7% of £21·56
37 $2\frac{3}{4}$% of £79·29	[**38**] $3\frac{1}{2}$% of £16·48

39 What premium is charged to insure property worth £2870 against fire, if the premium is $\frac{1}{5}$ per cent of the value?

40 Find the premium for insuring warehouse goods, valued at £350, against fire, at the rate of 12p per cent.

[**41**] Find the premium for insuring the hull and machinery of a steamer, valued at £45 000, at the rate of £4·20 per cent.

[**42**] The window glass in a house can be insured at the rate of 75p per £100 of rent. Find the premium to the nearest penny if the rent is £65.

43 The premium for fire insurance for the contents of a house, at the rate of 8p per cent, is £3. Find the value of the contents.

44 A man takes out a life insurance policy for £1000 on his 35th birthday; the annual premium is £1·75 per cent. How much altogether has he paid when he is $60\frac{1}{2}$ years old?

Proportional Parts The statement that the weights of 3 boys, A, B, C are **proportional** to the numbers 9, 11, 12, means that

the ratio, A's weight : B's weight = 9 : 11,

and the ratio, B's weight : C's weight = 11 : 12;
also the ratio, A's weight : C's weight = 9 : 12 = 3 : 4.

This is written concisely in the form,

A's weight : B's weight : C's weight = **9 : 11 : 12.**

Example 9 A rod 1 m long is broken into 3 pieces whose lengths are in the ratios 5 : 2 : 3. Find the length of each piece.

If the rod is divided into (5 + 2 + 3) equal parts, the first piece contains 5 of these parts and is therefore $\frac{5}{10}$ of the whole rod.

∴ the length of the first piece = $\frac{5}{10}$ of 1 m = **50 cm.**

Similarly the length of the second piece = $\frac{2}{10}$ of 1 m = **20 cm**

and the length of the third piece = $\frac{3}{10}$ of 1 m = **30 cm.**

Check: 50 cm + 20 cm + 30 cm = 100 cm = 1 m.

Example 10 Find four whole numbers proportional to 90p, 63p, 135p, 108p.

90p : 63p : 135p : 108p = **10 : 7 : 15 : 12**

(dividing each number by 9).

Example 11 Divide £108 between A, B, C so that B has 5 times as much as C, and A has half as much again as B.

If C's share is taken as 1 unit, B's share is 5 units, and therefore A's share is (5 + $\frac{1}{2}$ of 5) units = $7\frac{1}{2}$ units.

∴ A's share : B's share : C's share = $7\frac{1}{2}$: 5 : 1 = 15 : 10 : 2.

∴ A's share = $\frac{15}{15+10+2}$ of £108 = $\frac{15}{27}$ of £108 = **£60;**

B's share = $\frac{10}{27}$ of £108 = **£40;** C's share = $\frac{2}{27}$ of £108 = **£8.**

Check: £60 + £40 + £8 = £108.

EXERCISE 24

[*Nos. 1–8 are intended for oral work or class discussion*]

1 Divide £60 in the ratio (i) 2 : 3; (ii) 5 : 7; (iii) 4 : 1.

[**2**] Divide 48 cm in the ratio (i) 3 : 5; (ii) 1 : 5; (iii) 9 : 7.

3 Divide £24 into three shares in the ratios 1 : 2 : 3.

[**4**] Divide 96p into three shares in the ratios 5 : 2 : 9.

5 Divide 10 kg into three parts in the ratios 7 : 4 : 9.

6 Divide 22 m into three parts in the ratios $\frac{1}{2} : \frac{1}{4} : \frac{1}{6}$.

[**7**] Divide £7 into four parts in the ratios 2 : 4 : 5 : 9.

8 Find as simply as possible three whole numbers proportional to: (i) £2, £1·50, £3; (ii) $2\frac{1}{2}$ kg, 4 kg, $1\frac{1}{2}$ kg.

9 Share 210p between 3 boys in the ratios 9 : 4 : 1.

[**10**] 20 tonnes of coal are shared between three families in the ratios 2 : 5 : 9. How much does each receive?

11 The sides of a triangle are in the ratios 1:1·5:2, and its perimeter is 36 cm. Find the length of each side.

[**12**] The profits, £750, of a business are divided between 3 men so that their shares are proportional to 3, 4, 8. Find each share.

13 A legacy of £450 is divided among 3 sons in the ratios $1\frac{1}{2}:2\frac{1}{4}:3$. How much does each receive?

14 A, B, C provide £250, £500, £750 respectively to buy a business; their shares of profits are proportional to the capital they provide. If the profits are £1470, what does each receive?

15 Find the three smallest whole numbers proportional to a, b, c, if (i) $a:b = 5:6$; $b:c = 9:4$; (ii) $a = 5b$, $b = 3c$.

[**16**] Divide 35p into two parts such that one is half as much again as the other.

17 Divide £66 between A, B, C so that A has twice as much as B, and B has half as much again as C.

[**18**] A load of 95 kg is distributed between A, B, C, so that A carries $\frac{1}{3}$ of what B carries, and B carries 4 times as much as C. How much does B carry?

19 A bankrupt owes B £2144, C £2130, D £771, E £315. His assets (what he possesses) are £1200. How much does B receive?

[**20**] Three ingredients costing 3p, 4p, 1p, per kg, are mixed so that their weights are proportional to 2, 3, 7 respectively. Find the cost of 60 kg of the mixture.

21 A puts £900 into a business for 1 year, and B puts into it £400 for 9 months. The profits shared between A and B are £180. How much should each receive? (Divide in the ratio $900 \times 12:400 \times 9$.)

[**22**] A man borrows £300 for 10 months from B, £500 for 9 months from C, and £700 for 3 months from D, and pays altogether £72 interest. How is this shared between B, C and D?

23 Three firms A, B, C undertake a piece of work for which £160 is paid. In carrying out the work, A provides 10 men for 15 days, B 8 men for 20 days, C 15 men for 6 days. If all the men are paid at the same rate, find the amount each firm receives.

***24** Profits amounting to £2415 are divided between A, B, C, so that for every £4 A gets, B gets £5, and for every £9 B gets, C gets £16. Find the shares.

***25** A takes 10 days to build a fence which B can build in 8 days. If £4·50 is paid for the job and if both work together, how much should A receive?

***26** One employer has 40 workmen and another 75 workmen, and their weekly wage-bills are in the ratio 3:5. Find the ratio of the average wages per man paid by the two employers.

Use of Ratio in Problems

Example 12 If $\frac{3}{7}$ of a number is 42, what is the number?

If the number is divided into 7 equal parts, 3 of these parts make up 42. Therefore the ratio of the number to 42 equals 7:3.

$$\therefore \text{ the number } = 42 \times \tfrac{7}{3} = \mathbf{98}.$$

Algebraically, if the number is x, $\frac{3}{7}x = 42$; $\therefore x = 42 \times \frac{7}{3}$.

Example 13 A boy, after spending $\frac{2}{9}$ of his money, has 56p left. How much had he at first?

After spending $\frac{2}{9}$ of his money, $\frac{7}{9}$ of his money remains;

$\therefore$ the ratio of what he had at first to what remains equals 9:7.

$$\therefore \text{ he had at first } 56\text{p} \times \tfrac{9}{7} = 8\text{p} \times 9 = \mathbf{72p}.$$

Example 14 A picture-dealer deducts 10p in the £ for cash from the price on the ticket. If the cash price of a picture is £2·25, find the price on the ticket.

For each 100p marked on the ticket, the cash price is 90p.

$\therefore$ the ratio of the marked price to the cash price = 100:90 = 10:9.

$$\therefore \text{ the marked price } = £2{\cdot}25 \times \tfrac{10}{9} = £\tfrac{22{\cdot}5}{9} = \mathbf{£2{\cdot}50}.$$

EXERCISE 25

1 If $\frac{2}{5}$ of a number is 30, find the number.

2 When a man spends $\frac{5}{9}$ of his income, he saves £720 a year; find his income.

[**3**] When $\frac{5}{12}$ of a trench has been dug, 42 m still remain to be done; find the length of the trench.

4 If 20p in the £ is deducted for cash from the marked price of a chair, the cash price is £2·80; find the marked price.

[**5**] After paying tax at the rate of 40p in the £, a man has £604·80 left; what had he at first?

6 To each 10p a boy saves, his father adds 5p; how much has the boy saved when he possesses £2·34?

7 25p in the £ is added to all the prices in a shop. Find how much has been added in the case of an article whose new price is £1·05.

8 A shopkeeper makes a profit of 20p in the £ on the cost price of his goods. What does he pay for an article which he sells for £1·50?

[**9**] By selling my car for £588, I lose at the rate of 30p in the £ on the price I paid for it. What did I pay for it?

10 A retailer makes a profit of 40p in the £ on the cost price of his goods. What is his profit on an article he sells for £5·60?

11 In Réaumur's thermometer, freezing-point and boiling-point of water are marked 0, 80 respectively; in Fahrenheit's thermometer they are marked 32, 212. What is the Fahrenheit reading for 12 Réaumur?

12 A watch was 2 min slow at 8 a.m., and 1 min fast at 6 p.m. on the same day. Find when it was right if it gained time uniformly.

13 Coal at £9·60 a tonne is mixed with coke at £7·20 a tonne in the ratio 5:3 by weight. Find how much money is saved by using 10 tonnes of the mixture instead of 11 tonnes of coal.

14 After paying tax at an average rate of 40p in the £, a man's income was £1476. Find his income before tax was paid.

[**15**] In a sale 20p in the £ is deducted from the marked price. The sale price of a chair is £3·76, find the marked price.

16 A machine is constructed so that the ratio of the effort to the load raised is 2:7. Find in kg the effort required to raise a load of 140 kg.

17 P, Q are points on AB such that AP:PB = 2:7 and AQ:QB = 5:4; AB = 6 cm. Find PQ.

[**18**] X, Y are points on AB produced and BA produced such that AX:BX = 11:3 and AY:BY = 3:5; AB = 4 cm. Find XY.

[**19**] A and B invest £252 and £208 at the same date in Savings Certificates. What is the value of B's holding when A's holding is worth £315?

20 If a railway climbs h metres for a length s metres of track, the ratio $h:s$ is called the *average gradient* of the track. If the average gradient between 2 stations, 6 km apart, is 1:288, *called* 1 in 288, find in metres to the nearest metre how much one station is higher than the other above sea-level.

Numbers in a given Ratio If the ratio of two numbers, x and y, is given, either number can be expressed in terms of the other.

For example, if $x:y = 4:9$, then

$$\frac{x}{y} = \frac{4}{9}, \qquad \therefore 9x = 4y;$$

$$\therefore x = \frac{4y}{9} \quad \text{and} \quad y = \frac{9x}{4}.$$

Further, we may write $\frac{x}{4} = \frac{y}{9}, = k$, say;

$$\therefore x = 4k \quad \text{and} \quad y = 9k.$$

This shows that any two numbers in the ratio 4:9 can be expressed in the form, $4k$ and $9k$.

In general, two numbers in the given ratio $a:b$ can be expressed in the form, ak and bk, and it is often convenient to use this fact.

Example 15 If $x:y = 3:4$, find the value of $\dfrac{4x^2-y^2}{x^2+y^2}$.

The numbers x, y can be expressed in the form $3k$, $4k$;

$$\therefore \frac{4x^2-y^2}{x^2+y^2} = \frac{36k^2-16k^2}{9k^2+16k^2} = \frac{20k^2}{25k^2} = \mathbf{\frac{4}{5}}.$$

Example 16 If $\dfrac{4x+y}{x+3y} = \dfrac{3}{2}$, find the ratio $x:y$.

Multiply each side of the given equation by $2(x+3y)$,

$$\therefore 2(4x+y) = 3(x+3y),$$

$$\therefore 8x+2y = 3x+9y, \qquad \therefore 5x = 7y;$$

$$\therefore x = \frac{7y}{5}, \qquad \therefore \frac{x}{y} = \frac{7}{5}, \qquad \therefore x:y = \mathbf{7:5}.$$

Example 17 The weekly wages of two men are in the ratio 6:5. If each receives an increase of £1·50 a week, the ratio of their wages becomes 13:11. Find their original weekly wages.

The original weekly wages of the two men can be written

$$£6k, \quad £5k;$$

$\therefore$ the increased weekly wages are $£(6k+1\tfrac{1}{2})$, $£(5k+1\tfrac{1}{2})$;

$$\therefore \frac{6k+1\frac{1}{2}}{5k+1\frac{1}{2}} = \frac{13}{11}, \qquad \therefore 11(6k+1\tfrac{1}{2}) = 13(5k+1\tfrac{1}{2}),$$

$$\therefore 66k+16\tfrac{1}{2} = 65k+19\tfrac{1}{2}, \qquad \therefore k = 3.$$

$\therefore$ the original weekly wages were **£18, £15.**

EXERCISE 26

1 If $x:y = 7:3$, find the value of $(x-y):(x+y)$.

2 If $\dfrac{3x-y}{x+9y} = \dfrac{3}{5}$, find the values of $\dfrac{x}{y}$ and $\dfrac{x^2+4y^2}{(x-y)^2}$.

[**3**] If $\dfrac{3x+2y}{5x+6y} = \dfrac{3}{7}$, find x in terms of y and the value of the ratio of $\dfrac{1}{x}+\dfrac{1}{y}$ to $\dfrac{1}{x}-\dfrac{1}{y}$.

4 If $2x-3y = 8$ and $3x+5y = 20$, find the ratio $x:y$ *without* finding the separate values of x and y.

5 If $(a-b)$ is equal to 60% of $(a+b)$, express b as a percentage (i) of a, (ii) of $a+b$.

[**6**] If x is equal to n per cent of $(x+y)$, find the ratio $x:y$.

7 A man spends £11·70 on a week's holiday. The ratio of travelling expenses to hotel expenses is 2:11 and his other expenditure is 20% of the total of these two items. Find what he spends on each item.

8 Full marks for the 12 questions in an examination paper are 240. The ratio of the marks for each of the first 7 questions to the marks for each of the last 5 questions is 5:9. Find the marks for the first and for the last question.

[**9**] Children under 14 years old travel half-price with a half-ticket. The ratio of the number of whole tickets to the number of half-tickets for the passengers on an excursion is 2:3. Find the number of passengers if they take the equivalent of 252 full tickets.

10 A man employs 2 men, each for 8 hours a day, and 3 boys, each for 6 hours a day, at a total daily cost of £20·24. The ratio of the pay per hour of a man to that of a boy is 7:4. How much does he pay each man and each boy per day?

11 The full speed of a steamer in still water is u km per hour. Find its full speed (i) upstream, (ii) downstream, if the stream flows at v km per hour. If the ratio of the full speed upstream to that downstream is 5:9, find the ratio of the full speed of the steamer in still water to the speed of the stream.

[**12**] The ratio of the speed of a steamer in still water to the speed of the current of a river is 13:3. Find the time taken to go upstream as far as it takes 15 minutes to go downstream.

13 The purchasing price of £1 in 1947 was equal to that of 74p in 1938; in 1954 it was equal to that of 52p in 1938. A man's salary was £600 a year in 1947 and £900 a year in 1954. Taking 1938 as the standard, find the ratio of the purchasing power of his 1954 salary to that of his 1947 salary.

14 A bottle weighs 14 g when empty, and 82 g when full of water and 99 g when full of glycerine. Find the ratio of the weights of equal volumes of glycerine and water.

[**15**] An empty bottle weighs 330 g; when full of water it weighs 1080 g and when full of alcohol it weighs 930 g. Find the ratio of the weights of equal volumes of alcohol and water.

16 A train X passes three stations A, B, C at 9 a.m., 9.40 a.m., 10.32 a.m. respectively. Another train Y passes A, B at 9.50 a.m., 10.15 a.m. respectively. If both trains travel uniformly, find the time at which Y will pass C. Find also the speed of Y if X is travelling at 50 km per hour.

CHAPTER 6

ISOSCELES TRIANGLES AND POLYGONS

Isosceles Triangle The point of intersection of the equal sides of an isosceles triangle is called **the vertex,** the angle included by the equal sides is called the **vertical angle,** and the third side is called **the base.**

THEOREM

If two sides of a triangle are equal, then the angles opposite to those sides are equal.

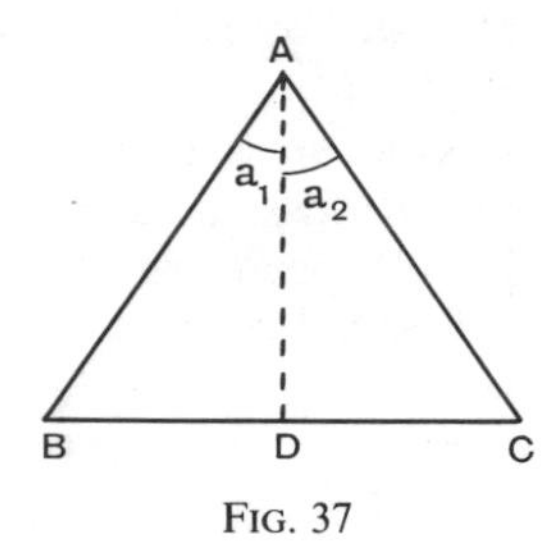

FIG. 37

Given a triangle ABC in which AB = AC.

To prove that $\angle B = \angle C$.

Construction Let AD be the **bisector of** $\angle BAC$ and let it meet BC at D.

Proof In the $\triangle$s ABD, ACD,

AB = AC *given,*

AD = AD

$a_1 = a_2$ *constr.,*

$\therefore \triangle s \begin{matrix} ABD \\ ACD \end{matrix}$ are congruent SAS.

$\therefore \angle B = \angle C$.

Abbreviation for reference: base $\angle$s, isos. $\triangle$.

This theorem is often stated in the form:
The angles at the base of an isosceles triangle are equal.

Converse Theorems The statement,

If, in $\triangle ABC$, $\angle B = \angle C$, then AB = AC, is called the **converse** of the statement,

If, in △ABC, AB = AC, then ∠B = ∠C,

because what is given and what is to be proved have been interchanged.

When a theorem has been proved, it does not necessarily follow that its converse is also true, although it often is so.

For example, the following statement is *true*:

If two angles are right angles, then they are equal.

But the converse statement,

If two angles are equal, then they are right angles,

is **not** true.

THEOREM

If two angles of a triangle are equal, then the sides opposite to those two angles are equal.

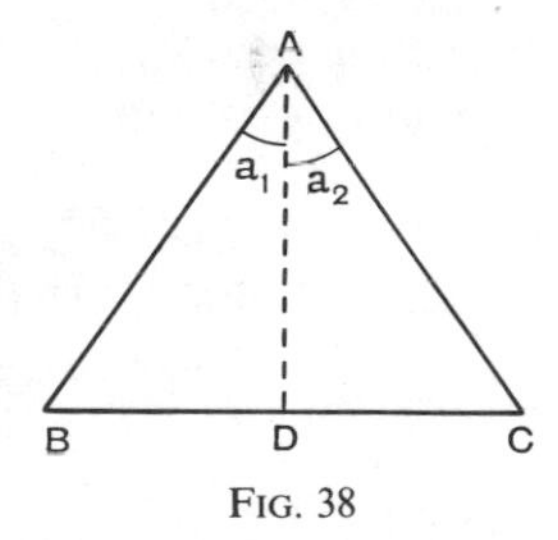

FIG. 38

Given a triangle ABC in which ∠B = ∠C.

To prove that AB = AC.

Construction Let AD be the **bisector of** ∠BAC and let it meet BC at D.

Proof In the △s ABD, ACD,

∠B = ∠C *given,*

$a_1 = a_2$ *constr.*

AD = AD.

∴ △s ABD, ACD are congruent AAS.

∴ AB = AC.

Abbreviation for reference: sides opp. equal ∠s.

EXERCISE 27 (Class Discussion)

1 If ABC is an equilateral triangle, prove that

∠A = ∠B = ∠C = 60°.

Explain why ∠A = ∠B and ∠A = ∠C. What follows?

2 If ABC is a triangle in which AB = AC and ∠BAC = 60°, prove that △ABC is equilateral.

3 If ABC is a triangle in which ∠ACB = 90° and ∠ABC = 60°, prove BC = $\frac{1}{2}$BA.
Produce BC to D making BC = CD.
Join AD.

(i) Explain why ∠ACD = 90°.
(ii) Prove that △ABD ≡ △ACB.
(iii) What is the size of ∠D? Prove △ABD is equilateral.

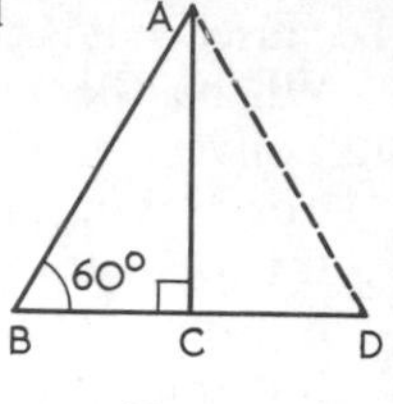

FIG. 39

Note This result is best recognised and remembered by thinking of △ACB as **'half an equilateral triangle'.**

The statement in each of Nos. 4–6 is true. Write down each converse and state whether it is true or untrue.

4 If ABCD is a parallelogram, then AB = DC.

5 If $x°$, $y°$, $z°$ are the angles of some triangle, then $x° + y° + z° = 180°$.

6 If, in △ABC, ∠BAC is obtuse, then BC is the longest side.

7 If AB = AC and ∠BAC = $t°$, find ∠ABC in terms of t.

8 In Fig. 40, AOB is a diameter of a circle, centre O. Find in terms of x, ∠POB, ∠OBP. What can you now say about ∠APB?

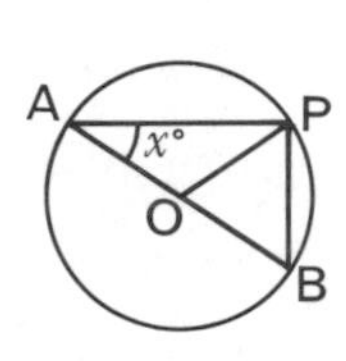

FIG. 40

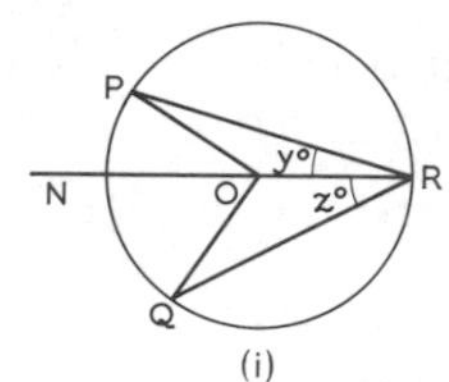

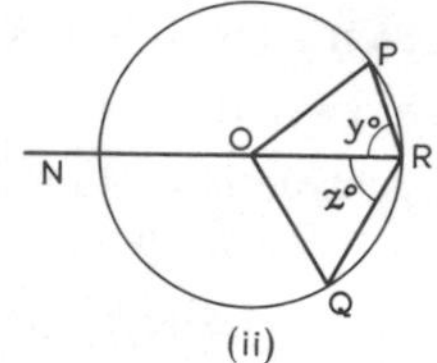

FIG. 41

9 In Fig. 41, the line RON passes through the centre O of the circle. Find in terms of y, z, ∠PON and ∠QON. What can you say about ∠POQ in Fig. 41 (i) and reflex ∠POQ in Fig. 41 (ii)?

EXERCISE 28

[*Arrows mean that lines are given parallel*]

1 One of the angles of an isosceles triangle is $x°$; find the other angles in terms of x. [Two sets of answers.]

2 Find the angles of an isosceles triangle if a base angle is double the vertical angle.

[3] Find the angles of an isosceles triangle if the vertical angle is three times a base angle.

4 In Fig. 42, AB = AC and CX = CB, find ∠ACX.

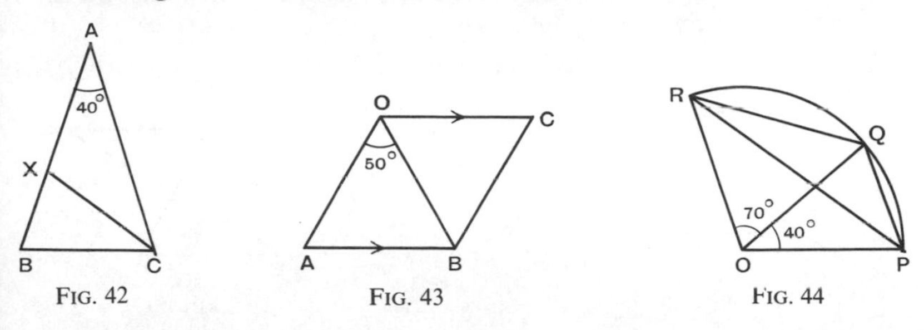

Fig. 42 Fig. 43 Fig. 44

5 In Fig. 43, OA = OB = OC, find ∠OCB.

6 In Fig. 44, O is the centre of the circular arc PQR. Find the angles of △PQR. Verify that reflex ∠POR = 2∠PQR.

[7] In △ABC, the bisector of ∠A cuts BC at D. If AD = DB and if ∠C = 66°, find ∠B.

8 In Fig. 45, AN = AC, find ∠NAB.

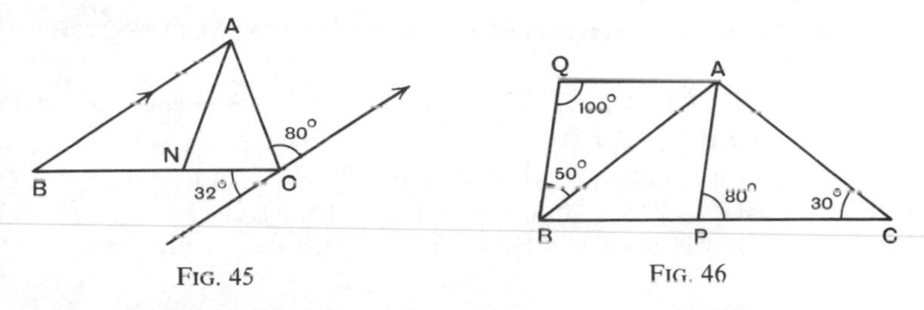

Fig. 45 Fig. 46

9 In Fig. 46, AB = AC and BPC is a straight line; prove that BP∥QA and BQ∥PA.

10 In Fig. 47, ABCD is a straight line and BE = BC, prove that CE = CD.

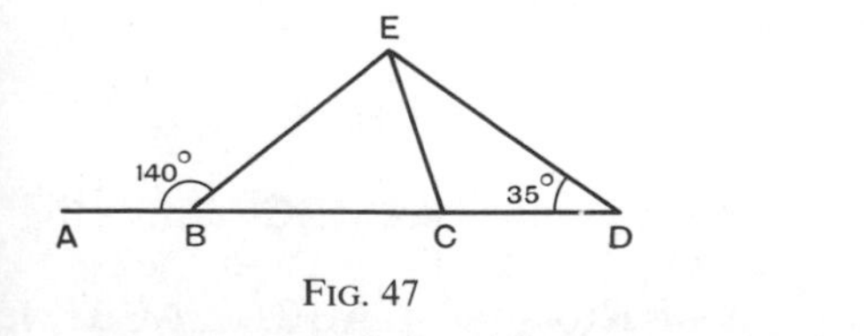

Fig. 47

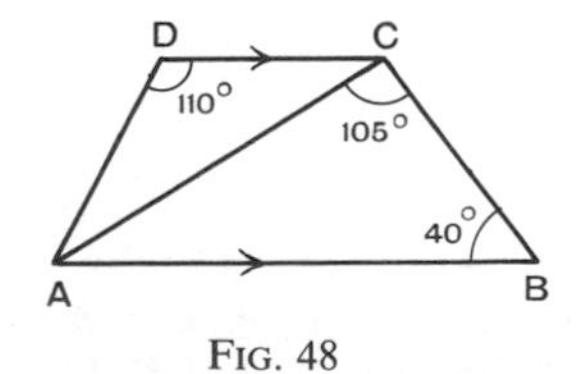

Fig. 48

11 In Fig. 48, prove that DA = DC.

[12] The side BC of △ABC is produced to D; ∠BAC = 40°, ∠ACD = 75°. P is a point on AB such that AP = AC, prove PB = PC.

13 In $\triangle$ABC, AB = AC; D is a point on AC produced such that BD = BA. If $\angle$CBD = 36°, prove that BC = CD.

In Nos. 14, 15, work throughout in terms of the small letter which is to appear in the answer.

14 In Fig. 49, if AB = AC, find x in terms of y.

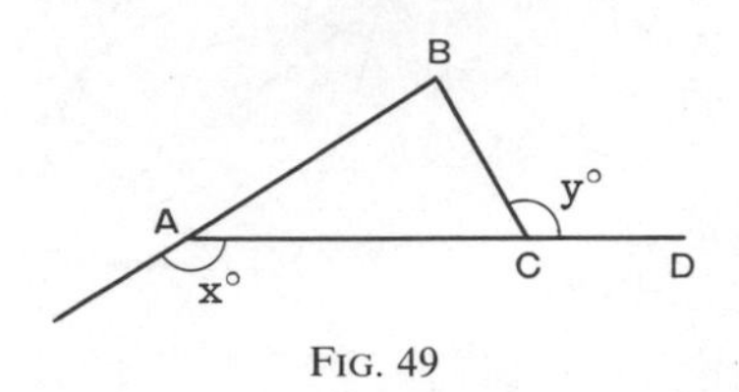

FIG. 49

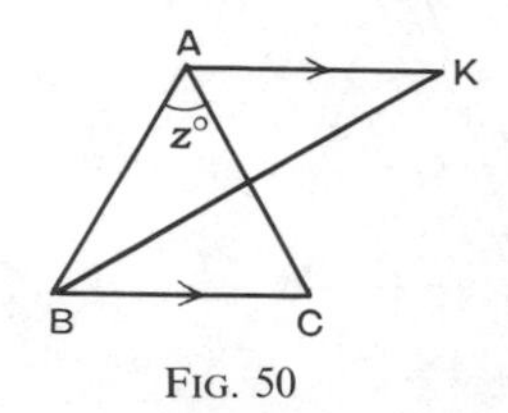

FIG. 50

15 In Fig. 50, AB = AC and BK bisects $\angle$ABC. Find $\angle$AKB in terms of z.

EXERCISE 29

[*Arrows mean that lines are given parallel*]

1 The side BC of $\triangle$ABC is produced to D. If AB = AC, prove that $\angle$ABD + $\angle$ACD = 2 rt. $\angle$s.

[**2**] The side BC of $\triangle$ABC is produced to D. If $\angle$ACD = 2$\angle$BAC, prove that CA = CB.

3 In $\triangle$ABC, AB = AC. If the bisector of $\angle$B meets AC at P, prove that $\angle$APB = 3$\angle$PBC.

4 ABCD is a quadrilateral in which AB = AD and $\angle$ABC = $\angle$ADC. Prove that CB = CD. [Join BD.]

5 In Fig. 51, AP bisects $\angle$BAC. Prove that AQ = QP.

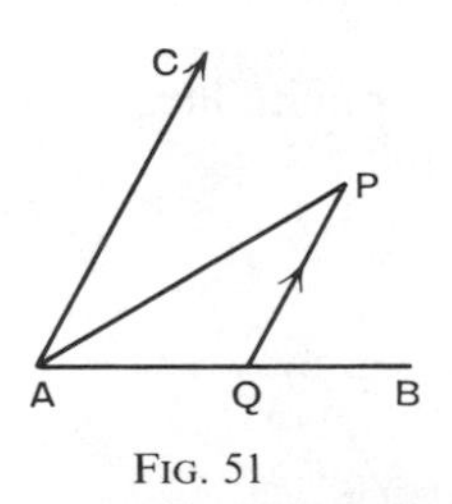

FIG. 51

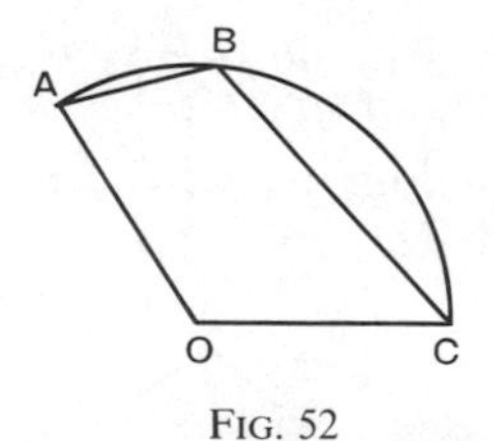

FIG. 52

6 In Fig. 52, O is the centre of the circular arc ABC. Prove that $\angle$OAB + $\angle$OCB = $\angle$ABC. [Join OB.]

7 In Fig. 53, IB, IC are the bisectors of $\angle$ABC, $\angle$ACB. If AB = AC, prove that IB = IC.

8 In Fig. 54, AB = AC, BP = QC, and BPQC is a straight line. Prove $\angle$APQ = $\angle$AQP. [Prove two triangles are congruent.]

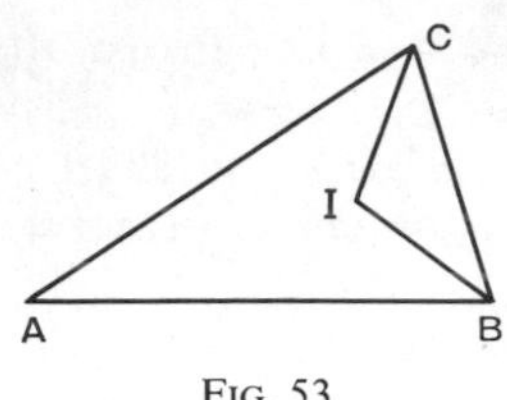

FIG. 53

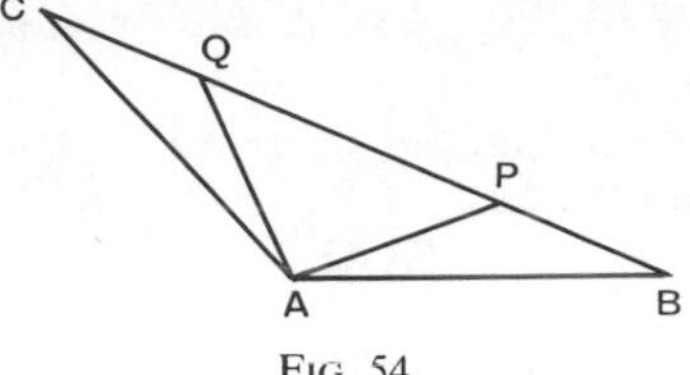

FIG. 54

[9] ABCD is a quadrilateral in which ∠B and ∠C are equal acute angles. If AB = CD, prove that ∠A = ∠D. [Produce BA and CD to meet at K.]

10 In △ABC, AB = AC. If AC is produced to any point X and if BX is joined, prove that ∠ABX + ∠AXB = 2∠ACB.

11 IB, IC are the bisectors of ∠ABC, ∠ACB of △ABC, see Fig. 53. If AB = AC and if BC is produced to D, prove that ∠ACD = ∠BIC.

[12] D is the mid-point of the side BC of △ABC. If AD = BD, prove that ∠BAC is a right angle.

[13] In △ABC, AB = AC. BA is produced to E and AX is drawn bisecting ∠CAE. Prove that AX is parallel to BC.

14 In △ABC, AB = AC. If CN is the perpendicular from C to AB, prove that ∠NCB = $\frac{1}{2}$∠A.

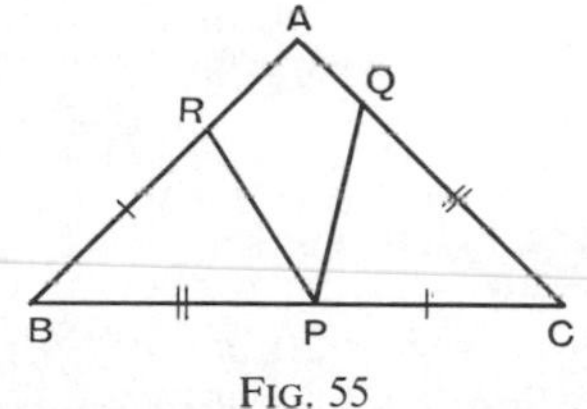

FIG. 55

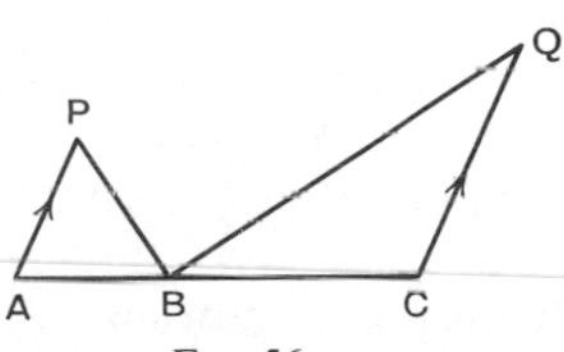

FIG. 56

15 In Fig. 55, P, Q, R are points on the sides of △ABC such that BR = PC and QC = PB. If AB = AC, prove that (i) PQ = PR, (ii) ∠RPQ = ∠B.

[16] In Fig. 56, AP = AB, CB = CQ, and ABC is a straight line. Prove that ∠PBQ is a right angle.

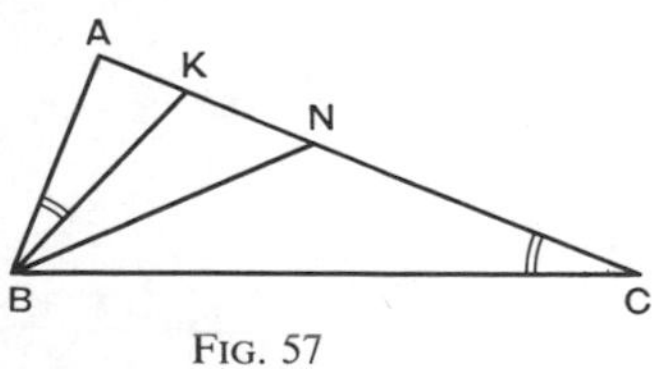

FIG. 57

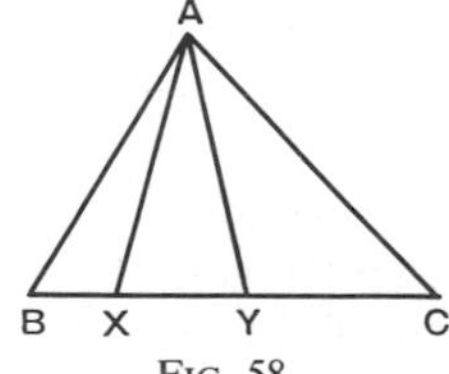

FIG. 58

17 In Fig. 57, ∠ABK = ∠C and BN bisects ∠KBC. Prove that AN = AB.

18 In Fig. 58, $\angle XAC = \angle B$ and $\angle YAB = \angle C$. Prove that $AX = AY$.

19 In Fig. 59, O is the centre of the circle; PBOA and PQR are straight lines. If PQ is equal to the radius of the circle, prove that $\angle AOR = 3\angle BOQ$.

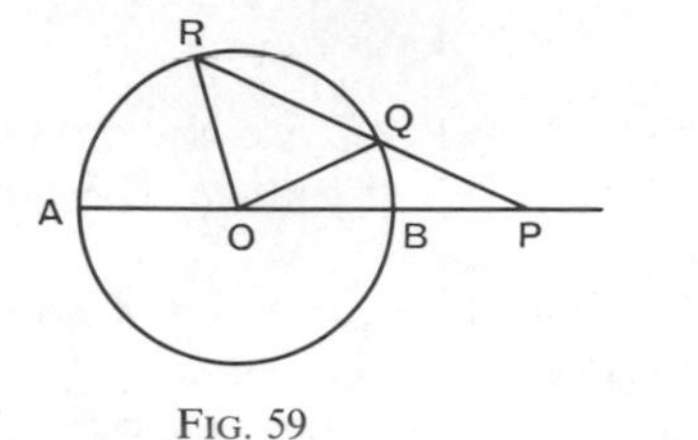

FIG. 59

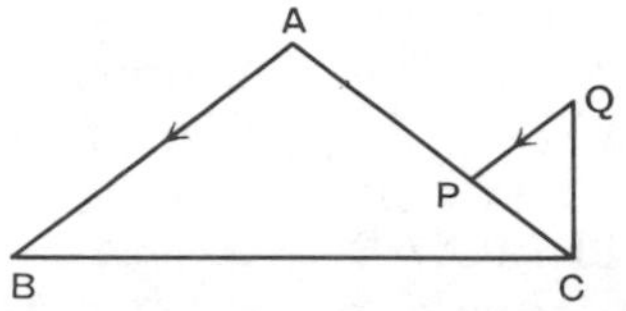

FIG. 60

20 In Fig. 60, $AB = AC$ and $PQ = PC$. Prove $\angle BCQ = 1$ rt. $\angle$.

Angles of a Quadrilateral The simplest way of finding the sum of the angles of a quadrilateral ABCD is to draw a diagonal AC, thus forming the two triangles ABC, ACD.

With the notation of Fig. 61,

$m+n+p = 2$ rt. $\angle$s, $\angle$ *sum of* $\triangle$,

$r+s+t = 2$ rt. $\angle$s, $\angle$ *sum of* $\triangle$.

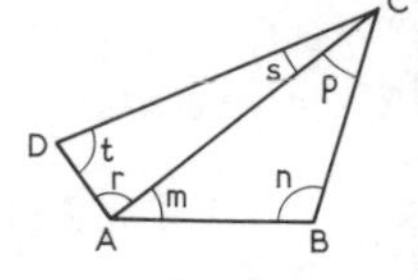

FIG. 61

But the angles of ABCD are $r+m$, n, $p+s$, t;
$\therefore$ by addition,

the sum of the angles of a quadrilateral is 4 right angles.

Angles of a Polygon If a polygon has a large number of sides, it is simpler to use a different method for finding the sum of its angles.

Examples for Oral Work

1 Find the sum of the angles of the hexagon ABCDEF.

Take any point O inside the hexagon; join O to each vertex.

(i) What is the *total* sum of all the angles of the 6 triangles?

(ii) What is the sum of the 6 angles at O?

(iii) Subtract. What follows?

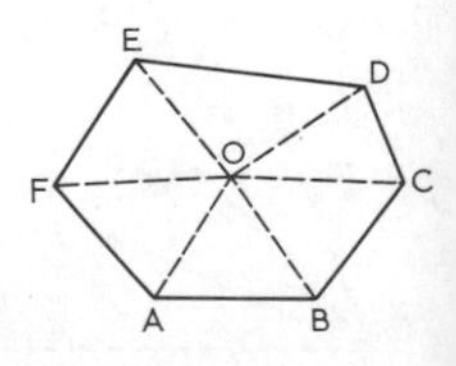

FIG. 62

2 Draw a 7-sided polygon and use the *method* of No. 1 to find the sum of its interior angles.

3 Use the *argument* of No. 1 to find the sum of the interior angles of (i) a 10-sided polygon, (ii) a 100-sided polygon.

4 A point O inside a polygon with n sides is joined to each vertex.
(i) How many triangles are so formed?
(ii) What is the *total* sum of all the angles of all the triangles?
(iii) Hence prove the general statement:

The sum of the interior angles of an n-sided polygon is (2n—4) right angles.

Exterior Angles of a Polygon A polygon is called **convex** if each of its interior angles is less than two right angles.

If a side of a convex polygon is produced, the angle so formed outside the polygon is called an **exterior angle of the polygon.**

Thus in Fig. 63, ABCDE is a convex pentagon and the marked angles formed by producing the sides are *exterior* angles.

A polygon is called **re-entrant** if at least one of its interior angles is greater than two right angles.

For example, the pentagon QPSCD in Fig. 64, is re-entrant, and in this case the angle which QP produced makes with PS is *inside* the pentagon QPSCD; it is *not* an exterior angle.

The phrase, **exterior angle of a polygon,** is used only if the polygon is convex.

Examples for Oral Work

1 If the sides of a convex pentagon ABCDE are **produced in order,** see Fig. 63, find the sum of the exterior angles so formed.

Copy and complete the following:
The interior $\angle$+the exterior $\angle$ at each vertex = 2 rt. $\angle$s,...;
$\therefore$ the sum of the 5 int. $\angle$s+the sum of the 5 ext. $\angle$s = ...;
but the sum of the 5 int. $\angle$s = ...;
$\therefore$ the sum of the 5 ext. $\angle$s = ...

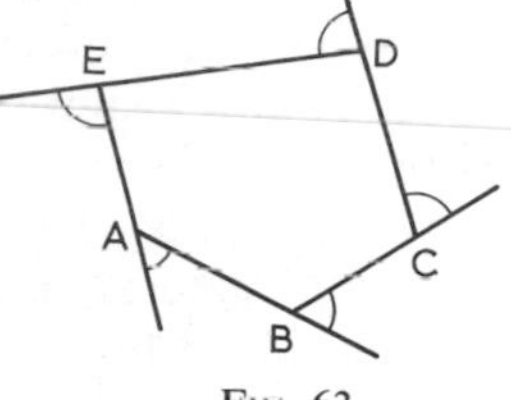

FIG. 63

2 Draw a convex 7-sided polygon and produce the sides in order. Use the *method* of No. 1 to calculate the sum of the exterior angles so formed.

3 Use the *argument* of No. 1 to find the sum of the exterior angles of a 10-sided convex polygon if the sides are produced in order.

4 The sides of an n-sided convex polygon are produced in order.
(i) What is the *total* sum of the n interior angles and the n exterior angles?
(ii) What is the sum of the n interior angles?
(iii) Hence prove the general statement:

The sum of the exterior angles of an n-sided convex polygon formed by producing the sides in order is 4 right angles.

Regular Polygon A polygon is called **regular** if all its sides and all its angles are equal.

EXERCISE 30

1 Three of the angles of a quadrilateral are 88°, 100°, 110°; find the other angle.

[**2**] ABCD is a quadrilateral in which $\angle A = \angle B = \angle C$ and $\angle D = 120°$; find $\angle A$.

3 The angles of a quadrilateral taken in order are $2x°, 3x°, 7x°, 8x°$. Find x and prove that two opposite sides are parallel.

[**4**] The angles of a pentagon are $x°, 2x°, 2x°, 2x°, 3x°$; find x.

5 ABCDE is a regular pentagon; AB, DC are produced to meet at P; find $\angle BPC$.

6 Find the sum of the interior angles of a polygon which has (i) 30 sides, (ii) 40 sides.

[**7**] Prove that the sum of the interior angles of an octagon (8 sides) is twice the sum of the interior angles of a pentagon.

8 Find the size of an exterior angle of a regular polygon having (i) 15 sides, (ii) n sides.

9 Find the number of sides of a polygon (i) if each exterior angle is 40°, (ii) if each interior angle is 144°.

[**10**] Find the number of sides of a polygon (i) if each exterior angle is 15°, (ii) if each interior angle is 160°.

11 If the sum of the interior angles of a polygon is 30 right angles, find the number of its sides.

12 Find the sum of the interior angles of the re-entrant pentagon APBCD in Fig. 64.

13 In Fig. 64, PA, PB, RC, RD are the bisectors of the angles of the quadrilateral ABCD. If $\angle DAB = 80°$, $\angle ABC = 70°$, $\angle BCD = 150°$, prove that $\angle QPS + \angle QRS = 180°$.

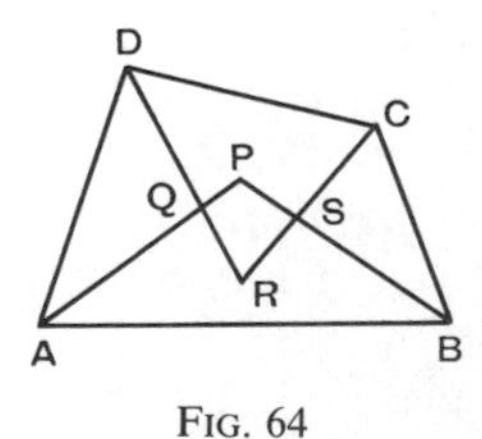

FIG. 64

FIG. 65

14 Find the sum of the interior angles of the re-entrant polygon in Fig. 65.

EXERCISE 31

1 ABCD is a quadrilateral in which ∠A = ∠C and ∠B = ∠D. Prove that (i) ∠A + ∠B = 2 rt. ∠s; (ii) ABCD is a parallelogram.

[2] ABCD is a quadrilateral in which ∠A + ∠C = 2 rt. ∠s. If AB is produced to E, prove that ∠CBE = ∠D.

3 In Fig. 66, NA, NB, NC, ND are the bisectors of the angles of the quadrilateral ABCD. Prove ∠ANB + ∠CND = 2 rt. ∠s.

[4] If, in Fig. 66, NA and NB are the bisectors of ∠DAB, ∠CBA, prove that ∠ADC + ∠BCD = 2∠ANB.

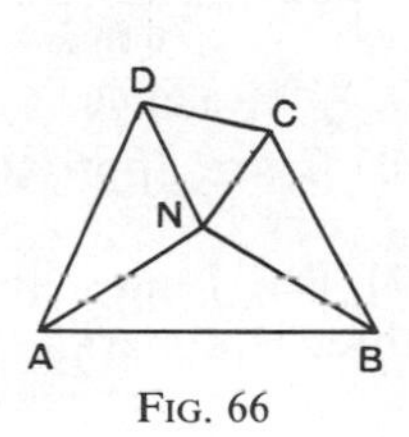

FIG. 66

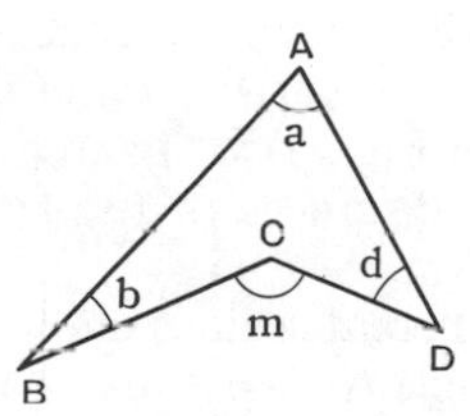

FIG. 67

5 In Fig. 67, prove that $m = a + b + d$.

[6] If, in Fig. 67, $m = 2a$ and $a = 2b$, prove that $b = d$.

7 If, in Fig. 67, the bisectors of ∠ABC, ∠ADC meet at K, prove that $a + m = 2$∠BKD.

[8] ABCD is a quadrilateral in which ∠A and ∠C are each right angles. Prove that the bisectors of ∠B and ∠D are parallel.

9 Find the sum of the marked angles in Fig. 68.

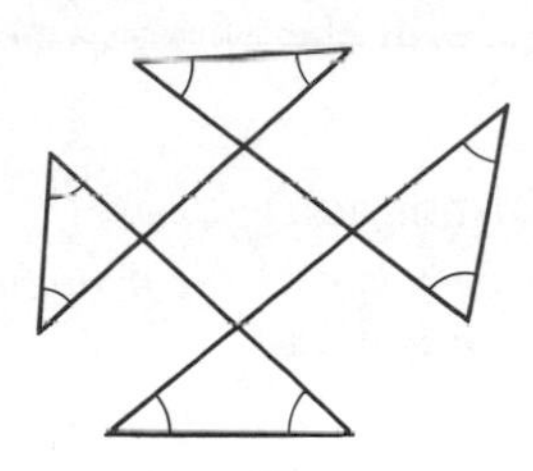
FIG. 68

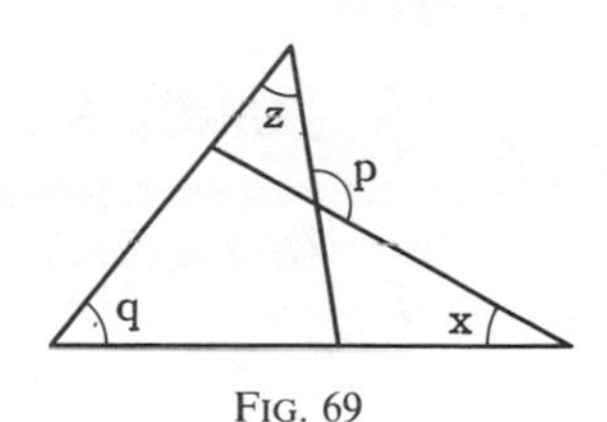

FIG. 69

***10** If, in Fig. 69, p and q are supplementary angles, prove that

$$p = 90° + \tfrac{1}{2}(x + z).$$

CHAPTER 7

THE TANGENT OF AN ANGLE

Similar Triangles A map of three landmarks A, B, C on level ground can be drawn when the distance BC, say 6 km, and the angles ABC, ACB, say 56°, 41°, have been measured. With the scale, 1 cm: 1 km, the map is $\triangle$ABC in Fig. 70; *draw accurately* $\triangle$ABC and mark on your figure the lengths of AB, AC; check your measurements by comparison with Fig. 71. *Use your figure* (or Fig. 71) for Exercise 32. It follows that the distances of the landmark A from B, C are approximately 4 km, 5 km.

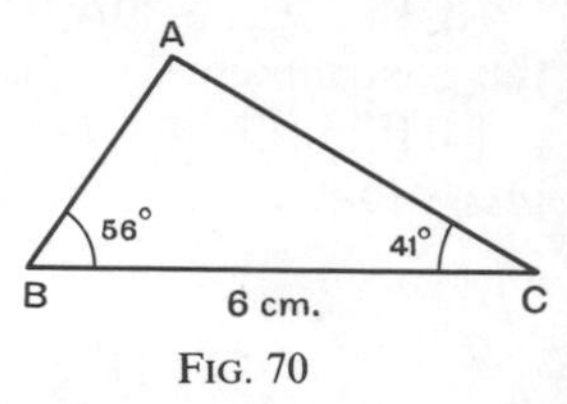

FIG. 70

If a different scale is chosen so that, in Fig. 71, B′C′ represents 6 km, the triangle A′B′C′ is also a map of the same three landmarks.

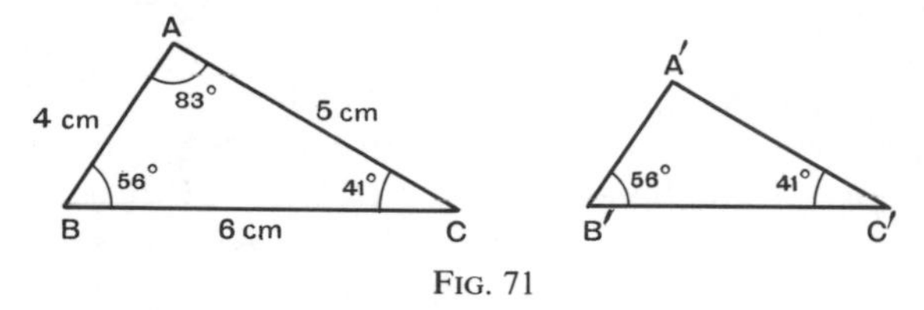

FIG. 71

The sizes of the angles of $\triangle$A′B′C′ are the same as those of $\triangle$ABC, but the lengths of the sides are all decreased (or increased) in the same ratio. The triangles ABC, A′B′C′ are of the *same shape* and are called **similar.**

EXERCISE 32 (Class Discussion)

[*Throughout this exercise,* $\triangle$ABC *is to be taken as the triangle whose dimensions are shown in Fig.* 71.]

1 Draw $\triangle$A′B′C′ so that $\angle$B′ = $\angle$B = 56°, $\angle$C′ = $\angle$C = 41°, but with B′C′ = 4·5 cm. Measure A′B′, A′C′ and complete the following:

$$\frac{B'C'}{BC} = \frac{4{\cdot}5}{6} = 0{\cdot}75; \qquad \frac{C'A'}{CA} = \frac{\cdots}{5} = 0{\cdot}\ldots; \qquad \frac{A'B'}{AB} = \frac{\cdots}{4} = 0{\cdot}\ldots$$

$\triangle$A′B′C′ is a *scale drawing* of $\triangle$ABC, scale 0·75 : 1 or 3 : 4.

2 Draw $\triangle$PQR so that $\angle$Q = $\angle$B = 56°, $\angle$R = $\angle$C = 41°, and QR = $\frac{4}{5}$BC = $\frac{4}{5}$ of 6 cm = 4·8 cm. Measure PQ, PR and use the data of Fig. 71 to find the values of $\frac{PQ}{AB}$ and $\frac{PR}{AC}$.

$\triangle$PQR is a *scale drawing* of $\triangle$ABC, scale 4:5. This is stated

$$\triangle\text{s}\ \begin{matrix}\text{PQR}\\ \text{ABC}\end{matrix}\ \textbf{are similar,}\ \text{scale } 4:5.$$

Use this fact to *calculate* the lengths of PQ, PR.

3 If the $\triangle$s $\begin{matrix}\text{XYZ}\\ \text{ABC}\end{matrix}$ are similar and the scale is 7:10, calculate the lengths of YZ, ZX, XY, see Fig. 71. What are $\angle$X, $\angle$Y, $\angle$Z?

4 If the $\triangle$s $\begin{matrix}\text{DEF}\\ \text{ABC}\end{matrix}$ are similar and if EF = 9 cm, find the remaining sides and the angles of $\triangle$DEF, see Fig. 71. What is the scale?

The general statement may be expressed as follows:

If the $\triangle$s $\begin{matrix}\text{XYZ}\\ \text{ABC}\end{matrix}$ are such that

$$\angle \mathbf{Y} = \angle \mathbf{B} \quad \text{and} \quad \angle \mathbf{Z} = \angle \mathbf{C}$$

then the $\triangle$s $\begin{matrix}\text{XYZ}\\ \text{ABC}\end{matrix}$ are **similar,** that is

$$\frac{\mathbf{YZ}}{\mathbf{BC}} = \frac{\mathbf{ZX}}{\mathbf{CA}} = \frac{\mathbf{XY}}{\mathbf{AB}} \quad \text{and} \quad \angle \mathbf{X} = \angle \mathbf{A}.$$

If the lengths of the three sides of a triangle are given, the triangle is fixed in shape *and* size. For example, suppose Z is a church 6 km from Y, and that X is a church 4 km from Y and 5 km from Z; then the data fix the shape and size of $\triangle$XYZ. $\triangle$ABC, in Fig. 71, is a scale-drawing or map of $\triangle$XYZ, scale 1 cm:1 km, and the angles of $\triangle$ABC are equal to the corresponding angles of $\triangle$XYZ. Thus, if we draw $\triangle$ABC so that

$$\frac{\mathbf{BC}}{\mathbf{YZ}} = \frac{\mathbf{CA}}{\mathbf{ZX}} = \frac{\mathbf{AB}}{\mathbf{XY}}$$

then the $\triangle$s $\begin{matrix}\text{ABC}\\ \text{XYZ}\end{matrix}$ are **similar,** that is

$$\angle \mathbf{A} = \angle \mathbf{X}, \quad \angle \mathbf{B} = \angle \mathbf{Y}, \quad \angle \mathbf{C} = \angle \mathbf{Z}.$$

5 Using the lengths of the sides of $\triangle$ABC in Fig. 71, draw $\triangle$EFG so that FG = $\frac{4}{5}$BC, GE = $\frac{4}{5}$CA, EF = $\frac{4}{5}$AB.

Measure $\angle$E, $\angle$F, $\angle$G. What are the errors in your answers?

EXERCISE 33

[*Arrows mean that lines are given parallel*]

1 Are the triangles in Fig. 72 similar? If so, give the reason and express the fact in the form $\triangle$s $\begin{matrix}\text{ABC}\\ \dots\end{matrix}$ are similar. Find also the lengths of the remaining sides.

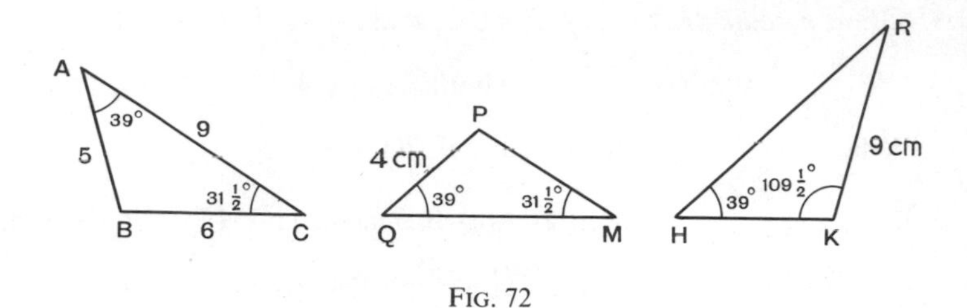

FIG. 72

2 In Fig. 73, (i) complete △s $\frac{\text{ABC}}{\dots}$ are similar because …
(ii) Find the lengths of BP, BQ.

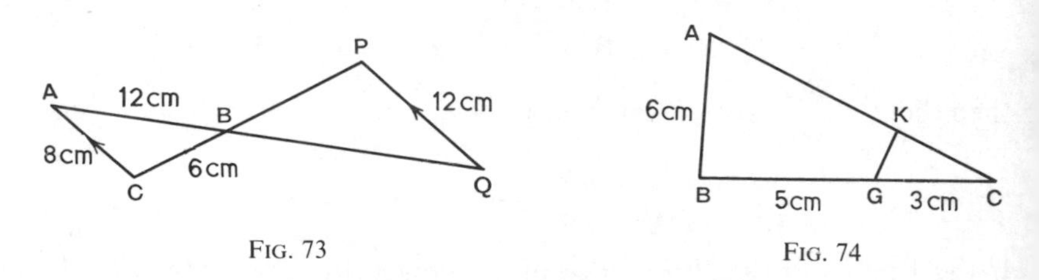

FIG. 73

FIG. 74

3 In Fig. 74, AC = 10 cm and ∠GKC = ∠B.

(i) Complete △s $\frac{\text{ABC}}{\dots}$ are similar because …

(ii) Find the lengths of KG, KA.

4 In Fig. 75, find the lengths of AB and QC.

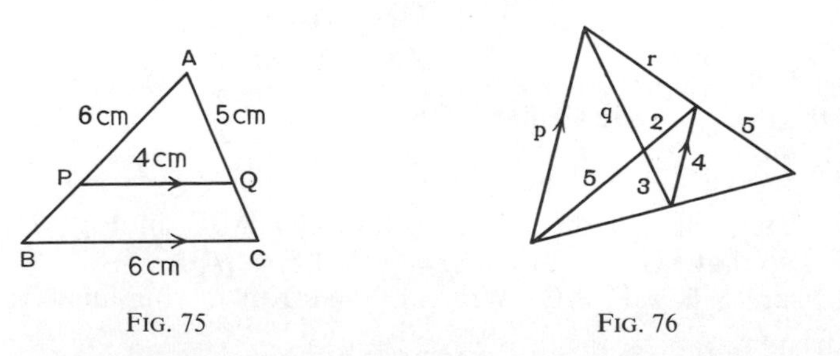

FIG. 75

FIG. 76

5 In Fig. 76, calculate the values of p, q, r.

The Tangent of an angle Draw two angles A, B each 38° and from points P, Q on one arm of each angle draw the *perpendiculars* PM, QN to the other arm.

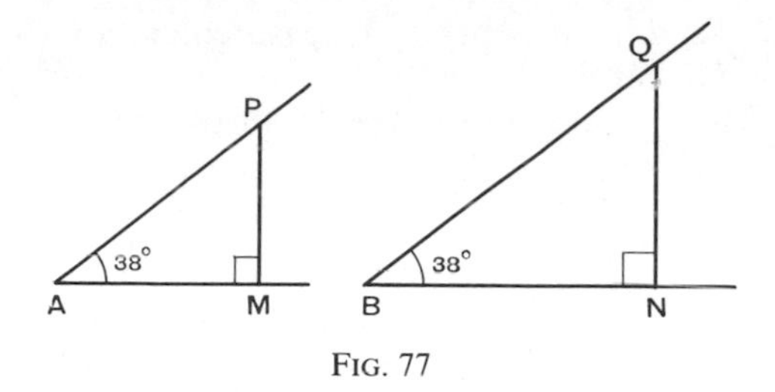

FIG. 77

Then $\angle A = \angle B$ and $\angle M = \angle N$,

$$\therefore \Delta s \begin{matrix} APM \\ BQN \end{matrix} \text{ are similar}$$

$$\therefore \text{ the ratio } \frac{MP}{AM} \text{ equals the ratio } \frac{NQ}{BN}.$$

Thus the value of the ratio $\frac{MP}{AM}$ does not depend on the distance of P from the point A, but only on the size of $\angle A$.

The ratio $\frac{MP}{AM}$ is called the **tangent of the angle** PAM and is written tan PAM or tan 38° since $\angle PAM = 38°$.

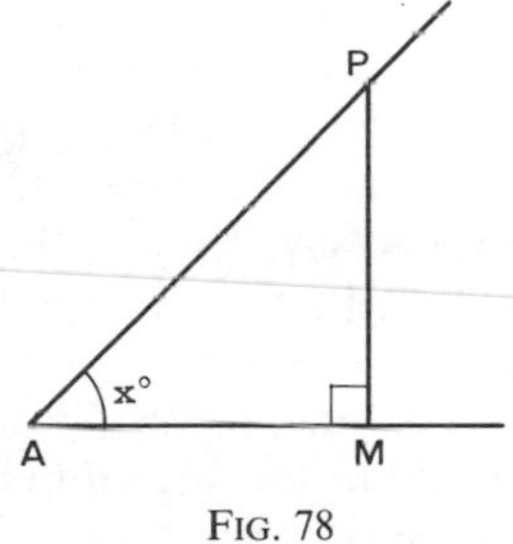

FIG. 78

If P is any point on one arm of an angle $x°$ at A, and if PM is the perpendicular from P to the other arm, AMP is a right-angled triangle whose **hypotenuse** is AP. MP is the side *opposite* to $\angle A$, AM is that side *adjacent* to $\angle A$ which is not the hypotenuse. For brevity, in relation to $\angle A$, we call MP the **opp.** side and AM the **adj.** side.

With this notation,

$$\textbf{tan } x° = \frac{MP}{AM} = \frac{\textbf{opp.}}{\textbf{adj.}}$$

An approximate value of the tangent of an angle can be found by measurement and computation.

If Fig. 77 is drawn so that AM = 5 cm, measurement shows that MP = 3·9 cm, approximately.

Hence $\tan 38° = \frac{3{\cdot}9}{5} = 0{\cdot}78$, approximately.

Example 1 Find by drawing and measurement approximate values of tan 25°, tan 50°, tan $12\frac{1}{2}°$, tan $37\frac{1}{2}°$.

In Fig. 79, OA = 10 cm and APQRS is drawn perpendicular to OA;

$$\angle AOQ = 25°,\ \angle AOS = 50°,\ \angle AOP = 12\tfrac{1}{2}°,\ \angle AOR = 37\tfrac{1}{2}°.$$

By measurement, AP = 2·2 cm, AQ = 4·7 cm, AR = 7·7 cm, AS = 11·9 cm.

$$\therefore \tan 25° = \frac{AQ}{OA} = \frac{4\cdot7}{10} = \mathbf{0\cdot47}, \text{ approximately.}$$

The reader should now write down approximate values of tan 50°, tan $12\frac{1}{2}°$, tan $37\frac{1}{2}°$, by measuring AS, AP, AR in his own figure.

In Fig. 79, AS is drawn perpendicular to OA; if we draw a circle, centre O, radius OA, we see that it meets AS at only one point and we say that AS *touches* the circle at A and we call AS the **tangent** to the circle at A. In Example 1, approximate values of the tangents of the given angles have been found by measuring the lengths from A cut off along the tangent to the circle, and this explains the choice of name for this particular ratio. If the radius had contained 1 unit of length, the number of units of length cut off on the tangent, shown in Fig. 79, would represent the tangent of the corresponding angle. But it is important to realise that the tangent of an angle is a *ratio*, not a length. If we draw a circle of radius, say, 5 units, then the lengths cut off along the tangent must be divided by 5 to give the required ratios.

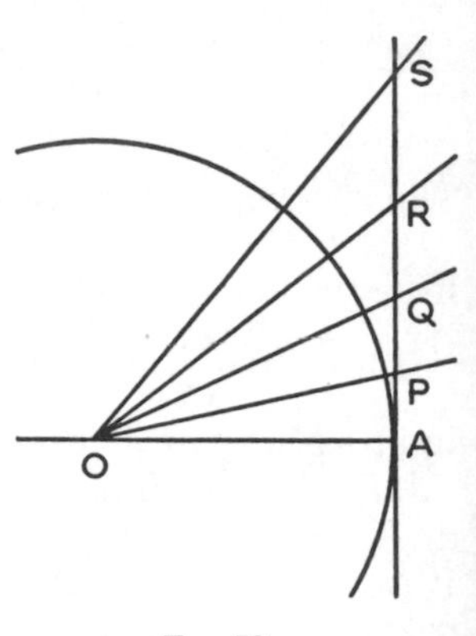

FIG. 79

The reader should notice that the tangent of an angle is *not* proportional to the size of the angle, thus tan 50° is more than twice tan 25°, and more than four times tan $12\frac{1}{2}°$.

Tangent Tables The value of the tangent of an angle can be found by drawing and measurement to 2 or, at most, 3 significant figures. For a higher degree of accuracy, calculation is necessary; the results are given in books of tables.

In most practical work it is customary to use four-figure tables.

These give the tangents of angles from 0° to 90° at intervals of tenths of a degree (1° = 60′; 0·1° = 6′). By means of the *difference-columns* at the side, it is possible to find the values for intervals of 1 minute.

The figures in the middle of the pages of the tables are figures after the decimal point which is omitted; the whole number, if any, is printed *only* at the beginning of the line. Thus the extract below shows, that, to 4 places.

tan 50° = 1·1918, tan 50° 6′ = 1·1960, tan 50° 12′ = 1·2002, etc.

The narrow columns at the end of the line, called *difference* columns, show what compensations must be made for intermediate angles.

Extract from Table of Natural Tangents

	0′	6′	12′	18′	24′	30′	36′	42′	48′	43′	1′	2′	3′	4′	5′
50	1·1917	1960	2002	2045	2088	2131	2174	2218	2261	2305	7	14	22	29	36

Example 2 Find tan 50° 44′.

tan 50° 42′	1·2218
Add diff. for 2′	14
tan 50° 44′	1·2232

tan 50° 44′ = **1·2232**

The difference columns are obtained by finding the *average increase* per minute taken over an interval of 1°; the fourth decimal place is not therefore reliable, and for angles over 70° it is necessary to take the average over smaller intervals to avoid large errors. For example, from the tables,

tan 75° 12′	3·7848
tan 75°	3·7321
∴ Diff. for 12′	527
∴ Diff. for 1′	44

tan 76°	4·0108
tan 75° 48′	3·9520
∴ Diff. for 12′	588
∴ Diff. for 1′	49

Accordingly where necessary the tables give the average difference for 1′, calculated over 12′ intervals.

Example 3 Find 75° 8′.

Diff. for 1′, interval 1′ to 11′, is 44;

∴ Diff. for 2′ is 44 × 2, = 88.

tan 75° 6′	3·7583
Diff. for 2′	88
tan 75° 8′	**3·7671**

Example 4 Find tan 75° 51′.

Diff. for 1′, interval 49′ to 59′, is 49;

∴ Diff. for 3′ is 49 × 3, = 147.

tan 75° 48′	3·9520
Diff. for 3′	147
tan 75° 51′	**3·9667**

Example 5 If $\tan x^\circ = 0{\cdot}7068$, find x°.

From the tables, $\tan 35^\circ\ 12' = 0{\cdot}7054$, $\tan 35^\circ\ 18' = 0{\cdot}7080$,

$$\therefore x^\circ \text{ lies between } 35^\circ\ 12' \text{ and } 35^\circ\ 18'.$$

Since $0{\cdot}7068 - 0{\cdot}7054 = 0{\cdot}0014$, we read off from the difference columns the number of minutes which corresponds to a difference of 14. The nearest number is 13, and this is the compensation for 3′;

$$\therefore x^\circ = 35^\circ\ 12' + 3' = \mathbf{35^\circ\ 15'}.$$

Example 6 From the top A of a cliff AN, 120 m high, the angle of depression of a boat B is 49°.

Find the distance of B from N.

If AD is horizontal,

$$\angle ABN = \angle BAD = 49^\circ,$$

$$\therefore \angle BAN = 90^\circ - 49^\circ = 41^\circ;$$

$$\therefore \frac{BN}{AN} = \tan BAN = \tan 41^\circ,$$

$$\therefore BN = 120 \tan 41^\circ \text{ m} = 120 \times 0{\cdot}8693 \text{ m}.$$

$$\therefore BN = \mathbf{104\ m}, \text{ to nearest metre.}$$

Fig. 80

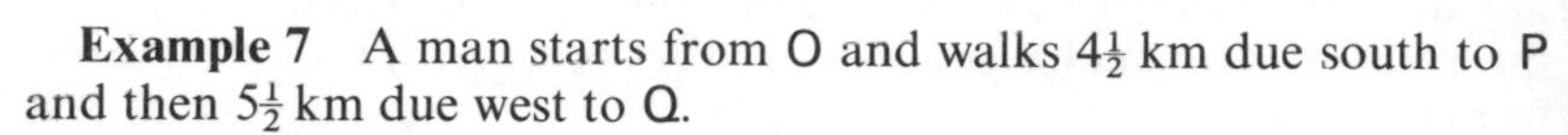

Example 7 A man starts from O and walks $4\frac{1}{2}$ km due south to P and then $5\frac{1}{2}$ km due west to Q.

Find the bearing of Q from O.

$$\tan QOP = \frac{QP}{OP} = \frac{5\frac{1}{2}}{4\frac{1}{2}},$$

$$\therefore \tan QOP = \tfrac{11}{9} = 1{\cdot}2222.$$

$$\therefore \angle QOP = 50^\circ\ 43'.$$

$\therefore$ the bearing of Q from O is **S. 50° 43′ W.**

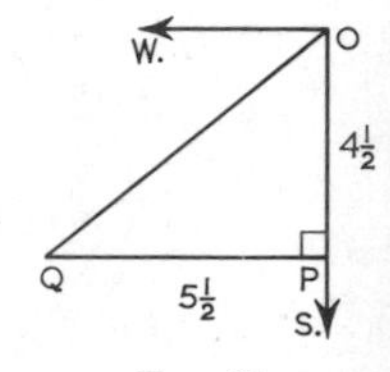

Fig. 81

EXERCISE 34

1. Draw two triangles ABC, PQR, in which

$$\angle A = \angle P = 35^\circ, \quad \angle C = \angle R = 90^\circ, \quad AC = 8 \text{ cm}, \quad PR = 6 \text{ cm}.$$

Measure BC and QR, and find the values of $\frac{BC}{AC}, \frac{QR}{PR}$. Find approximately $\tan 35^\circ$.

2 In Fig. 82, BC = 8 cm; measurement gives AC = 4·8 cm. Calculate PR if QR is (i) 6 cm, (ii) 5 cm. Find approximately tan 31°.

3 When the shadow of a pole 3 m high is 2 m long, the shadow

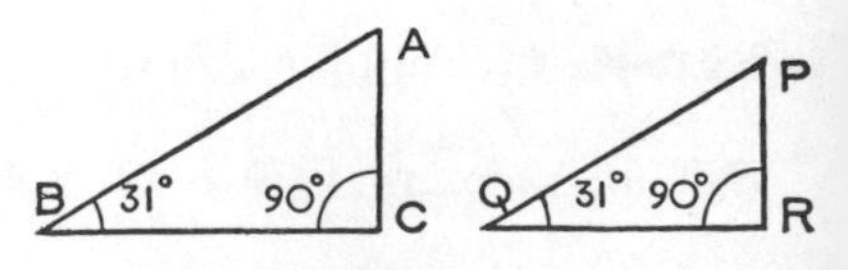

Fig. 82

of a church tower is 17 m long. What is the height of the tower? If the angle of elevation of the sun (often called the *altitude* of the sun) is $x°$, what is the value of tan $x°$?

4 Find by drawing the values of tan 20°, tan 40°, tan 45°, tan 50°, tan 60°, tan 75°. (It saves time to use squared paper.) Then find their values from the tables.

5 Find by drawing the angles whose tangents are $\frac{1}{4}$, 0·9, 1·6, 2·3. Then find these angles from the tables.

6 Use tables to write down the tangents of: 62°, 32° 24′, 50° 18′, 63° 24′, 63° 30′, 17° 38′, 25° 40′, 50° 15′, 16° 41′, 53° 47′, 79° 2′, 79° 55′, 74° 27′, 76° 20′, 77° 41′, 82° 14′, 81° 33′.

7 Use tables to write down the angles whose tangents are: 0·4245, 2·9042, 0·2754, 1·0283, 1·7532, 3·0061, 2·0057, 28·64, 0·2536, 0·2431, 0·5004, 1·2692, 1·7079, 0·6883, 1·0022, 1·1193, 1·6354, 4·0209, 4·5992, 5·2175, 10·45.

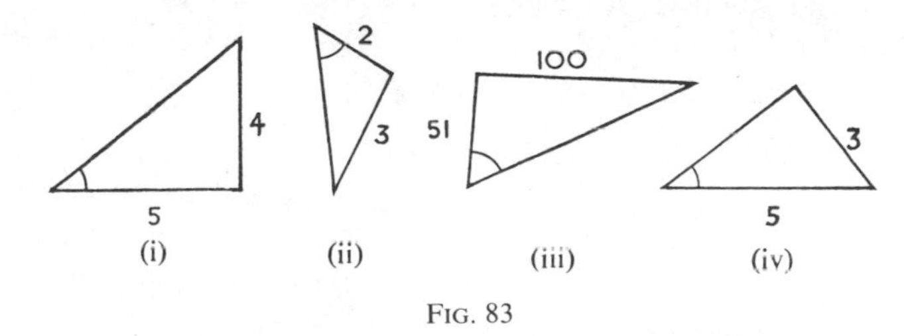

FIG. 83

8 Find the marked angles in Fig. 83, (i), (ii), (iii), (iv), given that the triangles are right-angled.

9 Find the lengths of marked sides in Fig. 84, (i), (ii), (iii), (iv), given that the triangles are right-angled.

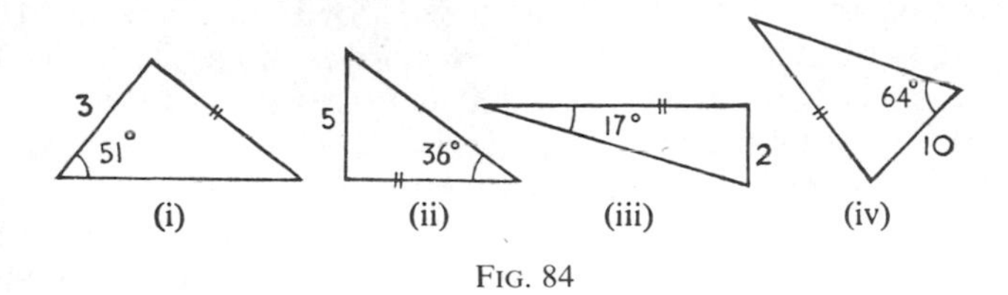

FIG. 84

10 A ladder leaning against a vertical wall makes an angle of 21° with the wall; the foot of the ladder is 2m from the wall. How high up the wall does the ladder reach?

11 A man starts from O and walks 2 km north and then $\frac{1}{2}$ km east. What is his bearing from O? What is his new bearing from O when he walks another $\frac{1}{2}$ km east?

[12] From a point on the ground, 100 m away from a tower, the angle of elevation of the top of the tower is 33° 30′. Find the height of the tower.

13 From the top of a cliff 100 m high the angle of depression of a boat is 17°. Find the distance of the boat from the foot of the cliff below the observer.

14 What is the angle of elevation of the top of a spire 80 m high from a point on the ground 200 m from the foot of it?

15 The steps of a staircase are 20 cm deep and 12 cm high; what angle does the bannister rail make with the horizontal?

[16] The shadow of a vertical pole 4 m high is 5·5 m long. What is the altitude of the sun?

Example 8 AB, AC are the equal sides of an isosceles triangle ABC; BC = 6 cm and ∠BAC = 48°. Calculate the length of the perpendicular AD from A to BC.

Since AB = AC,

$$\angle B = \angle C = \tfrac{1}{2}(180^\circ - \angle BAC) = 66^\circ,$$

and

$$DB = DC = \tfrac{1}{2}BC = 3\text{ cm};$$

$$\therefore \frac{AD}{BD} = \tan ABD = \tan 66^\circ,$$

$$\therefore AD = 3 \tan 66^\circ \text{ cm}$$

$$= 3 \times 2{\cdot}2460 \text{ cm}$$

$$= \mathbf{6{\cdot}74\ cm}, \text{ to 3 figures.}$$

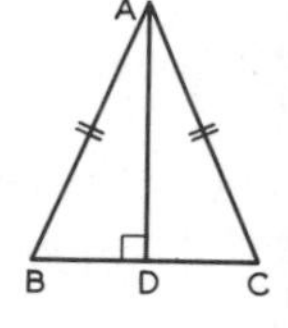

FIG. 85

Contour Lines In a map of a district, the features are represented by supposing that they are all flattened down to sea-level. If you look at an ordnance map you will see a series of curves on it. Each point on the curve marked, say, 200 m represents a place which is 200 m above sea-level and this curve is called a **200 m contour line.** On a map whose Representative Fraction is 1:50 000, the contour lines are usually at 10 m intervals: 0, 10, 20, 30, ... m above sea-level. In addition, 'spot heights' of peaks of hills are recorded on the map.

Example 9 A map, scale 5 cm to 1 km, shows a straight road crossing the contour lines 70 m, 100 m at P, Q. The length of PQ on the map is 1·9 cm. If P′, Q′ are the points on the ground represented by P, Q, find the angle of elevation of Q′ from P′.

The perpendiculars P′P, Q′Q from P′, Q′ to a horizontal plane at sea-level are 70 m, 100 m long, see Fig. 86 (ii), not drawn to scale.

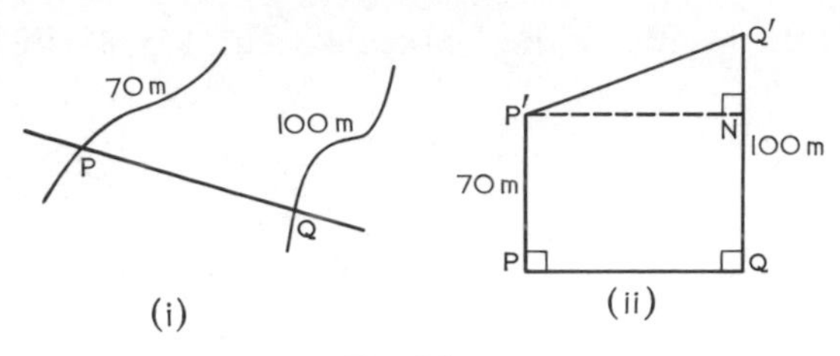

FIG. 86

Since the scale of the map is 5 cm to 1 km, the line PQ on the map represents 0·38 km, or 380 m. Draw a horizontal line P′N through P′ to meet the vertical line Q′Q at right angles at N, see Fig. 86 (ii); then

$$Q'N = Q'Q - NQ = Q'Q - P'P = (100 - 70)\text{ m} = 30\text{ m};$$

$$\therefore \tan Q'P'N = \tfrac{30}{380} = 0{\cdot}079.$$

$$\therefore \angle Q'P'N = 4^\circ\ 30';$$

∴ the angle of elevation of Q′ from P′ is **4° 30′**.

Notice that the distance P′Q′ direct from the place P′ to the place Q′ is greater than the distance PQ obtained from the map. P and Q are called the *projections* of P′ and Q′ on a horizontal plane at sea-level.

EXERCISE 35

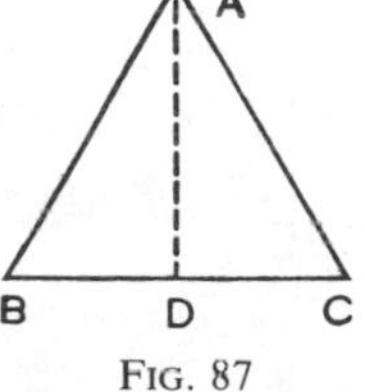

FIG. 87

1 ABC is an isosceles triangle with AB = AC; AD is drawn perpendicular to BC.
(i) Find AD if ∠B = 37°, BC = 6 cm.
(ii) Find BC if ∠B = 42°, AD = 5 cm.
(iii) Find AD if ∠A = 96°, BC = 4 cm.

2 The vertical angle of a cone is 102° and the diameter of its base is 5 cm. What is its height?

[3] The pole of a bell-tent is 2·4 m high, and the diameter of the base of the tent is 4·2 m. What angle does the slant side of the tent make with the ground?

4 PQ is a chord of a circle, centre O; PQ = 6 cm, ∠POQ = 103° 30′. Calculate the length of the perpendicular from O to PQ.

5 Using tables, find the angles A and B if

(i) $\tan A = 2$, $\tan B = \frac{1}{2}$, (ii) $\tan A = \frac{3}{5}$, $\tan B = \frac{5}{3}$.

Find the value of A + B in each case and make sketches to explain the two answers.

6 The sides of a rectangle are 4, 5 cm long. What is the angle between the diagonals?

[7] Find a value of θ if $\tan \theta^\circ = 3 \tan 20^\circ$.

8 Find the lengths of the marked sides in Fig. 88, (i), (ii), (iii).

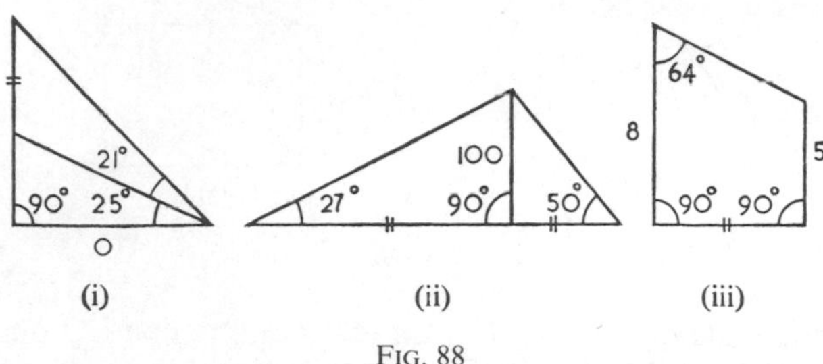

FIG. 88

[**9**] The greatest and least heights of a lean-to shed are 3·1 m and 2 m; the floor is 4·8 m wide. Find the slope of the roof.

10 The diagonals of the quadrilateral ABCD cut at right angles at N; AN = 2 cm, NC = 4 cm, BN = 3 cm, ND = 5 cm. Find the angles of the quadrilateral.

11 ABC is a triangle in which AB = 3 cm, AC = 7 cm, $\angle$BAC = 90°, P is a point on AC such that $\angle PBA = \frac{1}{4}\angle CBA$. Find (i) $\angle$PBA, (ii) PC.

12 Fig. 89 represents the section of a railway cutting; the base BC is horizontal and 5 m wide; the tops A, D are each 6 m above BC. Find AD. [Draw BH, CK perpendicular to AD.]

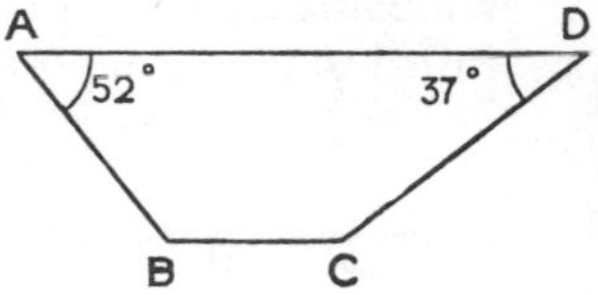

FIG. 89

[**13**] In $\triangle$ABC, $\angle$B = 132°; $\angle$C = 28°; AD is the perpendicular from A to CB produced; AD = 7 cm. Find BC.

14 The triangles $\begin{matrix}\text{ABP}\\ \text{ABQ}\end{matrix}$ are congruent; PQ = 8 cm, $\angle$PAQ = 47°, $\angle$PBQ = 75°. Find AB.

15 A and B are points 0·9 cm apart on a map, scale 1 cm to the kilometre, which lie on the 50 m and 130 m contour lines. Calculate the angle of elevation of B from A.

[**16**] The angle of elevation of a peak P of a mountain from a point A is 17° 30′; A lies on the 660 m contour line and the spot height of P is 1060 m. Find correct to the nearest millimetre the distance on a map, scale 4 cm to 1 km, between the points representing A and P.

17 A map, scale 10 cm to 1 km, shows a straight line ABC which represents a path running up to the peak C of a steep hill. A and B lie on the 30 m and 130 m contour lines and the spot height of C is 205 m. On the map, AB = 3·2 cm, BC = 2·9 cm. Find the angles of elevation of B from A and of C from B. Is C visible from A?

[18] Two cones have as common base a circle, radius 8 cm; their vertical angles are 114° and 78°. Find the length of the line joining their vertices. [Two answers.]

19 N is a point on the side AB of the rectangle ABCD; AN = 3 cm, NB = 7 cm, BC = 8 cm. Find ∠CND.

[20] BCD is a road running west; A is 7 km due north of B and bears N. 32° E. from C, N. 50° E. from D. Find BC and CD.

21 The shadow of a vertical pole is 3 m long when the sun's elevation is 35°. What is the length of the shadow when the sun's elevation is 25°?

[22] The angle of elevation of the top of a tower from a point on the ground 120 m from its foot is 21° 48′. What will it be from a point on the ground 20 m nearer the tower?

23 From the top of a cliff 100 m high the angles of depression of two boats in a vertical plane with the observer are 25° 24′, 37° 52′. Find the distance between the boats.

[24] ABC is a triangle in which AB = 3 cm, BC = 7 cm, ∠B = 90°. The line bisecting ∠BAC cuts BC at P. Find (i) ∠BAC, (ii) PC.

25 In Fig. 90, ∠PQS = ∠QSR = 90°, ∠QPS = 48° 33′, ∠QRS = 57° 16′, QS = 10 cm. Calculate PQ, RS, ∠PKQ.

26 ABCD is a quadrilateral in which AB = 6 cm, CD = 5 cm, BD = 8 cm, ∠ABD = ∠BDC = 90°. Find the acute angle at which AC cuts BD, see Fig. 91.

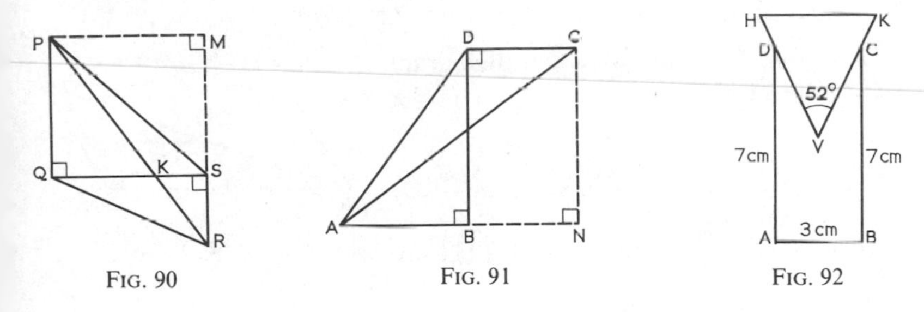

FIG. 90 FIG. 91 FIG. 92

27 Fig. 92 represents a vertical section of an inverted cone, vertex V, horizontal base HK, resting on the upper rim of an open circular cylinder whose base AB is horizontal. Find the height of V above AB.

28 ABC is a triangle in which ∠B = 37° 15′, ∠C = 59° 40′, BC = 8 cm; the line which bisects BC at right angles cuts BA, CA produced at P, Q. Find PQ.

29 From a window 10 m above the ground, the angle of depression of the foot of a tower is 23° 30′, and the angle of elevation of the top of the tower is 50° 12′. Find the height of the tower.

CHAPTER 8

PARALLELOGRAMS

Definition A quadrilateral which has both pairs of opposite sides parallel, see Fig. 93, is called a **parallelogram.**

Abbreviation: ‖gram *or* parm.

The properties of a parallelogram and the tests for a quadrilateral to be a parallelogram are discussed in Exercise 36 and give useful practice in rider-work. Formal proofs of Nos. 1–3 are given on pp. 269–271, but at this stage familiarity with the facts and an understanding of their proofs is more important than ability to reproduce the proofs.

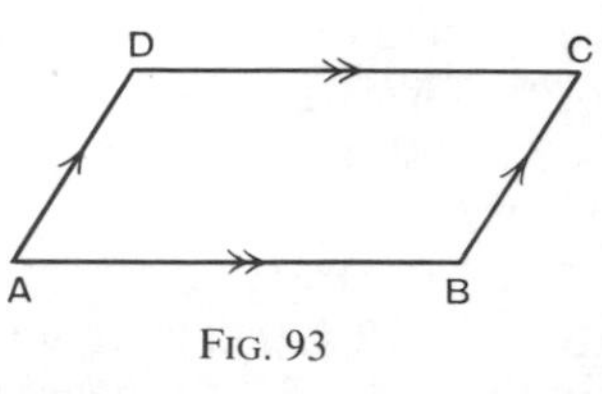

FIG. 93

EXERCISE 36 (Class Discussion)

Properties of a Parallelogram

1 ABCD is a parallelogram. Prove that

(i) $AB = DC, AD = BC$; (ii) $\angle A = \angle C, \angle B = \angle D$;

(iii) **Each diagonal bisects the area of the parallelogram.**

Draw your own figure. Insert arrows to show what is given.
Join BD and give the reasons why $\triangle ABD \equiv \triangle CDB$.
Reference: (i) opp, sides ‖gram; (ii) opp. $\angle$s ‖gram.

2 If the diagonals of a parallelogram ABCD cut at K, prove

$AK = KC$ **and** $BK = KD$.

Give the reasons why $\triangle AKB \equiv \triangle CKD$. (Use No. 1.)

Abbreviation for reference: diags ‖gram.

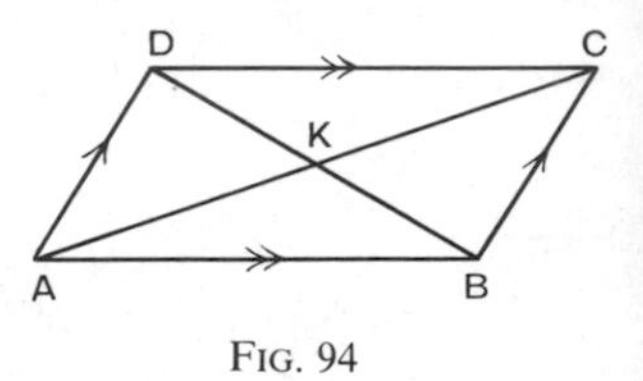

FIG. 94

Tests for a Parallelogram

3 ABCD is a quadrilateral in which $AB = DC$ and AB is parallel to DC. Prove that ABCD is a parallelogram.

Draw your own figure and show on it by suitable markings what is given.

Join BD. Give the reasons why $\triangle ABD \equiv \triangle CDB$.

Hence prove AD is parallel to BC. Why is this sufficient?

Abbreviation for reference: 2 sides equal and ‖.

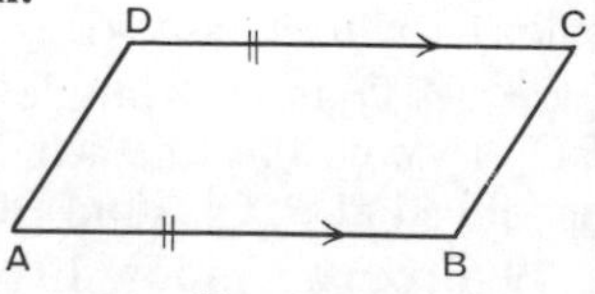

FIG. 95

4 ABCD **is a quadrilateral in which** $\angle A = \angle C$ **and** $\angle B = \angle D$. **Prove that** ABCD **is a parallelogram.**

Draw your own figure and show on it by suitable markings what is given. No construction.

What is the sum of the angles of any quadrilateral?

Use the data to prove that AB is parallel to DC and that AD is parallel to BC.

5 ABCD **is a quadrilateral in which** AB = DC **and** AD = BC. **Prove that** ABCD **is a parallelogram.**

Draw your own figure and show on it by suitable markings what is given. Join BD. Give the reasons why $\triangle ABD \equiv \triangle CDB$.

Hence prove AB is parallel to DC and AD is parallel to BC.

6 ABCD **is a quadrilateral in which the diagonals bisect each other at** K, **i.e.** AK = KC **and** BK = KD.

Prove that ABCD **is a parallelogram.**

Use $\triangle$s AKB, CKD to prove that AB is parallel to DC.

How can you prove that AD is parallel to BC?

7 (*Discovery Example.*) ABCD is a quadrilateral in which AB = CD and AD is parallel to BC. Does it follow that ABCD *must* be a parallelogram? *If it does*, give reasons fully. *If it does not*, explain why not.

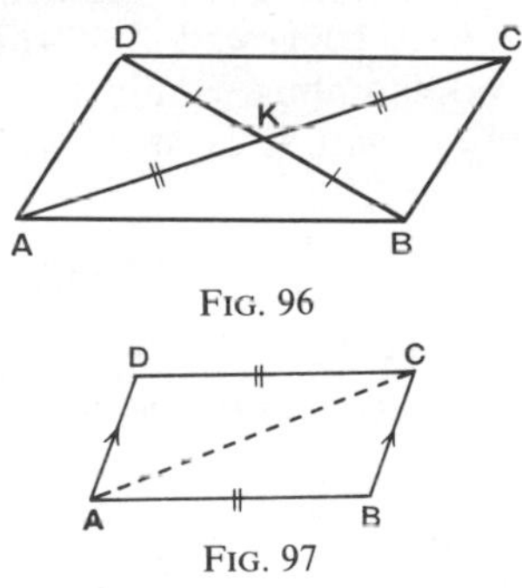

FIG. 96

FIG. 97

Rectangle, Square, Rhombus and Trapezium

A parallelogram in which *one* angle is a right angle is called a **rectangle** (*abbreviation*, rect.).

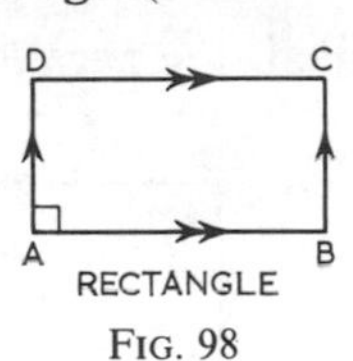

FIG. 98

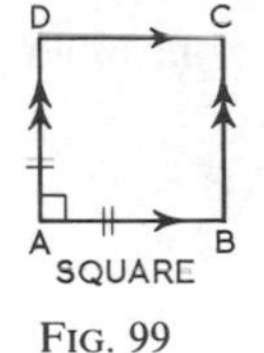

FIG. 99

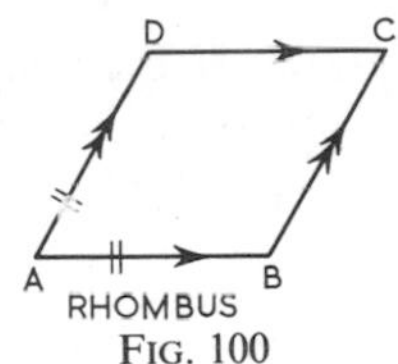

FIG. 100

A rectangle which has *two adjacent* sides equal is called a **square** (*abbreviation*, sq.). The square ABCD is often called *the square on* AB, or *sq. on* AB.

A parallelogram which has *two adjacent* sides equal is called a **rhombus.**

A quadrilateral which has one, and only one, pair of sides parallel is called a **trapezium.** (See Fig. 101 on the next page.) If the other two sides are equal, the trapezium is called **isosceles.**

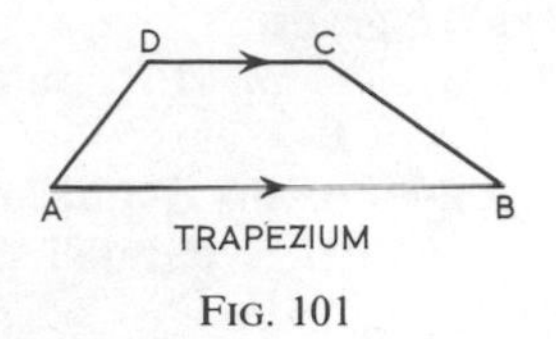

Fig. 101

These definitions must be noted carefully. A definition must include sufficient to be free from ambiguity, but must not include facts that can be deduced from the definition. For example, a rectangle is *defined* as a parallelogram in which *one* angle is a right angle; it can then be **proved** that the remaining three angles are also right angles; it is wrong to include this fact in the definition.

It is suggested that the chief properties of the rectangle, square and rhombus should be taken orally as riders, see Exercise 37. *The results may be assumed in rider-work unless a proof is explicitly asked for.*

EXERCISE 37 (Class Discussion)

I. Properties of a Rectangle

(i) **All the angles of a rectangle are right angles.**

If $\angle A = 1$ rt. $\angle$,

then

$$\angle B = \angle C = \angle D = 1 \text{ rt. } \angle.$$

[*No construction:* use parallels.]

(ii) **The diagonals of a rectangle are equal.**

[Prove that $\triangle DAB \equiv \triangle CBA$.]

Fig. 102

(iii) **If the diagonals of a parallelogram are equal, the parallelogram is a rectangle.**

[Prove that $\triangle DAB \equiv \triangle CBA$; why does $\angle A + \angle B = 2$ rt. $\angle$s?]

II. Properties of a Square

(i) **All the sides of a square are equal.**

If $AB = AD$

then $AB = BC = CD$.

Fig. 103

[*No construction*: opp. sides ||gram are equal.]

(ii) **The diagonals of a square are equal and cut at right angles.**

(iii) **The angle which a diagonal makes with a side of the square is 45°.**

(iv) **If the diagonals of a parallelogram are equal and cut at right angles, the parallelogram is a square.**

III. Properties of a Rhombus

(i) **All the sides of a rhombus are equal.**

If AB = AD,

then AB = BC = CD.

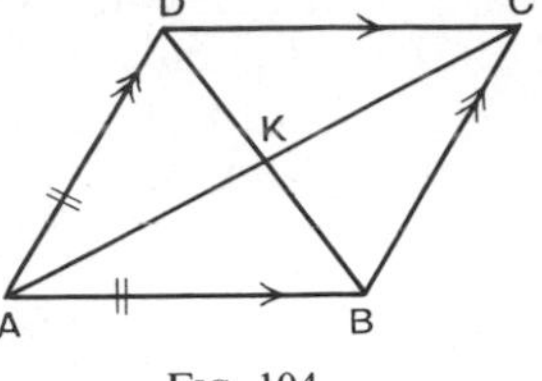

FIG. 104

(ii) **The angles of a rhombus are bisected by the diagonals.**

(iii) **The diagonals of a rhombus cut at right angles.**

(iv) **If the diagonals of a parallelogram cut at right angles, the parallelogram is a rhombus.**

Summary Familiarity with the properties which have been discussed is gained by working numerical examples, see Exercise 38, and is essential both for the rider-work in Exercise 39 and for the constructions in Exercise 40. The reader may find the following summary useful when tackling these exercises.

(1) **In any parallelogram**
- (a) opposite sides are parallel (this is really the definition);
- (b) opposite angles are equal;
- (c) opposite sides are equal;
- (d) the diagonals bisect each other;
- (e) each diagonal bisects the area.

(2) **A quadrilateral is a parallelogram, if**
- (a) both pairs of opposite sides are parallel;
- (b) both pairs of opposite angles are equal;
- (c) both pairs of opposite sides are equal;
- (d) one pair of opposite sides are equal and parallel;
- (e) the diagonals bisect each other.

(3) **In any rectangle,**
- (a) all the angles are right angles;
- (b) the diagonals are equal.

(4) **In any square,**
- (a) the diagonals are equal;
- (b) the diagonals cut at right angles;
- (c) the angle which each diagonal makes with each side of the square is 45°.

(5) **In any rhombus,**
- (a) all the sides are equal;
- (b) the diagonals cut at right angles;
- (c) the angles are bisected by the diagonals.

When constructing a figure from numerical data,

(1) make a neat sketch of the required figure;

(2) mark on your sketch the given measurements;

(3) try to find or draw some *triangle* in the figure which can be constructed from the data, or by deductions from the data, see Examples 1, 2, p. 78.

EXERCISE 38

1 ABCD is a rectangle. If $\angle BAC = 32°$, find $\angle DBC$.

[**2**] ABCD is a rectangle. If $\angle ACD = 67°$, find $\angle ADB$.

3 ABCD is a rhombus. If $\angle ABC = 56°$, find $\angle ACD$.

[**4**] ABCD is a rhombus. If $\angle BAC = 35°$, find $\angle ADC$.

5 The diagonals of a rectangle ABCD cut at K. If $\angle AKB = 110°$, find $\angle ACB$ and $\angle ACD$.

[**6**] The diagonals of a rectangle ABCD cut at N. If $\angle ABD = 74°$, find $\angle DNC$.

7 ABCD is a square; a straight line CPQ cuts BD at P and cuts BA at Q. If $\angle CPD = 80°$, find $\angle CQA$.

8 ABCDE is a regular pentagon; ABPQ is a square inside the pentagon. Find $\angle CBP$ and $\angle DBQ$.

[**9**] ABC is an equilateral triangle; BCPQ and BCHK are the two squares on BC. Find $\angle APB$ and $\angle AHB$.

10 The diagonals of a square ABCD cut at K. From AB a part AQ is cut off equal to AK. Prove that $\angle AKQ = 3\angle BKQ$.

[**11**] The side AD of the square ABCD is produced to E, and the bisector of $\angle EDB$ meets AC produced at X. Find $\angle AXD$ and prove that CD = CX.

12 ABCD is a square; H is a point on AC such that AH = AB; K is a point on CB such that CK = CH. Prove that $\angle BHK = 45°$.

[**13**] ABCDE is a regular pentagon; ABP is an equilateral triangle inside the pentagon. Find the angles of $\triangle PEB$.

14 The diagonals of a rectangle ABCD cut at K; KAP is an equilateral triangle drawn so that B and P are on the same side of AC. If $\angle ACD = 25°$, find the angles of $\triangle APB$.

***15** ABCD is a square; ABX is an equilateral triangle inside the square. Find $\angle DXC$.

***16** ABCDE is a regular pentagon and EDCP is a parallelogram inside the pentagon. Find $\angle EPA$ and prove that APC is a straight line.

***17** The diagonals of the rectangle ABCD cut at K, and AK is greater than AB. The circle, centre A, radius AK, cuts AB produced at E. If $\angle AKB = 4\angle BKE$, find $\angle BAC$.

***18** ABCD is a square; P is a point on CA produced such that the parallelogram DAPQ is a rhombus. If QC cuts PD at R, find the angles of △DRQ and prove that RP = RC.

EXERCISE 39

1 State, without proof, what you know about the parallelogram ABCD (i) if AC bisects ∠BAD, (ii) if AC = BD, (iii) if AC is perpendicular to BD.

2 In Fig. 105, ABCD and ABXY are parallelograms such that DCYX is a straight line. Use the **SAA** test to prove that △ADY ≡ △BCX.

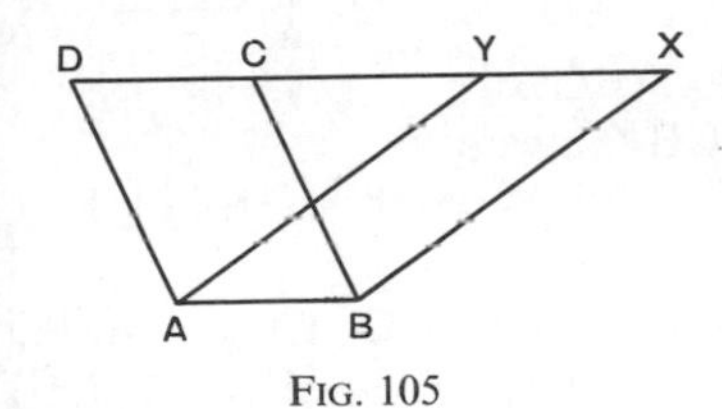

FIG. 105

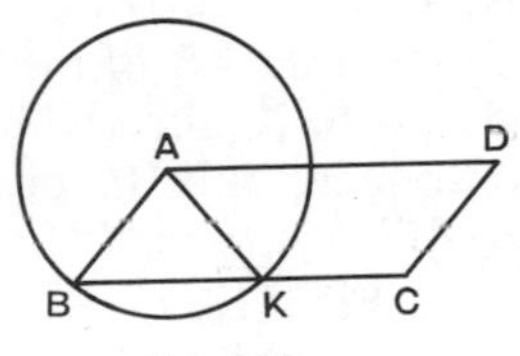

FIG. 106

3 In Fig. 106, ABCD is a parallelogram and A is the centre of the circle. Prove that ∠KAD = ∠CDA. [No construction.]

4 The diagonals of the parallelogram ABCD cut at K; any line through K cuts AB, CD at X, Y. Prove that KX = KY.

[5] P is the mid-point of the side BC of a parallelogram ABCD; DP and AB meet, when produced, at Q. Prove that AB = BQ.

[6] ABCD is a parallelogram. Prove that the perpendiculars from B and D to AC are equal.

7 In Fig. 107, ABCD and ABPQ are parallelograms. Prove that CDQP is a parallelogram.

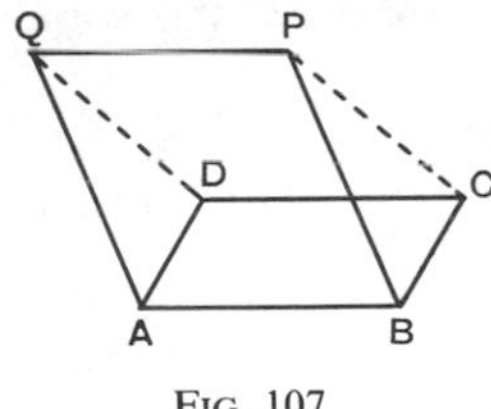

FIG. 107

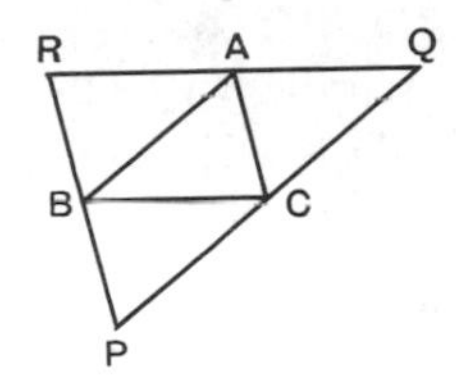

FIG. 108

8 In Fig. 108, △PQR is formed by drawing lines through A, B, C parallel to BC, CA, AB respectively. Prove that A, B, C are the mid-points of the sides of △PQR.

[9] Two equal circles, centres A, B, cut at C, D. Prove that ACBD is a rhombus.

10 ABCD is a rhombus. If the bisector of ∠DAC cuts CD at P, prove that ∠DPA = 3∠DAP.

[**11**] ABCD is a parallelogram such that the bisectors of ∠A and ∠B meet on CD. Prove that AB = 2BC.

12 In △ABC, ∠A is a right angle; ABPQ and ACXY are squares lying outside △ABC. Prove that PAX is a straight line.

13 In Fig. 109, ABCD and APQR are squares. Prove that BP = DR.

[**14**] ABCD is a parallelogram; ABLM and BCHK are squares outside the parallelogram. Prove that (i) ∠LBK = ∠BCD, (ii) KL = BD.

FIG. 109

[**15**] The side AB of the parallelogram ABCD is produced to X, and the bisector of ∠CBX meets DA produced and DC produced at E and F. Prove that DE = DF = BA+BC.

16 In Fig. 110, H, K are the mid-points of AB, AC; CP is parallel to BA, and HKP is a straight line. Prove that (i) CP = AH, (ii) CPHB is a parallelogram, (iii) HK = $\frac{1}{2}$BC.

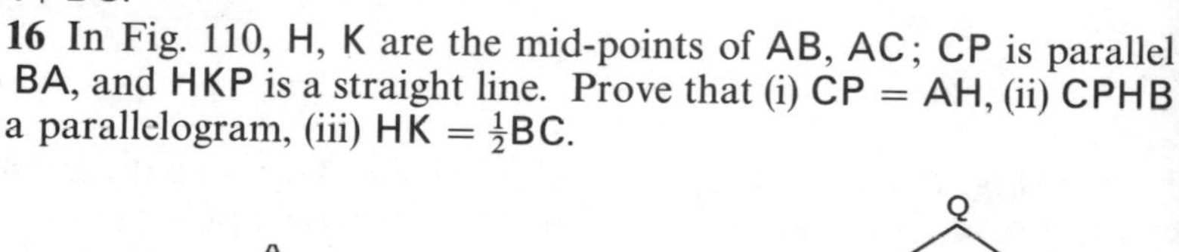

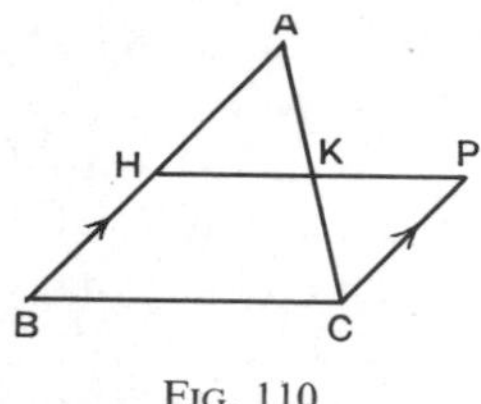

FIG. 110

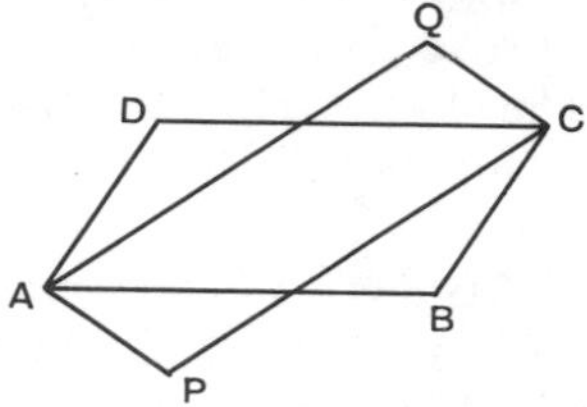

FIG. 111

17 In Fig. 111, ABCD and APCQ are parallelograms. Prove that (i) AC, BD, PQ are concurrent, *i.e.* pass through the same point; (ii) PB is parallel to DQ.

[**18**] ABCD is a rhombus; PABQ is a straight line such that PA = AB = BQ. Prove that PD and QC when produced cut at right angles.

Example 1 (Class Discussion) **Construct** a parallelogram ABCD, given AB = 6 cm, AC = 10 cm, BD = 8 cm. Measure AD.

Sketch a parallelogram ABCD and let AC cut BD at K, see Fig. 112. Complete: AK = KC = ..., BK = KD = ...; then construct △AKB and use it to construct the points C, D. Measure AD.

Example 2 (Class Discussion) **Construct** a trapezium ABCD, in which DC is parallel to AB, given AB = 8·5 cm, BC = 3·5 cm, CD = 4·5 cm, DA = 3 cm. Measure BD.

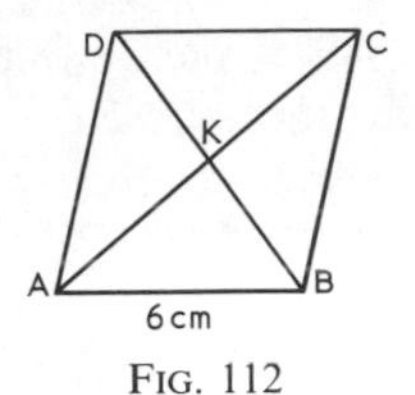

FIG. 112

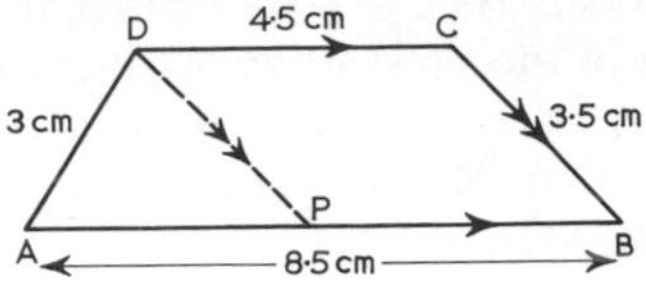

FIG. 113

Mark the data on a *sketch* of the trapezium ABCD, see Fig. 113. In your sketch, draw DP parallel to CB to meet AB at P. Complete: PB = ..., ∴ AP = ..., and DP = ...; then construct △APD and use it to construct the points B, C. Measure BD.

EXERCISE 40

[*Use only ruler and compasses for the constructions in Nos. 1–12*]

1 Construct a square ABCD, side 3 cm; measure AC.

2 Construct a rectangle ABCD, given AB = 4 cm, AC = 6 cm; measure AD.

[**3**] Construct a rhombus ABCD, given AB = 5 cm, ∠BAD = 60°; measure AC.

4 Construct a parallelogram ABCD, given AB = 4 cm, AD = 5 cm, ∠BAD = 45°; measure AC.

5 Construct a triangle ABC, given that BC = 6 cm, ∠ABC = 75°, ∠ACB = 45°. Construct and measure the perpendicular from A to BC.

[**6**] Construct a square ABCD, given AC = 5 cm. Measure AB.

7 Construct a triangle ABC, given AB = AC, ∠BAC = 30°, BC = 6 cm. Construct the perpendicular bisector of AB, meeting AC at K; measure AK.

8 Construct a rhombus ABCD, given AC = 6 cm, BD = 9 cm; measure AB.

[**9**] Construct a parallelogram ABCD, given that AC = 6 cm, BD = 8 cm, and that AC makes $67\frac{1}{2}$° with BD. Measure the sides.

10 Construct a parallelogram ABCD, given AC = 7 cm, BD = 9 cm, BC = 5 cm. Measure AB.

11 Construct a regular hexagon ABCDEF, side 5 cm; measure AC.

12 Construct a regular octagon (8 sides), side 4 cm; measure a diagonal.

Set-squares and protractors may be used in the following constructions:

13 Construct a rectangle ABCD given that BD = 8 cm and that AC cuts BD at an angle of 54°. Measure the sides.

[**14**] Construct △ABC given that ∠A = 70°, ∠C = 35°, and the length of the perpendicular from A to BC is 4 cm; measure BC.

15 Construct $\triangle ABC$ given that $\angle C = 68°$, $AB = 6$ cm, and the length of the perpendicular from A to BC is 4 cm; measure BC.

16 D is a point on the side BC of an equilateral triangle ABC. Given that $BD = 3$ cm and $\angle DAC = 40°$, construct $\triangle ABC$ and measure BC.

[**17**] K is a point between two parallel lines AB, CD and at distances 2 cm, 3 cm from AB, CD. Draw the figure and construct a line PKQ meeting AB, CD in P, Q so that $PQ = 6$ cm.

18 The perpendicular distances between the opposite sides of a parallelogram are 3 cm, 4 cm, and one angle is 70°. Construct the parallelogram and measure one of its longer sides.

19 Construct a trapezium ABCD, given AB is parallel to DC, $AB = 5$ cm, $BC = 6$ cm, $CD = 2$ cm, $DA = 4$ cm; measure $\angle A$.

[**20**] Construct a trapezium ABCD, given AB is parallel to DC, $AB = 8$ cm, $CD = 5$ cm, $\angle A = 72°$, $\angle B = 40°$; measure BC.

***21** Construct a trapezium ABCD, given AB is parallel to DC, $AB = 6{\cdot}5$ cm, $CD = 3$ cm, $AC = 7$ cm, $BD = 5$ cm. Describe your method. [In your sketch, complete parallelogram CDBP.]

***22** Construct a triangle ABC in which $\angle B = 60°$, $\angle C = 40°$, perimeter $= 9$ cm. Measure BC. [In your sketch, produce CB to P and BC to Q so that $BP = BA$, $CQ = CA$; what are the sizes of $\angle APQ$, $\angle AQP$? Hence construct $\triangle APQ$.]

CHAPTER 9

SIMULTANEOUS EQUATIONS AND PROBLEMS

Example 1 Two tanks, A and B, contain respectively 120 litres and 300 litres of water. If water is allowed to run into A at the rate of 10 litres per minute, and to run out of B at the rate of 20 litres per minute, find a formula for the number of litres n in tank A after t minutes and a similar formula for tank B.

Water runs into A at 10 litres per minute;

$\therefore$ in t minutes, $10t$ litres run into A;

$\therefore$ after t minutes, there are $(120+10t)$ litres in A.

$$\therefore \text{ for tank A, } \boldsymbol{n = 120+10t}.$$

Water runs out of B at 20 litres per minute;

$\therefore$ in t minutes, $20t$ litres run out of B;

$\therefore$ after t minutes, there are $(300-20t)$ litres in B.

$$\therefore \text{ for tank B, } \boldsymbol{n = 300-20t}.$$

Since the amount of water in A steadily increases, and that in B steadily decreases, there must come a single moment when A and B contain equal amounts of water. This means that there is one value of t which makes the value of n for tank A the same as the value of n for tank B.

At this moment, *and only at this moment*, the equations

$$n = 120+10t; \quad n = 300-20t$$

are true for each tank, and we call them **simultaneous equations.**

If the equations $\begin{cases} n = 120+10t \\ n = 300-20t \end{cases}$ are simultaneous,

then $\quad 120+10t = 300-20t; \quad \therefore 10t+20t = 300-120;$

$$\therefore 30t = 180; \quad \therefore t = \mathbf{6}.$$

Also $n = 120+10t = 120+60 = \mathbf{180}.$

Check: Take the other equation,

$$n = 300-20t = 300-20\times 6 = 300-120 = 180, \text{ as before.}$$

This shows that in 6 minutes' time there will be equal amounts of water, namely 180 litres, in the two tanks. At no other time will this be true.

EXERCISE 41 (Oral)

1 A and B have £110 and £600 respectively in the bank. If A increases his bank balance by £30 every year and if his balance is £P after n years, find P in terms of n. If B decreases his bank balance by

£40 every year and if his balance is £P after n years, find P in terms of n.

Find also the time at which their balances are equal and the amount of the balance at this time.

2 A joins a club for which he has to pay an entrance fee of £10 and a subscription of £9 a year. If the total cost for t years is £n, find n in terms of t. B joins another club for which he has to pay an entrance fee of £25, and a subscription of £6 a year. If the total cost for t years is £n, find n in terms of t.

Find also the number of years for which the total costs of the two clubs are the same and what this cost is.

3 Find the formula for the tax £T on an income of £P, calculated by each of the following rules:

Rule I. No tax on the first £200 of a man's income, and 50p in the £ on the rest. [Assume P is greater than 200.]

Rule II. No tax on the first £160 of a man's income, and 40p in the £ on the rest.

For what income do the two rules give the same tax, and how much is this tax?

4 When a load is attached to the hook of the spring A (see Fig. 114), the spring stretches 2 cm for each 100 g weight attached. If the *total* length of spring A, when a load of weight W grammes is attached, is l cm, find l in terms of W.

When a load is attached to the spring B, the spring stretches 5 cm for each 100 g weight attached. If the *total* length of spring B, when a load of weight W grammes is attached, is l cm, find l in terms of W.

Find also what load makes the total lengths equal, and what this length is.

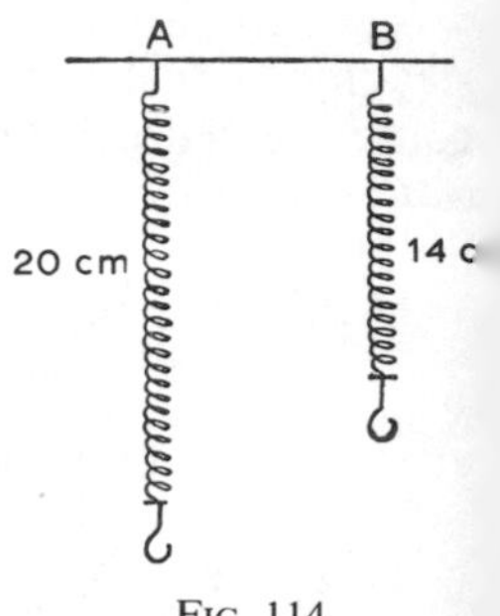

FIG. 114

There are two methods for solving simultaneous equations.

Method I. Solution by Substitution

Example 2 Solve the simultaneous equations:

$$3x+2y = 21 \qquad \text{(i)}$$

$$2x+5y = 3 \qquad \text{(ii)}$$

From (i),

$$2y = 21-3x,$$

$$\therefore y = \frac{21-3x}{2} \qquad \text{(iii)}$$

Substitute this value of y in (ii), then

$$2x+\frac{5(21-3x)}{2} = 3,$$

$$\therefore\ 4x+105-15x = 6; \qquad \therefore\ -11x = -99,$$

$$\therefore\ x = 9.$$

Put $x = 9$ in (iii), $\qquad \therefore\ y = \dfrac{21-27}{2} = -3,$

$$\therefore\ \boldsymbol{x = 9, \quad y = -3.}$$

Check: If $x = 9$ and $y = -3$, $2x+5y = 18-15 = 3$, as in (ii). The value of y was found by substituting $x = 9$ in (iii), which is equivalent to (i). We therefore use (ii) for the check, *not* (i).

EXERCISE 42 (Oral)

Find y in terms of x in Nos. 1–3:

1 $3x-y = 11$ **2** $5x+2y = 12$ **3** $\frac{1}{3}x-4y = 16$

Find x in terms of y in Nos. 4–6:

4 $x+5y = 3$ **5** $4x+9y = 0$ **6** $\frac{1}{3}x = \frac{1}{5}(y-1)$

Is it better to find x in terms of y or to find y in terms of x when solving the following? *Do not solve them.*

7 $7x-2y = 4$, $3x-y = 1$ **8** $2x-5y+3 = 0$, $3x+4y-1 = 0$ **9** $\frac{1}{2}y+x = 1$, $\frac{1}{3}y-7x = 4$

Method II. Solution by Addition or Subtraction

If the method by substitution involves awkward fractions, it is easier to use the method by addition or subtraction.

Example 3 Solve the simultaneous equations:

$$3x-2y = 11 \qquad . \quad . \quad . \quad . \qquad \text{(i)}$$

$$5x+2y = 29 \qquad . \quad . \quad . \quad . \qquad \text{(ii)}$$

The result of adding the left side of (i) to the left side of (ii) equals the result of adding the right sides; but if we do this, the term in y disappears, leaving a simple equation in x.

Thus, $\qquad 3x+5x = 11+29; \qquad \therefore\ 8x = 40;$

$$\therefore\ x = 5.$$

Put $x = 5$ in (i), then $\quad 15-2y = 11; \qquad \therefore\ -2y = -4;$

$$\therefore\ y = 2.$$

$\therefore$ the solution is $\boldsymbol{x = 5, \quad y = 2.}$

Check: If $x = 5$ and $y = 2$, $5x+2y = 25+4 = 29$, as in (ii).

Example 4 Solve the simultaneous equations:

$$2x-5y = 27 \qquad . \quad . \quad . \quad . \qquad \text{(i)}$$
$$2x+3y = 3 \qquad . \quad . \quad . \quad . \qquad \text{(ii)}$$

If we subtract, thc term in x disappears.

Subtracting, $-5y-3y = 27-3$; $\therefore -8y = 24$;

$$\therefore y = -3.$$

Put $y = -3$ in (ii), then $2x-9 = 3$; $\therefore 2x = 12$;

$$\therefore x = 6.$$

$\therefore$ the solution is $\boldsymbol{x = 6}$, $\boldsymbol{y = -3}$.

Check: If $x = 6$ and $y = -3$, $2x-5y = 12+15 = 27$, as in (i).

The process of getting rid of one of the unknowns is called **elimination.** In Example 3 we eliminated y, and in Example 4 we eliminated x. **It does not matter which unknown is eliminated, always do whichever is easier.**

Example 5 Solve the simultancous equations:

$$7x-6y = 20 \qquad . \quad . \quad . \quad . \qquad \text{(i)}$$
$$3x+4y = 2 \qquad . \quad . \quad . \quad . \qquad \text{(ii)}$$

The L.C.M. of 6 and 4 is 12; therefore if we multiply each side of (i) by 2, and each side of (ii) by 3, we shall obtain equations in which the coefficients of y are *numerically* equal.

It is easier to do this than to make the coefficients of x equal, because in this case we should have to multiply by 3 and 7 respectively.

Multiply each side of (i) by 2, $\therefore 14x-12y = 40$ . . (iii)

Multiply each side of (ii) by 3, $\therefore 9x+12y = 6$ . . (iv)

From (iii) and (iv) by adding, $23x = 46$,

$$\therefore x = 2.$$

Put $x = 2$ in (ii), $\therefore 6+4y = 2$;

$$\therefore 4y = -4;$$
$$\therefore y = -1.$$

$\therefore$ the solution is $\boldsymbol{x = 2}$, $\boldsymbol{y = -1}$.

Check: If $x = 2$ and $y = -1$, $7x-6y = 14+6 = 20$, as in (i).

We must use (i) for the check because the value of y was found by substituting in (ii).

General Instructions

(i) First decide which unknown it is easier to eliminate.

(ii) When one unknown has been found, obtain the other by substituting the EASIEST equation you have containing it.

(iii) When checking, use that one of the ORIGINAL equations which is not equivalent to the equation used for substitution.

(iv) If the answers involve awkward fractions or decimals, it is often better to look over your working again or solve by an independent method, instead of checking by substitution.

(v) Number the chief equations, and use the numbering to explain your method. A way of abbreviating the explanation is indicated on **p. 89.**

EXERCISE 43

Solve the following pairs of simultaneous equations and check the answers. In each case state *at the start* which unknown can be climinated the more easily, or whether it makes no difference.

1 $x+y = 13$, $x-y = 1$

2 $u = 3v$, $u = 15-2v$

3 $3p+q = 11$, $p+q = 7$

4 $2r-s = 3$, $r+s = 9$

5 $2y-3z = 13$, $2y-4z = 10$

6 $b = 2a-1$, $b = 9-2a$

[7] $c-5d = 4$, $c-2d = 16$

[8] $x+y = 11$, $x-y = 5$

[9] $r+3s = 8$, $r-2s = 3$

[10] $x-5y = 1$, $x+4y = 28$

[11] $p = 2+q$, $q = 8-p$

[12] $y-2z = 5$, $2z+y = 1$

13 $2a = 3b-6$, $a = 2b-5$

14 $5c-d = 3$, $3d-8c = 5$

15 $c-3d = 0$, $2c-d = 20$

[16] $l+2m = 8$, $2l+m = 7$

[17] $3s-t = 1$, $5s+2t = 20$

[18] $a-2b = 3$, $3a+b = 72$

19 $x+2y = 11$, $2x-y = 2$

20 $3y-z = 11$, $2y-3z = 5$

[21] $3r+5s = 21$, $r+2s = 7$

22 $2m+5n = 8$, $3m+4n = 5$

[23] $7p-q = 2$, $6p = q$

24 $4s = 3r+2$, $3s+r+1 = 0$

[25] $x-2y = 27$, $7x+y = 9$

[26] $b = 3a-2$, $a = 1-2b$

[27] $2c+d = 10$, $3c-2d = 1$

28 $3x+4y+11 = 0$, $5x+6y+7 = 0$

29 $2(r-2s) = 3s$, $3s-2r = 1$

[30] $2(y+2) = 6(z+1)$, $2y-5z = 4$

31 $0{\cdot}2x+0{\cdot}2y = 1$, $0{\cdot}3x+0{\cdot}5y = 2{\cdot}1$

32 $1{\cdot}2x+y = 0{\cdot}6$, $0{\cdot}7x+0{\cdot}8y = 1$

33 $0{\cdot}2x+0{\cdot}9y = 3$, $0{\cdot}3x+0{\cdot}6y = 1{\cdot}5$

[34] $4a-2b = 4b-5a = a+b-3$

[35] $3l-8 = l-2m-1 = 40$

36 $3p-1 = 10p+q = 1\frac{1}{2}p$

37 $2y-3z = 6y+5z = 70$

[38] $5a-3b = 7a-6b = 9$

[39] $x-y = y-x+2 = 3y$

40 $2c-d-3 = 3d-c+4 = 5c-8d-4$

[41] $4x-6y-3 = 7x+2y-4 = 3y-2x+24$

42 $\frac{4}{x}-\frac{1}{y} = 17$ **43** $\frac{2}{a}+\frac{5}{b} = 5$ [**44**] $\frac{3}{x}+\frac{5}{y} = 5$

$\frac{2}{x}+\frac{3}{y} = 19$ $\frac{1}{a}-\frac{7}{b} = 12$ $\frac{15}{y}-\frac{12}{x} = 1$

45 If $y = mx+b$ is satisfied by $x = 2$, $y = 3$, and also by $x = -4$, $y = 1$, find the values of m and b.

[**46**] If $x = 3$, $y = -2$ satisfy simultaneously the equations $x+ay = 5$, $bx+y = 7$, find the values of a and b.

47 Find *two pairs* of numbers satisfying $7x-2y = 38$, and such that one number is three times the other.

[**48**] If $2a+3b = 9$ and $3a+2b = 16$, find the value of $3a-2b$.

49 If $x = my+a$ is satisfied by $x = 2$, $y = 3$, and also by $x = -5$, $y = 4$, find the value of y when $x = 9$.

50 If $y = ax(x+1)+b$ is satisfied by $x = -2$, $y = 1$, and also by $x = 3$, $y = 21$, find the value of y when $x = 2$.

Problems

Example 6 A certain number is formed of two digits; its value equals four times the sum of its digits. If 27 is added to it, the sum is the number obtained by interchanging the digits. What is the number?

Let x be the tens-digit and y the unit-digit. Then the value of the number is $10x+y$, and the sum of the digits is $x+y$.

$$\therefore\ 10x+y = 4(x+y);$$

$$\therefore\ 10x+y = 4x+4y; \qquad \therefore\ 6x-3y = 0;$$

$$\therefore\ 2x-y = 0 \quad . \quad . \quad . \quad . \quad (i)$$

The value of the number formed by interchanging the digits is $10y+x$.

$$\therefore\ 10x+y+27 = 10y+x; \qquad \therefore\ 9x-9y = -27;$$

$$\therefore\ x-y = -3 \quad . \quad . \quad . \quad . \quad (ii)$$

From (i) and (ii) by subtracting, $x = 3$.

$$\therefore\ \text{from (i)},\ 6-y = 0; \qquad \therefore\ y = 6.$$

$\therefore$ the original number is **36.**

Check by using the data of the problem:

$4\times(\text{sum of digits}) = 4\times(3+6) = 4\times 9 = 36 =$ the number.

$36+27 = 63 =$ number obtained by interchanging the digits.

EXERCISE 44

Use **simultaneous** *equations to solve the following problems*:

1 Find two numbers whose sum is 39 and difference is 11.

2 A jug and basin cost 144p; the jug cost 6p more than the basin. Find the cost of each.

3 2 knives and 5 forks cost £2; 5 knives and 6 forks of the same kind cost £3·05. Find the cost of a knife and of a fork.

4 A farmer can buy 3 cows and 5 sheep for £135, and he can buy 4 cows and 10 sheep for £210. Find the price of a cow and of a sheep.

[**5**] 5 kg of salt and 2 kg of sugar cost 48p; 1 kg of salt and 2 kg of sugar cost 24p. Find the cost of 1 kg of sugar.

6 Fig. 115 shows the lengths of the sides of a triangle in centimetres. If the triangle is equilateral, find the values of x and y and the perimeter of the triangle.

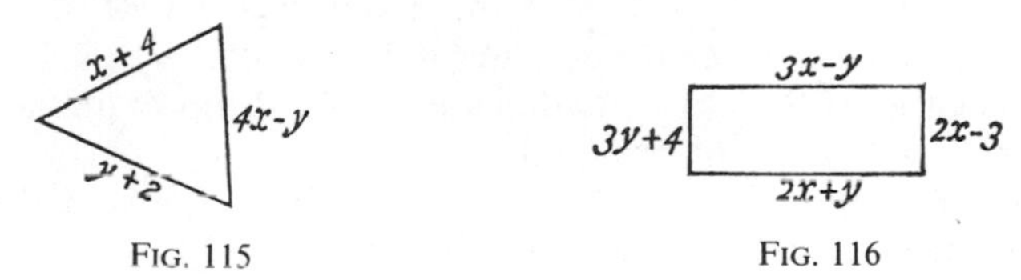

FIG. 115 FIG. 116

7 Fig. 116 shows the lengths of the sides of a rectangle in centimetres. Find the values of x and y and the area of the rectangle.

8 I am thinking of a pair of numbers. If I add 11 to the first, I obtain twice the second; if I add 20 to the second, I obtain twice the first. What are the numbers?

9 Find two numbers whose sum is 90 and such that one-third of the smaller is equal to one-seventh of the larger.

[**10**] A builder requires 3 lorry-loads and 8 cart-loads to fetch 15 tonnes of gravel. He would require 2 lorry-loads and 20 cart-loads for 21 tonnes. How much is a lorry-load and a cart-load?

11 In 4 years' time a father will be 3 times as old as his son will be; 4 years ago he was 5 times as old as his son was. Find their present ages.

[**12**] In $\triangle ABC$, $\angle B = x°$, $\angle C = y°$. If $\angle A = 2\angle B$ and $\angle C - \angle B = 36°$, find the angles of the triangle.

13 In $\triangle ABC$, $\angle B = q°$, $\angle C = r°$. If $\angle A$ can be expressed either as $(2r-q)$ degrees or as $(2q+r)$ degrees, find $\angle A$.

[**14**] If A gives B £3, B will have twice as much as A will have. If B gives A £5, A will have twice as much as B will have. How much has each?

15 A owes £4, B owes £5; A could just pay his debt if he borrowed from B one-eighth of what B has; B could just pay his debt if he borrowed from A two-sevenths of what A has. How much has each?

16 A number of two digits is equal to 7 times the sum of the digits and exceeds by 36 the number formed by interchanging the digits. What is the original number?

[**17**] A number of two digits is equal to 6 times the sum of the digits, and the number formed by reversing the digits exceeds four times the sum of the digits by 9. What is the original number?

18 A number of 3 digits is such that the tens-digit exceeds the unit-digit by 2. If this number exceeds the number formed by reversing the digits by 297 and if the sum of its digits is 14, find the number.

[**19**] If the larger of two numbers is divided by the smaller, the quotient and remainder are each 2. If 5 times the smaller is divided by the larger, the quotient and remainder are again each 2. Find the two numbers.

20 A man travels 22 km in $2\frac{3}{4}$ hours, part of the time at 16 km per hour and the rest at 5 km per hour. How far did he go at the faster speed?

21 Find a fraction which reduces to $\frac{2}{3}$ if the numerator and denominator are each increased by 1, and reduces to $\frac{3}{5}$ if the numerator and denominator are each diminished by 2.

[**22**] Find a fraction which reduces to $\frac{3}{4}$ either if 1 is added to the numerator or if 11 is subtracted from the numerator and the denominator is halved.

23 A heap of 10p coins and 5p coins is worth £2. If there were 3 times as many 10p coins and half the number of 5p coins, it would be worth 25p more. How many coins of each kind are there?

24 A delivers 50 litres of milk, some of it in litre bottles and the rest in $\frac{1}{2}$-litre bottles. B delivers one-third of the number of litre bottles but three times as many $\frac{1}{2}$-litre bottles as A, and this amounts to 70 litres. How many litre bottles did A deliver?

[**25**] In parallelogram ABCD, AB $= (3x-y+1)$ cm, BC $= (x-2y+12)$ cm, CD $= (x-3y+19)$ cm and DA $= 3y$ cm. Find x and y.

26 A skeleton cuboid (see Fig. 117) has a square base ABCD; the wire composing the cuboid is 132 cm long and of this 84 cm is used for the outside edges, *i.e.* all the rims. Find the dimensions of the cuboid.

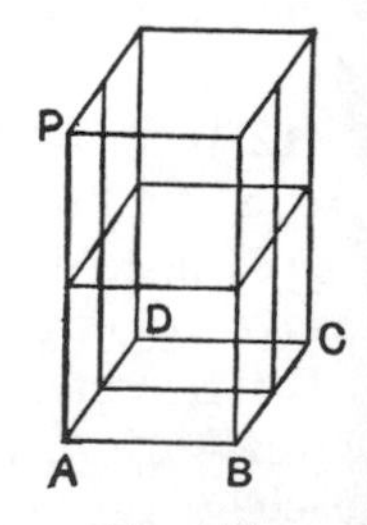

FIG. 117

[**27**] APQB is a straight line; PQ $= 2$ cm. If AQ $= 1\frac{1}{2}$QB and AP $= \frac{1}{2}$PB, find the lengths of AP and QB.

[**28**] x 10p coins and y 5p coins have the same value as y 10p coins and $(x-3)$ 5p coins. If this value is £3, find x and y.

29 A and B live 20 km apart. A leaves home at 10 a.m. and walks to meet B who started at 9.30 a.m. to meet A. The speeds of A and B are in the ratio 3:4. If they meet at 11.30 a.m., find their speeds.

30 If $t°$ Celsius corresponds to $T°$ Fahrenheit, there is an equation of the form $T = at+b$, where a, b are constants. If 10°, 20°C.

correspond to 50°, 68°F., find a and b and the value of T when $t = 100$.

[31] The resistance R kg to a train running at V km per h is given by the formula $R = a+bV^2$, where a, b are independent of the speed. At 20 km per h the resistance is 960 kg; at 50 km per h it is 2850 kg. Find the values of a, b and the resistance at 30 km per h.

32 The distance s metres in which a train running at v km per h can be stopped is given by the formula $s = av+bv^2$, where a, b are constants. At speeds of 20 km per h, 40 km per h, the distances are 100 m, 280 m. Find the distance for a speed of 50 km per h.

33 A number of two digits is equal to 7 times the sum of its digits. Prove that the number formed by reversing the digits is equal to 4 times the sum of its digits.

Fractional Equations When x or y have fractional (or decimal) coefficients, it is usually best to start by clearing these terms of fractions; *it is not necessary to clear of fractions the terms which do not contain* $\boldsymbol{x}$ *or* $\boldsymbol{y}$.

Number the chief equations: a short method of indicating that each side of equation (i) is to be multiplied by 15, say, is to write '(i) × 15' (see Example 7).

Example 7 Solve the simultaneous equations:

$$\frac{2x-1}{3}-\frac{3(y+1)}{5} = 1\tfrac{1}{2}. \qquad \text{(i)}$$

$$\frac{2x-y}{2}-\frac{y-1}{3} = 4\tfrac{1}{12}. \qquad \text{(ii)}$$

(i) × 15 gives $\quad 5(2x-1)-9(y+1) = 1\tfrac{1}{2}\times 15;$

$$\therefore\ 10x-5-9y-9 = 22\tfrac{1}{2};$$

$$\therefore\ 10x-9y = 36\tfrac{1}{2}. \qquad \text{(iii)}$$

(ii) × 6 gives $\quad 3(2x-y)-2(y-1) = 4\tfrac{1}{12}\times 6;$

$$\therefore\ 6x-3y-2y+2 = 24\tfrac{1}{2};$$

$$\therefore\ 6x-5y = 22\tfrac{1}{2}. \qquad \text{(iv)}$$

(iii) × 3 gives $\quad 30x-27y = 36\tfrac{1}{2}\times 3 = 109\tfrac{1}{2}. \qquad \text{(v)}$

(iv) × 5 gives $\quad 30x-25y = 22\tfrac{1}{2}\times 5 = 112\tfrac{1}{2}. \qquad \text{(vi)}$

(v)–(vi) gives $\quad -2y = -3; \quad \therefore\ y = 1\tfrac{1}{2}.$

Substituting in (iv) for y, $6x-5\times 1\tfrac{1}{2} = 22\tfrac{1}{2}$;

$$\therefore\ 6x = 22\tfrac{1}{2}+7\tfrac{1}{2} = 30; \quad \therefore\ x = 5.$$

$$\therefore \text{ the solution is } \boldsymbol{x = 5,\ y = 1\tfrac{1}{2}}.$$

The reader should now consider which of the equations, (i) or (ii), should be used for checking.

EXERCISE 45

Solve the following pairs of simultaneous equations:

1 $x+\frac{1}{2}y = 13$, $\frac{1}{3}x-y = 2$

2 $\frac{3}{4}p-\frac{1}{3}q = 9$, $2q-p = 2$

3 $\frac{1}{3}r+\frac{1}{5}t = 8$, $\frac{1}{9}r-\frac{1}{10}t = 1$

[4] $\frac{1}{2}a-2b = 5$, $\frac{1}{3}a+b = 1$

[5] $y-\frac{2}{7}x = \frac{1}{3}$, $x-y = 2\frac{1}{6}$

[6] $2c+5d = 1\frac{1}{2}$, $9c-7d = \frac{17}{20}$.

7 $Q+2 = \frac{1}{3}P$, $\frac{1}{5}P-1 = \frac{2}{3}Q$

[8] $\frac{1}{2}u+2v = 5$, $2u-\frac{1}{7}v = -8\frac{1}{2}$

[9] $\dfrac{2y}{5}-\dfrac{z}{3} = 2\frac{2}{3}$, $y = 2(z+1)$

10 $3a+\frac{1}{2}b = 8a+7b-9 = 2$

[11] $\dfrac{5p}{3}+\dfrac{2q}{3} = 3p+3q = 6p+9q+3$

12 $0{\cdot}5t-0{\cdot}7v = 2$, $1{\cdot}8t-2{\cdot}2v = 8{\cdot}8$

13 $1{\cdot}2x-0{\cdot}8y = 0{\cdot}4$, $y = 0{\cdot}3-0{\cdot}1x$

[14] $h+5{\cdot}2n = 12{\cdot}1$, $h+3{\cdot}8n = 10$

[15] $1\frac{1}{4}a-\frac{1}{2}b = \frac{1}{4}$, $0{\cdot}2a+0{\cdot}1b = 1{\cdot}3$

16 $\frac{1}{3}(r+2)-\frac{1}{4}(t-1) = 1,\ t = 1\frac{1}{4}r$

17 $l-5 = \dfrac{n-2}{7}$, $4n-3 = \dfrac{l+10}{3}$

18 $\dfrac{x+1}{2} = 8-\dfrac{y-1}{3}$, $\dfrac{y+1}{2} = 9-\dfrac{x-1}{3}$

19 $\dfrac{r}{10}+\dfrac{s}{8} = r-s$, $\dfrac{2r-s}{3}+2s = \frac{1}{2}$

20 $\dfrac{a-1}{2}+\dfrac{b+1}{5} = 4\frac{1}{5}$, $\dfrac{a+b}{3} = b-1$

[21] $2l-\dfrac{m-3}{5} = 9$, $3m+\dfrac{l-2}{3} = 25$

[22] $\dfrac{y+1}{3}-\dfrac{3z-1}{2} = 1$, $\dfrac{3-8z}{5}-\dfrac{7-3y}{4} = 1$

23 $\dfrac{x}{2}-\dfrac{y}{3} = \dfrac{2x}{5}-\dfrac{y}{4} = 3x-2y-5$

[24] $\dfrac{c-3}{5}+\dfrac{d-7}{3} = c-d = \dfrac{5c+d}{3}$

25 $\dfrac{4y+5z-9}{8} = 5y-5z,\quad \dfrac{2y-z-1}{3} = 2\frac{1}{2}-2z$

REVISION EXERCISE R 1 (Ch. 1–9)

Simplify:

1 $2(x^2-7x+3)-5(x^2-x+1)$ **[2]** $6d-3[b-2(d-c)]$

3 $\dfrac{2n-3}{6n}-\dfrac{n-2}{4n}+1$ **[4]** $\frac{1}{3}p\left(\dfrac{1}{6p}-\dfrac{1}{2p}\right)+\dfrac{4p}{9p}$

5 $[3x+y-(8y-11x)]\div 7$ **[6]** $[a-\frac{1}{3}(a-4b)]\div 2$

7 $\frac{1}{4}+[\frac{1}{3}\div(\frac{2}{3}-\frac{4}{7})]-\frac{3}{4}$ **[8]** $\frac{1}{4}+(\frac{1}{6}\div\frac{1}{7})-(\frac{1}{7}\times\frac{1}{8})$

9 $\dfrac{36{\cdot}8\times 0{\cdot}005}{0{\cdot}125}$ **10** $\dfrac{0{\cdot}4579}{24{\cdot}1}$ **[11]** $\dfrac{5{\cdot}5\times 0{\cdot}175}{0{\cdot}28\times 0{\cdot}275}$

12 Express as decimals: (i) $\dfrac{19}{40}$; (ii) $\dfrac{7}{16}$; (iii) $\dfrac{39}{2^2\times 5^4}$.

If $x = -3$, $y = -2$, $z = 0$, $t = 4$, evaluate the expressions in Nos. 13–16:

13 x^2+y^3 **[14]** y^2-2tz **15** $(x-y)(z-t)$ **[16]** $x(y-t)$

17 Add: $2x(3y-z+x)$; $x(x-7y+5z)$; $3xy-x(2x+z)$.

[18] Subtract $1-3x-x^2+2x^3$ from $4x^2-x+3$.

19 Multiply $4-6x+2x^2-4x^3$ by $-\frac{1}{2}x$.

20 What must be added to $m+n-p$ to make $m-n-p$?

Solve:

21 $(x+1)+(x+2)+(x+3) = 0$ **[22]** $\frac{3}{4}(1-y) = 1\frac{1}{2}$

23 $\dfrac{1}{t}+\dfrac{0{\cdot}2}{5t} = 0{\cdot}1$ **24** $\dfrac{z+2}{5} = 8-\dfrac{z-2}{4}$

25 Find p if $x = -1$ is a root of $x^3 = 5x-2x^2+p$.

[26] If $pq = p+q$ and $q = -1$, find p.

27 If $r+s+4 = 0$ and $s = -1$, find rs.

28 Find y if $(y-7)^2 = 25$.

29 Find the R.F. of a map which is 1 cm to 200 m.

[30] A man takes 21 min for a journey if he travels at 30 km per hour. How long will he take if he travels at 36 km per hour?

31 An article is reduced in price by 5p to 40p. In what ratio is the price reduced?

[32] If 25p in the £ is deducted from a bill, 99p is left. How much was the bill?

33 Purchase tax at the rate of 75p in the £ is added to the retail price. Find the retail price if the customer pays £1·05.

34 Increase £1·32 in the ratio 11:6.

35 Divide £4·20 between A, B, C so that A has *half as much again* as B, and B has twice as much as C.

[**36**] If $y = 3(2x+1)$, express x in terms of y.

37 If $x:y = \frac{1}{2}:\frac{1}{3}$, evaluate $(6x-5y):(2x+3y)$.

38 If $7x-2y = 1$ and $3x+5y = 2$, find $x:y$.

39 By how much per cent does $\frac{4}{5}n$ exceed $\frac{3}{4}n$?

40 If $a(5x-3)+b(x+5) = 7x$ for all values of x, where a, b are constants, find a, b.

[**41**] A man is y years older than his son whose present age is x years. In 5 years' time the father will be just twice as old as his son. Find y in terms of x.

42 If $(\frac{1}{5}n-1)$ is two-thirds of $(\frac{1}{3}n-1)$, find the value of n, and the value of each expression.

[**43**] From the formula $F = 32+\dfrac{9C}{5}$, find the value of C if $7C = 4F$.

[**44**] For what value of k does the fraction $\dfrac{3(k+1)}{4k+1}$ reduce to $\dfrac{4}{5}$?

45 If $2x$ 10p coins and $(3x+4)$ 5p coins are together worth £3, find the value of x.

[**46**] A has £a, B has £b. A owes £c to B, and B owes £10 to A. How many pounds will A have when both debts have been paid?

47 A man takes 10 minutes to go $1\frac{1}{4}$ km, running part of the way at 9 km per h and walking the rest at 6 km per h. How far does he walk?

48 If p is a whole number, find the least value of p for which $8(p-2)$ exceeds $5p+7$.

49 Use tables to obtain the values of the tangents of the angles, (i) 43° 50′, (ii) 67° 20′, (iii) 76° 9′, (iv) 76° 50′.

50 Use tables to find $x°$, if the values of tan $x°$ are

(i) 0·4397, (ii) 1·4415, (iii) 2·0381, (iv) 4·8150.

[**51**] In △ABC, AC = 3 cm, ∠A = 49°, ∠C = 90°, find BC.

52 In △FGH, GH = 9 cm, ∠F = 61°, ∠H = 90°, find FH.

[**53**] In △PQR, PQ = 7 cm, QR = 10 cm, ∠Q = 90°, find ∠P.

54 Find the marked angles in Fig. 118 (i), (ii), (iii).

[**55**] What is the angle of elevation of the top of a spire, 45 m high, from a point on the ground 100 m from the foot of it?

56 From a lighthouse window, 80 m above sea-level, the angle of depression of a boat is 23° 45′. Find the distance of the boat from the base of the cliff on which the lighthouse stands.

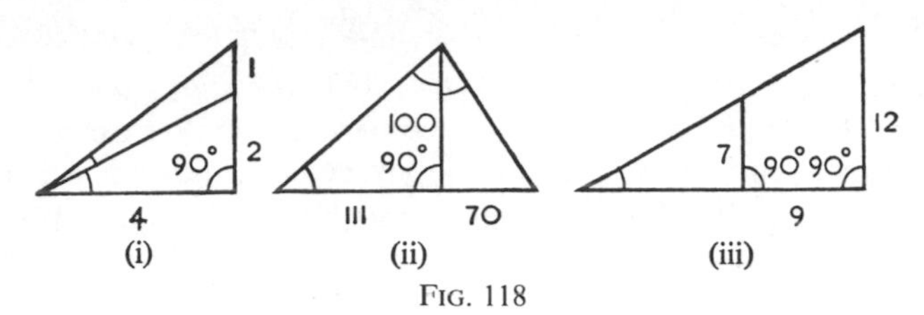

FIG. 118

[57] The vertical angle of a cone is 95° and the diameter of its base is 8 cm. Find its height.

58 The equal legs of a pair of steps stand at points 1·5 m apart on level ground and the top is 2·7 m above the ground. Find the angle which each leg makes with the ground.

[59] PQRS is a rectangle, ∠PRQ = 71°; calculate ∠QSR.

60 The diagonals of the rectangle EFGH cut at N; ∠ENF = 54°, calculate ∠NEF and ∠EHF.

[61] ABCD is a square and BCK is an equilateral triangle outside the square. Calculate ∠CKD and ∠KDA.

62 Construct the parallelogram ABCD given that AB = 6 cm, AC = 9 cm, BD = 7 cm. Measure BC.

[63] ABCD is a parallelogram; the line through C parallel to BD cuts AD produced at K. Prove AD = DK.

64 X, Y, Z are points on the sides BC, CA, AB respectively of △ABC such that BX = CY, CX = BZ. If XY = XZ, prove AB = AC.

65 ABCD is a rectangle; the line bisecting ∠BAC meets BC at K. If AK = KC, prove ∠ACB = 30° and AC = 2AB.

66 ABCD is a rhombus; DBXY is another rhombus such that X lies on BC produced. If ∠BCD = 116°, calculate ∠DBC and the acute angle which BY makes with CD.

[67] AOB and XOY are diameters of two concentric circles whose common centre is O; prove that AX is parallel to BY.

68 ABC and PQR are two concentric circles whose common centre is O; OP lies between OA, OB. If ∠POQ = ∠AOB, prove that *either* AP = BQ *or* AQ = BP.

69 The lines ANB, CND are equal and AD = BC. Prove that (i) ∠NAC = ∠NCA, (ii) AC is parallel to DB.

70 BCHK, CAPQ, ABMN are squares outside △ABC, right-angled at A; KB, HC are produced to meet MN, QP, produced if necessary, at D, E. Prove that (i) DB = BC = CE, (ii) BCED is a square.

Solve the equations:

71 $3a+2b = 4$
$a+3b = 13$

[72] $4c+3d = 1$
$5c+4d = 2$

73 $l = 2m+1$
$3l = 5(m+1)$

[**74**] If $x = 2y$ and $y+z = 1$, find xy when $z = 2$.

75 If $3p+4q = -119$ and $5p-11q = 102$, prove $p = q$.

[**76**] Find two numbers such that 3 times the smaller exceeds twice the larger by 3, and 7 times the smaller exceeds 5 times the larger by 2.

77 If A gives B 2p, B has 4 times as much as A then has. If C gives A and B 10p each, B will have twice as much as A. How much have A and B?

78 C is a point on the side AD of the triangle ABD such that AC = CB = BD; $\angle$ABC = $x°$, $\angle$CBD = $y°$; find x in terms of y. If also $\angle$ABD = 3$\angle$ADB, find x and y.

79 Solve $\dfrac{x+y}{5} = \dfrac{x-y}{3}$;

$$\frac{x+y}{15} - \frac{x-y}{12} = \frac{1}{4}.$$

[**80**] Solve $\dfrac{y+1}{3} - \dfrac{z+1}{2} = 2y+z+1$;

$$\frac{y-2}{3} = \frac{z+3}{2}.$$

CHAPTER 10

APPLICATIONS OF PERCENTAGE

Percentage Changes The importance of a change in the size or value of a quantity is often estimated by calculating what percentage the increase or decrease is of the **original** value. For example, if a car costs £800 when new and if its value is £720 at the end of 1 year, the decrease in value is £80; this is 10 per cent of the original value, £800, and we say that the car has depreciated in value by 10% after 1 year's use.

An *increase* of say 30 per cent means that for each 100 units in the original value there is an increase of 30 units, making the new value 130 units.

$\therefore$ the ratio of the increase to the original value is 30:100;

$\therefore$ the increase is $\frac{30}{100}$ times the original value.

Also the ratio of the new value to the original value is 130:100;

$\therefore$ the new value is $\frac{130}{100}$ times the original value.

Similarly the ratio of the new value to the increase is 130:30;

$\therefore$ the new value is $\frac{130}{30}$ times the increase; and so on.

A *decrease* of say 20 per cent means that for each 100 units in the original value there is a decrease of 20 units, making the new value 80 units.

$\therefore$ the ratio of the new value to the original value is 80:100;

$\therefore$ the new value is $\frac{80}{100}$ times the original value; and so on.

The ratios used as multipliers in the examples just given are called **multiplying factors;** *oral* practice of the form indicated in the next Exercise is intended to secure facility in their use.

EXERCISE 46 (Oral)

By what must a number be multiplied to increase it by:

1 17% **2** 83% **3** 70% **4** 20% **5** 139%

By what must a number be multiplied to decrease it by:

6 9% **7** 37% **8** 61% **9** 30% **10** 40%

11 Increase 300 by 8% **12** Decrease 400 by 20%

13 Decrease 80 by 10% **14** Increase 60 by 30%

15 If the price of an article is decreased by 7%, write down
(i) the ratio of the new price to the old price;
(ii) the ratio of the change in price to the new price;
(iii) the factor by which the change in price must be multiplied to give the old price.

16 (i) A exceeds B by 13%, write down the ratio of A to B.
(ii) C is 19% less than D, write down the ratio of C to D.

Complete the following:

17 If A exceeds B by 5%, $A = B \times \ldots$; $B = A \times \ldots$.
18 If C is 12% less than D, $C = D \times \ldots$; $D = C \times \ldots$.
19 If A exceeds B by 6%, $A - B = B \times \ldots$; $A - B = A \times \ldots$.
20 If C is 8% less than D, $D - C = C \times \ldots$; $D - C = D \times \ldots$.
21 If A exceeds B by x per cent, $B = A \times \ldots$.
22 If C is y per cent less than D, $D = C \times \ldots$.

Example 1 A man, whose salary is £1500 a year, receives an increase of 8 per cent. Find his new salary.

The ratio of the new salary to the old salary is 108:100;

$\therefore$ the new annual salary $= £1500 \times \frac{108}{100} =$ **£1620.**

Or as follows: the increase $= \frac{8}{100}$ of £1500 a year $=$ £120 a year;

$\therefore$ the new annual salary $= £1500 + £120 =$ **£1620.**

Example 2 If a man's salary is raised from £1500 a year to £1680 a year, find the increase per cent.

The increase is £180 a year; therefore the ratio of the increase to the first salary is 180:1500, or 3:25.

$\therefore$ the increase per cent $= \frac{3}{25} \times 100$ per cent $=$ **12%**.

Example 3 117 is 36% of a certain number. Find the number. The ratio of the number to 117 equals 100:36;

$\therefore$ the number $= 117 \times \frac{100}{36} =$ **325.**

Or algebraically, if x is the number, $\frac{36}{100}x = 117$;

$\therefore x = 117 \times \frac{100}{36} =$ **325.**

Example 4 A line, whose true length is known from calculation to be 7·5 cm, is found by drawing and measurement to be 7·2 cm. What is the error per cent?

The error $=$ 7·5 cm $-$ 7·2 cm $=$ 0·3 cm; therefore the ratio of the error to the true length is 0·3:7·5, that is 3:75 or 1:25;

$\therefore$ the error per cent $= \frac{1}{25} \times 100$ per cent $=$ **4%**.

Example 5 After 5% of a bill has been deducted from it, £57 remains to be paid. How much was the bill?

After 5% of the bill has been deducted, 95% of the bill remains.

$\therefore$ the ratio of the bill to £57 equals 100:95;

$\therefore$ the bill is £57 $\times \frac{100}{95}$, that is **£60.**

Or the working may be finished as follows:

95% of the bill is £57; $\therefore$ 100% of the bill is £57 $\times \frac{100}{95}$;

$\therefore$ the bill is **£60.**

EXERCISE 47

1 Increase 80 by 35%

[2] Decrease 75 by 40%

3 Decrease 216 by $37\frac{1}{2}$%

[4] Increase 416 by 125%

Find the number or quantity of which:

5 25% is 7

[6] $37\frac{1}{2}$% is 84

7. $7\frac{1}{2}$% is 150 p

8 What number when increased by 20% becomes 144?

[9] What number when decreased by 20% becomes 108?

[10] What sum of money when increased by 35% becomes £216?

11 What sum of money when decreased by 35% becomes £156?

12 A man, whose salary is £1520 a year, receives an increase of 15%; find his new salary.

13 A car costs £875 when new; after 1 year, its value is £735. By how much per cent has its value decreased?

[14] A man spends £1320 a year, and this is 80% of his income. What is his income?

15 55 per cent of the pupils in a school are girls. What percentage of the pupils are boys? Find the number of pupils if there are 216 boys.

16 What are full marks for a paper if a boy who gets 112 marks obtains 70 per cent of the total?

[17] A spends 88% of his income. Find his income if he saves £243 a year.

18 If 10% is deducted from a bill, £27 remains to be paid. How much is the bill?

19 My bank deposit has increased by 40% during the past year. It is now £504; what was it a year ago?

20 Calculation shows that the true length of a line is 3·2 cm; the length by drawing and measurement is 3·4 cm; find the error per cent.

21 Find to two significant figures the error per cent in taking the area of a field which is 165 m long, 120 m wide, as 20 000 m^2.

22 In a sale the price of an armchair was lowered by 30% to £3·15. How much was it reduced?

[23] If the price of an article is raised by 8%, the increase in the price is 50p. Find the new price.

24 24% by weight of an explosive mixture is saltpetre. Find the weight of a sample in which the other ingredients weigh 9·12 g.

25 15% of a sum of money is £27·70; find the value of $16\frac{1}{2}$% of the same sum.

[**26**] A rectangular enclosure is 80 m long, 25 m wide. If the length of each side is increased by 20%, find the percentage increase in the area.

27 Over a period of 18 years, wireless licences increased from 7 080 000 to 13 455 000. Find the increase per cent to the nearest whole number.

[**28**] The weight of a liquid was 3·75 g before heating and 3·25 g after heating. Find the loss per cent of its weight, correct to 2 figures.

Gain and Loss Per Cent If a dealer buys an article for £100 and sells it for £101, his profit is a small one for the transaction; but if he buys an article for £2 and sells it for £3, his profit is relatively large, although he gains £1 in both cases. In order to be in a position to perform the first transaction, he must first pay out £100 and his profit is only 1 per cent of this amount; but in order to be in a position to perform the second transaction, he must first pay out only £2, and his profit (£1) is 50 per cent of this amount. The fair way of comparing two sale-transactions is therefore to calculate the percentage that the profit (or loss) is of the **cost price,** because the cost price is what the dealer has had to invest in the goods he hopes to sell again.

Gain or loss per cent is *never* calculated on the number of articles sold. If a dealer sells 100 articles at a profit of £20, it is impossible to calculate his gain per cent unless the cost price of the articles can be found; if the 100 articles cost £200, he gains 10%; if they cost £50, he gains 40%; if they cost £2000, he gains 1%.

In calculating any percentage change, the increase or decrease is expressed as a percentage of the *first* value. Unless the contrary is stated, the phrase, **gain** or **loss per cent,** is taken to mean the percentage that the gain or loss is of the **cost price.** The commercial practice is different: a manufacturer finds it convenient to express his estimated profit as a percentage of his selling price.

Example 6 An article costing £1·60 is sold at a profit of 15%; find the selling price.

$$\text{The profit} = \tfrac{15}{100} \text{ of the } \textit{cost} \text{ price} = \tfrac{15}{100} \text{ of } £1{\cdot}60 = 24\text{p}.$$

$$\therefore \text{ the selling price} = £1{\cdot}60 + £0{\cdot}24 = \mathbf{£1{\cdot}84.}$$

Example 7 An article costing £55 is sold for £50; find the loss per cent.

The loss $= £55 - £50 = £5$; $\therefore$ the loss is $\frac{5}{55}$ of the *cost* price;

$$\therefore \text{ the loss per cent} = (\tfrac{5}{55} \times 100) \text{ per cent} = \mathbf{9\tfrac{1}{11}}\%.$$

Example 8 A dealer gained 40% by selling an article for 21p; find the cost price.

If the cost price is 100p, the gain is 40p, ∴ the selling price is 140p;

$$\therefore \text{cost price} = \tfrac{100}{140} \text{ of selling price}$$
$$= \tfrac{100}{140} \text{ of } 21\text{p} = \mathbf{15p.}$$

Example 9 A man buys hooks at 105p a hundred and sells them at 28p for a score. Find his gain per cent.

Choose any convenient number of hooks, and find the cost price and selling price of *this number*. Here, a convenient number is 100, because both the cost price and selling price of 100 hooks are a whole number of pence.

100 hooks cost 105p; 100 hooks are sold for 140p.

∴ the cost price is 105p, the selling price is 140p, and the gain is 35p.

$$\therefore \text{the gain is } \tfrac{35}{105}, \text{ or } \tfrac{1}{3}, \text{ of the cost price.}$$
$$\therefore \text{the gain per cent} = (\tfrac{1}{3} \times 100) \text{ per cent} = \mathbf{33\tfrac{1}{3}\%}.$$

Note It saves time to write **C.P.** for cost price, **S.P.** for selling price.

EXERCISE 48 (Oral)

Write down the factor by which the C.P. must be multiplied to give the S.P. in the following:

1 Gain 10% **2** Gain 50% **3** Loss 10% **4** Loss 20%
5 Gain 30% **6** Gain 5% **7** Loss 25% **8** Loss 8%

Write down the S.P. in the following:

9 C.P. 20p; gain 10% **10** C.P. 20p; loss 10%
11 C.P. £10; loss 20% **12** C.P. £60; gain 25%

Write down the factor by which the S.P. must be multiplied to give the C.P. in the following:

13 Gain 20% **14** Gain 80% **15** Loss 10% **16** Loss 40%
17 Gain 5% **18** Gain 8% **19** Loss 25% **20** Loss 12%

Write down the C.P. in the following:

21 S.P.£12; gain 20% **22** S.P. £15; gain 50%
23 S.P. £8; loss 20% **24** S.P. £14; loss 30%
25 S.P. £7; gain 40% **26** S.P. £42; gain 5%
27 S.P. £12; loss 60% **28** S.P. £12; loss 40%

Write down the gain or loss per cent in the following:

29 C.P. £50; gain £6 **30** C.P. £200; loss £40
31 C.P. 12p; loss 3p **32** C.P. 20p, gain 4p
33 S.P. £10; gain £2 **34** S.P. £60; loss £20
35 S.P. 12p; loss 3p **36** S.P. 12p; gain 3p
37 C.P. £50; S.P. £58 **38** C.P. 30p; S.P. 24p
39 C.P. £90; S.P. £100 **40** C.P. £1·25; S.P. 75p

EXERCISE 49

Find the S.P. in the following:

1 C.P. £15; gain 5%
[**2**] C.P. 6p; gain $33\frac{1}{3}$%
[**3**] C.P. £75; loss 8%
4 C.P. 40p; loss $17\frac{1}{2}$%
5 C.P. £54; gain $4\frac{1}{2}$%
[**6**] C.P. 150p; loss $13\frac{1}{3}$%

Find the gain or loss per cent in the following:

7 C.P. 80p; loss 24p
[**8**] C.P. £4·50; S.P. £4
9 C.P. 9p; S.P. $11\frac{1}{2}$p
[**10**] C.P. 198p; S.P. 231p
11 S.P. 21p; loss $3\frac{1}{2}$p
[**12**] S.P. £10·35; gain £1·35

Find the C.P. in the following:

13 S.P. £10·50; gain 5%
[**14**] S.P. 24p; loss 4%
15 S.P. 198p; loss 12%
[**16**] S.P. £32; gain $6\frac{2}{3}$%
17 S.P. £5·50; loss $31\frac{1}{4}$%
[**18**] S.P. 102p; gain $13\frac{1}{3}$%

19 A dealer buys a bicycle for £12 and sells it for £15; find his gain per cent.

20 I bought a house for £4800 and was forced to sell it for £4000; find my loss per cent.

[**21**] A man bought a chair for £4·20 and sold it at a gain of 15 per cent. How much profit did he make?

22 A man bought a car for £1125 and sold it one month later at a loss of 12 per cent. How much did he lose?

[**23**] By selling a mirror for £15, a dealer makes a profit of 25%; find what the dealer paid for it.

24 If I sell my house for £4200, I shall lose 20%; what did I pay for the house.

[**25**] Tea is bought at £450 per tonne and sold at 63p per kg: find the gain per cent.

[**26**] Steel screws costing 25p per gross (144) are retailed at 3p per dozen; find the gain per cent.

27 A grocer buys 6 dozen eggs for £1; 2 are broken and the rest are sold at 5 for 10p. Find his gain per cent.

[**28**] If I sell my wireless set for £9, I shall lose 64 per cent; what did I pay for it?

29 By selling a watch for £18, a jeweller makes a profit of 80 per cent; what did he pay for it?

30 A car is sold for £635 at a gain of 27 per cent. How much is the profit?

31 An ironmonger buys mops at 5 for £3 and sells them at 4 for £3; find his gain per cent.

[**32**] Christmas cards are bought at £4·20 per 100 and are sold at 7p each; find the gain per cent.

[**33**] A grocer buys 1500 bananas at 4p each; he sells 900 at 5p each and the rest at $4\frac{1}{2}$p each. Find his gain per cent.

34 1 tonne of tea costs £650; at what price per kg must it be sold to make a profit of 20 per cent?

35 Oranges are bought at the rate of 3 for 5p; at what price must they be sold to gain 80 per cent?

Miscellaneous Examples

EXERCISE 50

1 A man saves $\frac{3}{8}$ of his income. What percentage is this?

2 A man travels 8% of a journey by bus, 87% by train, and walks the rest. For what % of his journey does he walk?

[**3**] My insurance premium is £12·60 a year for my car which is valued at £525. What percentage is the premium of the value?

4 The profit on a wireless set, sold at a gain of $32\frac{1}{2}$%, is £6·50. Find the sale price.

[**5**] If *A* exceeds *B* by 80 per cent, find the factor by which *A* must be multiplied to give *B*.

6 A man measures the length of a road as 1934 m. It is really 1910 m; find his error per cent, to one place of decimals.

7 150 marks are assigned to each of two papers; if a boy obtains 63 marks for the first, how many must he get for the second to secure 36% on the whole?

8 If butter cost 40p per kg yesterday and has risen 25% today and will fall 25% tomorrow, what will be the price tomorrow?

[**9**] If a carpet 4 m by 3 m is laid in a room, there is a margin 50 cm wide all the way round. What percentage of the area of the floor is the area of the margin?

10 A coal merchant buys coal at £6·90 per tonne, and in addition pays 60p per tonne carriage. Find his gain per cent if he sells it at 94p per 100 kg.

[**11**] If wages are increased $12\frac{1}{2}$% all round, a firm's weekly wage-bill becomes £198. What is the increase per week in the bill?

12 2 litres of spirit containing 10% of water are mixed with 5 litres of spirit containing 8% of water, and 1 litre of water is added to the mixture. What is now the percentage of water?

13 A legacy of £4500 is left to 3 persons so that, after duties totalling 20% have been paid, their shares are proportional to 1, $1\frac{1}{2}$, 2. Find the shares.

[**14**] B is 20% heavier than A, and C is 25% heavier than B; by how much per cent is C heavier than A?

15 A is 20% older than B; by how much per cent is B younger than A?

16 One year a firm paid £7000 in wages, £2100 for other expenses, and made a profit of £1400. The next year it paid 75% more in wages and 45% more for other expenses. The receipts increased by 60 per cent. Find the increase per cent of the profits.

[**17**] On a journey across London, a taxi averages 32 km per h for 70% of the distance, 40 km per h for 10% of it, and 12·8 km per h for the remainder. Find the average speed for the whole journey.

***18** By what percentage must a motorist increase his average speed in order to reduce by 20% the time a particular journey takes?

***19** 20 kg of bronze contained 87% of copper and 13% of tin by weight. With how much copper must it be melted to obtain bronze containing 10% of tin by weight?

***20** A dealer obtains a commodity from the U.S.A. at a cost of 30 dollars per tonne. Taking £1 = 2·40 dollars, find the price per kg at which he must sell it to gain 20 per cent.

CHAPTER 11

PRODUCTS, QUOTIENTS AND SQUARES

Single Term Factors

If there is a factor common to each term of an expression, use **short division** to factorise the expression.

Example 1 Factorise $6a^2-12ax+3a$.

$$3a\overline{)6a^2-12ax+3a}$$
$$2a \;-\; 4x \;+\; 1$$

$$\therefore 6a^2-12ax+3a = \mathbf{3a(2a-4x+1)}.$$

EXERCISE 51

Simplify the following:

1 $(6c+3d)\div 3$ **[2]** $(8a^2-10a)\div 2a$ **3** $(y^2+y)\div y$
[4] $(6t^2+4t)\div 2t$ **5** $(p^3-p^2)\div p^2$ **[6]** $(b^2c-bc)\div bc$
7 $(10x^2y^2-15xy^3)\div 5xy^2$ **8** $(r+nr)\div r$

Express, *where possible*, the following in factors; *if there are no factors, say so.*

9 $6c-9$ **10** $4n+4$ **11** $xy+xz$ **12** p^2-2q
13 $6a^2-6ab$ **14** $20r^2+10r$ **15** $4+2s^2$ **16** b^2c-2bc^2
[17] $5r+10s$ **[18]** $bc-cd$ **[19]** $6y^2-2y$ **[20]** x^3z+z^3x
21 $ab+a$ **22** $3d^3-d$ **23** x^2+y^2 **24** p^2-rp^2
[25] $9pq-21rs$ **[26]** $r^2s^2+s^2$ **[27]** $4ab^2-4b$ **[28]** $y-cyz$
29 $ap+aq+ar$ **30** x^3-3x^2-3x **31** $ab+bc+ca$
[32] y^2+yz+y **[33]** $cd+c^2d+cd^2$ **[34]** $3r^2-rs-rt$
35 $a(r+s)+b(r+s)$ **36** $c(x+y)+d(x-y)$
37 $x(x+y)-y(y+x)$ **38** $(a+b)^2+c(a+b)$
39 $a(x+y)+b(y+z)$ **40** $(r-t)^2+(r-t)$
[41] $b(c-d)+x(c-d)$ **[42]** $a(p+q)+(p+q)b$

Geometrical Illustrations of Algebraic Identities

EXERCISE 52 (Class Discussion)

1 $$\mathbf{k(a+b+c+d) = ka+kb+kc+kd}$$

The rectangle P in Fig. 119 on the next page is k cm high, $(a+b+c+d)$ cm wide; it can be divided as shown into 4 rectangles A, B, C, D.

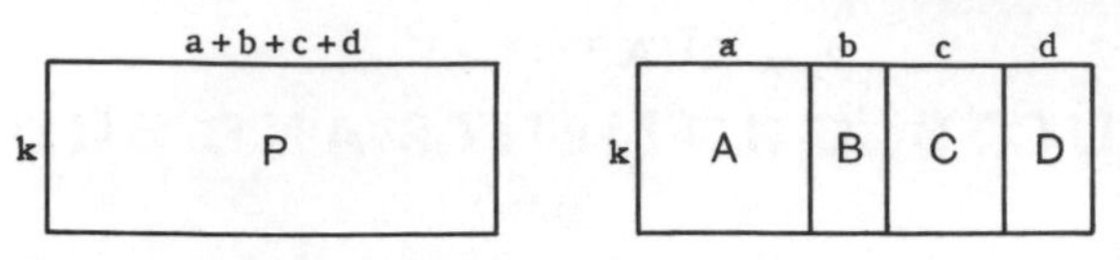

FIG. 119

The area of P is $k(a+b+c+d)$ cm^2;
the areas of A, B, C, D are ka, kb, kc, kd cm^2;
$\therefore\ k(a+b+c+d) = ka+kb+kc+kd.$

2 $$(a+b)^2 = a^2+2ab+b^2$$

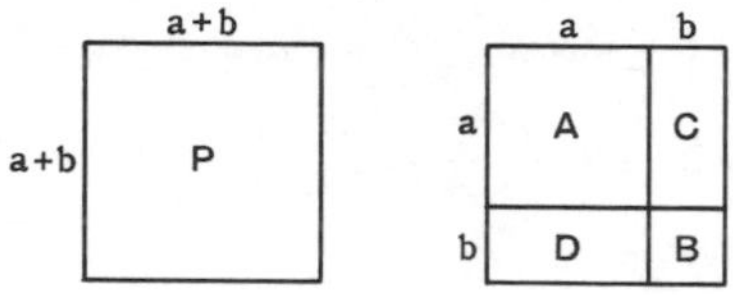

FIG. 120

The side of the square P in Fig. 120 is $(a+b)$ cm long; P can be divided as shown into 4 rectangles A, B, C, D.

The area of P is $(a+b)^2$ cm^2;
the areas of A, B, C, D are a^2, b^2, ab, ab cm^2;
$\therefore\ (a+b)^2 = a^2+2ab+b^2.$

3 What products are illustrated by (i) Fig. 121, (ii) Fig. 122?

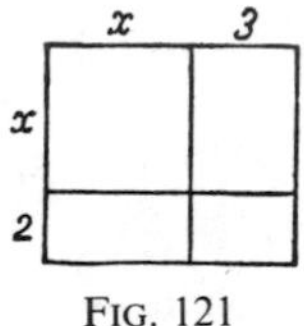

FIG. 121

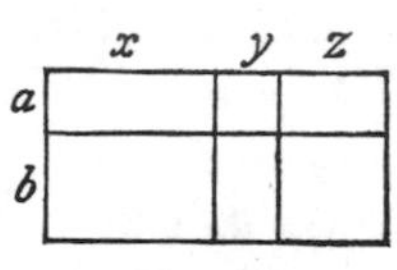

FIG. 122

4 $$(a-b)^2 = a^2-2ab+b^2$$

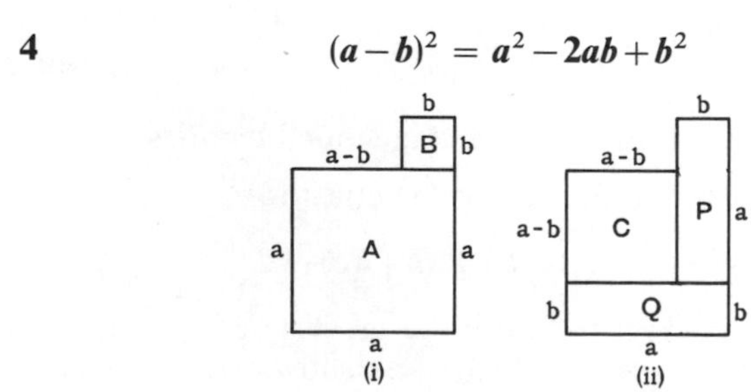

FIG. 123

Fig. 123 (i) is made up of the square A, side a cm and the square B, side b cm; and so its total area is (a^2+b^2) cm^2. The whole figure can be divided, as shown in Fig. 123 (ii), into 3 rectangles C, P, Q.

C is a square, side $(a-b)$ cm, area $(a-b)^2$ cm^2;
P is a rectangle, a cm high, b cm wide, area ab cm^2;
Q is a rectangle, b cm high, a cm wide, area ab cm^2;

$$\therefore\ a^2+b^2 = (a-b)^2+ab+ab,$$
$$\therefore\ a^2+b^2-2ab = (a-b)^2.$$

5 $$\boldsymbol{a^2-b^2 = (a+b)(a-b).}$$

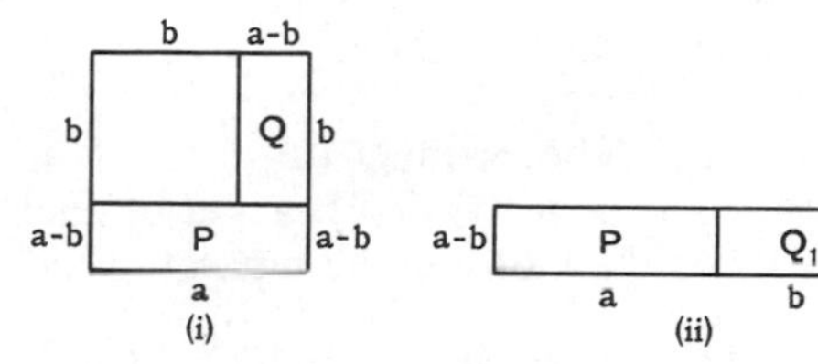

FIG. 124

Fig. 124 (i) is a square, side a cm, area a^2 cm^2, which is divided into a square, side b cm, area b^2 cm^2, and 2 rectangles P, Q.

$$\therefore\ \text{the total area of P and Q is } (a^2-b^2)\text{ cm}^2;$$

P and Q can be arranged in the positions, P, Q_1, in Fig. 124 (ii) and then make up a single rectangle, $(a-b)$ cm high, $(a+b)$ cm wide.

$$\therefore\ \text{the total area of P and Q is } (a+b)(a-b)\text{ cm}^2;$$
$$\therefore\ a^2-b^2 = (a+b)(a-b).$$

Expansions $\quad (a+b)k = k(a+b) = ka+kb$

and $\quad (a-b)k = k(a-b) = ka-kb;$

hence, if k is replaced by $c+d$,

$$(a-b)(c+d) = a(c+d)+b(c+d)$$

and $$(a-b)(c+d) = a(c+d)-b(c+d).$$

Example 2 Expand $(2x-3)(x-5)$.

$$(2x-3)(x-5) = 2x(x-5)-3(x-5) = (2x^2-10x)-(3x-15)$$
$$= 2x^2-10x-3x+15 = \mathbf{2x^2-13x+15.}$$

Example 3 Expand $(3a+5b)(3a-5b)$.

$$(3a+5b)(3a-5b) = 3a(3a-5b)+5b(3a-5b)$$
$$= 9a^2-15ab+15ab-25b^2$$
$$= \mathbf{9a^2-25b^2.}$$

EXERCISE 53

Expand the following expressions:

1 $(b+c)(y+z)$ **2** $(a-b)(x+y)$ **3** $(c+d)(y-z)$
4 $(a-d)(x-z)$ **5** $(a+b)(a-c)$ **6** $(x-a)(x-b)$
7 $(a+2)(a+3)$ **8** $(b+4)(b-1)$ **9** $(c-3)(c-5)$
10 $(3-m)(3-m)$ [**11**] $(4+n)(4-n)$ **12** $(5+p)(1-p)$
13 $(a-2b)(a-3b)$ [**14**] $(c+d)(c-4d)$ **15** $(x-y)(x+3y)$
16 $(3x+1)(3x-1)$ [**17**] $(5y+2)(5y-2)$ **18** $(2z-1)(4z-3)$
19 $(2r+3s)(4r-3s)$ [**20**] $(2x-5y)(3x+2y)$
21 $(a+b)^2$ **22** $(a-b)^2$ **23** $(a+b)(a-b)$ **24** $(3c-2d)^2$
[**25**] $(4x+y)^2$ [**26**] $(3y-4z)^2$ **27** $(5y+2t)^2$ **28** $(3z-5p)^2$

Products by Inspection The product $(a+b)(c+d)$ is obtained by multiplying each of the terms in the first bracket by each of the terms in the second bracket and then adding the separate products.

Example 4 Expand $(p-q)(r-s)$.

$$(p-q)(r-s) = pr+p(-s)-qr-q(-s)$$
$$= \mathbf{pr-ps-qr+qs}.$$

The middle step should be performed mentally.

Example 5 Expand $(3x-2)(4x+5)$.

$$(3x-2)(4x+5) = 12x^2+15x-8x-10$$
$$= 12x^2+7x-10.$$

After a little practice, *the middle step can be performed mentally.*

Inserting links is a help: $(3x-2)(4x+5)$.

For longer expressions the working may be arranged as in Arithmetic.

Example 6 Multiply $2x^2-5x-3$ by $3x+7$.

$$\begin{array}{l} 2x^2-5x-3 \\ 3x\ +7 \\ \hline 6x^3-15x^2-\ 9x \\ \qquad\ \ 14x^2-35x-21 \\ \hline \mathbf{6x^3-\ \ x^2-44x-21} \\ \hline \end{array}$$

It is useful to practice collecting coefficients of like terms mentally. It helps to insert links as before.

The links in $(3x+7)(2x^2-5x-3)$ show that the coefficient of x is

$7(-5)+3(-3)$, that is, $-35-9$, $=$ $\mathbf{-44}$.

EXERCISE 54

Write down the expansions of the following:

1 $(a+b)(x+y)$ **2** $(c-d)(y+z)$ **3** $(q-r)(a-b)$
[4] $(c+d)(x-y)$ **[5]** $(a+b)(a-c)$ **6** $(c-z)(d-z)$
[7] $(m-1)(n+2)$ **8** $(y+4)(y+7)$ **[9]** $(z+1)(z+10)$
10 $(a+5)(a-2)$ **11** $(b-7)(b+3)$ **12** $(c-3)(c-5)$
13 $(1-n)(5+n)$ **14** $(3-p)(3-p)$ **15** $(4+q)(4-q)$
[16] $(4+t)(3-t)$ **[17]** $(z-7)(z+7)$ **[18]** $(2\;\;s)(5-s)$
19 $(3+t)(7-t)$ **20** $(4-r)(10+r)$ **21** $(3b+2)(4b+5)$
[22] $(4y-1)(3y+1)$ **[23]** $(5z-1)(5z+1)$ **[24]** $(3c-2)(4c-7)$
25 $(x+y)^2$ **26** $(x-y)^2$ **27** $(x+y)(x-y)$
28 $(2c+5d)^2$ **29** $(7f-3g)^2$ **30** $(5y+7z)(5y-7z)$

Expand and simplify the following expressions:

31 $(x+3)(x^2+x+2)$ **32** $(a+b)(a^2-ab+b^2)$
[32] $(2c^2-3c-1)(3c-2)$ **[34]** $(1-7x-3x^2)(2+x)$
35 $(x^2+2xy+4y^2)(x-2y)$ **36** $(3r^2-rs-2s^2)(2r+3s)$
37 (i) $(p-1)^2+(p+1)^2-2p^2$; (ii) $(p+q)(p-q)-(p-q)^2$
[38] (i) $(r+s)^2-(r-s)^2$; (ii) $(x+\frac{1}{2}y)(4x-2y)$
39 $3(t-3)^2-2(t-1)^2+2(t+1)^2-3(t+3)^2$

Find the coefficient of x^2 and the coefficient of x in the expansions of the following products:

40 $(x-1)(5x^2-4x-3)$ **41** $(3x+5)(2x^2-x-2)$
42 $(4x^2+3x-7)(4x-3)$ **43** $(2x-3-5x^2)(3-2x)$
[44] $(1-2x)(3+7x-2x^2)$ **[45]** $(4-5x-8x^2)(4+5x)$

Solve the following equations:

46 $3(n-1)(n-2) = (1-n)(1-3n)$
47 $(y+1)^2+y^2+1 = (y-1)^2+y^2-1$
[48] $(x+1)(x-2)+5 = (1-x)(2-x)$

49 Expand the product of three consecutive whole numbers, the largest of which is $n+1$.

[50] Expand the product of three consecutive even numbers, the smallest of which is $2m-2$.

51 Expand the sum of the squares of three consecutive odd numbers, the smallest of which is $2p-1$.

52 By how much does the area of a square of side $(r+3)$ cm exceed the area of a rectangle $(r+5)$ cm long, $(r+1)$ cm high? Find also the perimeter of each figure.

[**53**] The length of each edge of a cubical block is $(l+1)$ cm. Find the total area of its surface in a form without brackets.

54 A rectangular lawn $(4a+b)$ m long, $(a+b)$ m wide, is surrounded by a path $(a-b)$ m wide. Find the area of the path.

Relation between $x-y$ and $y-x$

Statements such as $4+6 = 6+4, 2+7 = 7+2, 8+5 = 5+8$, etc., are all included in the single formula, $x+y = y+x$.

Since $6-4 = 2$ and $4-6 = -2$, $\therefore (6-4) = -(4-6)$.

Similarly $2-7 = -(7-2)$, $8-5 = -(5-8)$, etc.; and all these statements are included in the single formula,

$$x-y = -(y-x).$$

This equality follows from the ordinary rules for removing brackets, $-(y-x) = -y+x = x-y$;

or from short division,
$$-1\overline{)\;x-y\;}$$
$$\underline{-x+y}.$$

Example 7 Divide $2a^2-2ab$ by $-2a$.

$$-2a\overline{)\;2a^2-2ab\;}$$
$$\underline{-a+b}.$$

Quotient, **b − a.**

Example 8 Simplify $\dfrac{5r-5s}{6s-6r}$.

$$\frac{5r-5s}{6s-6r} = \frac{5(r-s)}{6(s-r)} = \frac{-5(s-r)}{6(s-r)} = -\frac{5}{6}.$$

EXERCISE 55

If $a = 7, b = 4$, write down the values of:

1 $a-b; b-a$ **2** $(a-b)^2; (b-a)^2$ **3** $(a-b)^3; (b-a)^3$

Simplify the following:

4 $\dfrac{b-a}{a-b}$ **5** $\dfrac{c-d}{3d-3c}$ [**6**] $\dfrac{2x-2y}{y-x}$ [**7**] $\dfrac{4(f+g)}{6(g-f)}$

8 $\dfrac{(r+s)(p-q)}{(s+r)(p+q)}$ [**9**] $\dfrac{(m-n)(m-n)}{(n-m)(n-m)}$ **10** $\dfrac{cxy-xyz}{abz-abc}$

Write more shortly:

11 $(p+q)(q+p)$ **12** $(c-d)(d-c)$ **13** $(x-y)^2 \div (y-x)$

14 $\dfrac{x}{x-y}+\dfrac{y}{y-x}$ **15** $\dfrac{x^2}{x-y}\times\dfrac{y-x}{xz}$ **16** $\dfrac{a-b}{c-d}\div\dfrac{b-a}{d-c}$

Important Expansions

The following expansions should be learnt by heart:

$$(A+B)^2 \equiv A^2+B^2+2AB;$$
$$(A-B)^2 \equiv A^2+B^2-2AB;$$
$$(A+B)(A-B) \equiv A^2-B^2.$$

In words,

The square of the sum of two numbers is equal to the sum of their squares **plus** *twice their product.*

The square of the difference of two numbers is equal to the sum of their squares **minus** *twice their product.*

The product of the sum and the difference of two numbers is equal to the difference of their squares.

Perfect Squares

Example 9 Write down the squares of $2x+3y$ and $5x-7y$.

$$(2x+3y)^2 = (2x)^2+(3y)^2+2(2x)(3y) = \mathbf{4x^2+9y^2+12xy.}$$
$$(5x-7y)^2 = (5x)^2+(7y)^2-2(5x)(7y) = \mathbf{25x^2+49y^2-70xy.}$$

After a little practice, the *middle step should be performed mentally*, not written down.

Example 10 Is $16y^2+25z^2-20yz$ a perfect square?

$16y^2 = (4y)^2$ and $25z^2 = (5z)^2$; therefore *if* the expression is a perfect square, it must be the square of $4y-5z$.

But $(4y-5z)^2 = 16y^2+25z^2-40yz$.

$$\therefore\ 16y^2+25z^2-20yz \text{ is not a perfect square.}$$

Square Roots Any positive number has two square roots; thus the square roots of 25 are $+5$ and -5.

Since $(B-A)^2 = B^2+A^2-2BA = (A-B)^2$, it follows that A^2+B^2-2AB has two square roots, namely $A-B$ and $B-A$; these may be written $A-B$ and $-(A-B)$ or, more shortly, $\pm(A-B)$.

Similarly, $(-A-B)^2 = A^2+B^2+2AB = (A+B)^2$; therefore the two square roots of A^2+B^2+2AB are $A+B$ and $-A-B$; these may be written $A+B$ and $-(A+B)$ or, more shortly, $\pm(A+B)$.

Completing the Square

Geometrical Illustration What must be added to x^2+6x to make the result a perfect square?

Figure 125 represents a rectangle $(x+6)$ cm long, x cm high; $\therefore$ its area $= x(x+6)$ cm^2 $= (x^2+6x)$ cm^2.

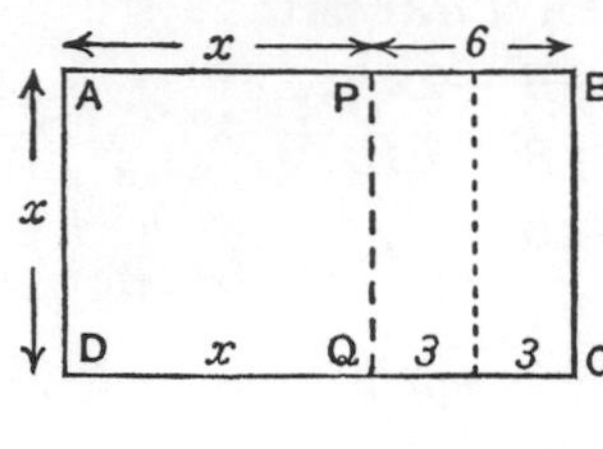

FIG. 125

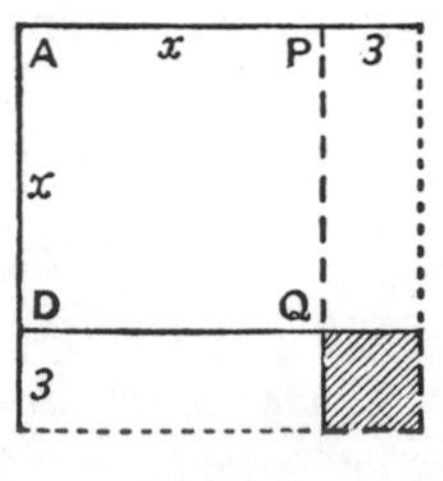

FIG. 126

Transpose half the rectangle PBCQ and fit it on to DQ, as in Fig. 126. This gives a square of side x cm, bordered with two rectangles, each of width, $\frac{6}{2}$ cm, $= 3$ cm. To complete the square, add the shaded area in Fig. 126; this is a square of side 3 cm, area 3^2 cm^2 $= 9$ cm^2. The result is a square of side $(x+3)$ cm, area $(x+3)^2$ cm^2, or (x^2+6x+9) cm^2.

Hence, to x^2+6x, add $(\frac{6}{2})^2 = 9$; then the sum is $(x+3)^2$.

To complete the square, start by looking for an expression whose square is of the required form.

Example 11 What must be added to a^2-5a to make the sum a perfect square? Of what expressions is the sum the square?

Since $a^2-2ab+b^2 = (a-b)^2$,

$$a^2-5a+? = a^2-2(\tfrac{5}{2})a+? = (a-\tfrac{5}{2})^2$$
$$= a^2-5a+(\tfrac{5}{2})^2.$$

$\therefore$ If $(\frac{5}{2})^2$ is added to a^2-5a, the sum equals $(a-\frac{5}{2})^2$, and this is the square of $a-\frac{5}{2}$, also of $-(a-\frac{5}{2})$, that is, $\frac{5}{2}-a$.

EXERCISE 56

Write down the squares of the following:

1 $a+5$ **2** $b-3$ **3** $3-b$ **4** $-a-5$

5 $4c+d$ **6** $4e-f$ **7** $2p+3q$ **8** $5s-7r$

[9] $x-y$ **[10]** $y-x$ **[11]** $-5r-3s$ **[12]** $4y-5z$

13 $x+\frac{1}{2}$ **14** $y-\dfrac{1}{y}$ **15** $3ab-4c$ **[16]** $a+\dfrac{1}{a}$

In Nos. 17–28, state whether the expression is a perfect square; if it is give the *two* square roots.

17 x^2-2x+1 **18** a^2+6a+9 **19** b^2+2b+4
20 c^2-4c+4 **21** $d^2-10d-25$ **22** $16-8k+k^2$
[23] n^2-25 **[24]** $1+2p+p^2$ **[25]** $4-12x+9x^2$
26 $9s^2+60s+100$ **27** $4a^2-40ab+25b^2$ **28** $t^2+t+\frac{1}{4}$

What term, added to the following, makes the sum a perfect square? Of what expressions is the sum the square?

29 x^2+10x **30** y^2-8y **31** a^2+6ab **32** c^2+3c
[33] y^2-14yz **[34]** d^2-7d **35** a^2+a **36** c^2-cd

The Difference of Two Squares

Since

$$A^2-B^2 = (A+B)(A-B),$$
$$x^2-9 = x^2-3^2 = (x+3)(x-3),$$

and

$$49a^2-25b^2 = (7a)^2-(5b)^2 = (7a+5b)(7a-5b),$$

and

$$(x+p)^2-q^2 = [(x+p)+q][(x+p)-q]$$
$$= (x+p+q)(x+p-q).$$

Example 12 Factorise $16a^2-9(b-c)^2$.

$$16a^2-9(b-c)^2 = [4a]^2-[3(b-c)]^2$$
$$= [4a+3(b-c)][4a-3(b-c)]$$
$$= \mathbf{(4a+3b-3c)(4a-3b+3c)}.$$

If there is a common factor, write it down **first.**

EXERCISE 57

Factorise the following:

1 y^2-z^2 **2** $9-x^2$ **3** c^2-16d^2
4 $4n^2-25$ **5** $1-4t^4$ **6** $16p^6-49q^8$
7 $(x+2y)^2-z^2$ **[8]** $(b-c)^2-9$ **9** $p^2-(q-r)^2$
10 $(a+3)^2-4$ **[11]** $(b-4)^2-25$ **12** $4c^2-(c-4d)^2$
13 ab^2-ac^2 **[14]** $3d^2-12$ **15** $5-45e^2$
16 $4a^2-9(b+c)^2$ **17** $c^2-16(c-d)^2$ **18** $4e^2-(e+2)^2$

Use factors to evaluate the following:

19 25^2-24^2 **20** 97^2-87^2 **21** $5{\cdot}4^2-4{\cdot}6^2$

22 Evaluate r^2-t^2 when $r = 3\frac{1}{4}$ and $t = 2\frac{1}{4}$.

23 Simplify (i) $(b^2-c^2)\div(b-c)$; (ii) $(1-9b^2)\div(1+3b)$.

CHAPTER 12

INTERCEPT THEOREMS

Definition If a transversal LM cuts two lines AB, CD at H, K, see Fig. 127, then HK is called the **intercept** made by AB and CD on LM.

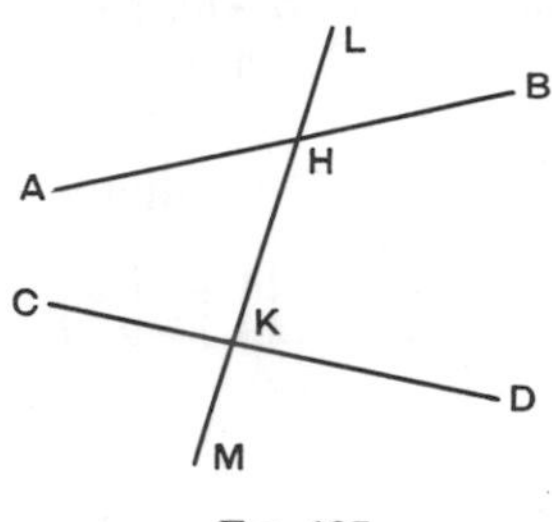

FIG. 127

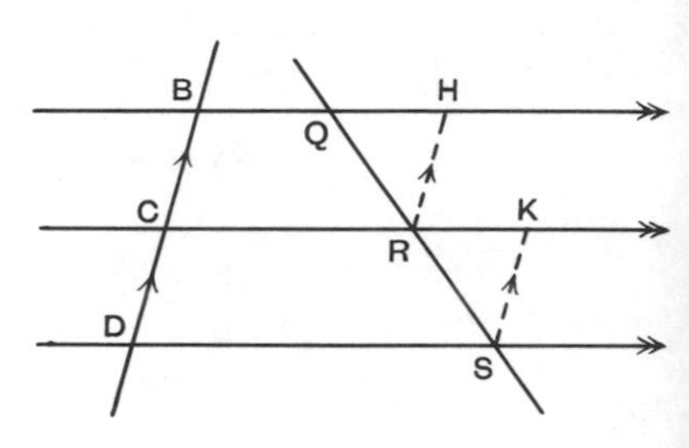

FIG. 128

EXERCISE 58 (Class Discussion)

1 In Fig. 128, BC = CD **and arrows indicate that lines are parallel. Prove that** QR = RS.

(i) Give the reasons why RH = SK.
(ii) Use the triangles RHQ, SKR to complete the proof.

2 (i) What figure is obtained from Fig. 128 by making Q the same point as B? Draw it.
(ii) Complete the sentence: the straight lines drawn through the mid-point of the side BD of △BDS, parallel to DS, . . .

3 Explain how to divide a given line AB into 3 equal parts without measuring it. Does Fig. 129 suggest a way of doing so? Give reasons.

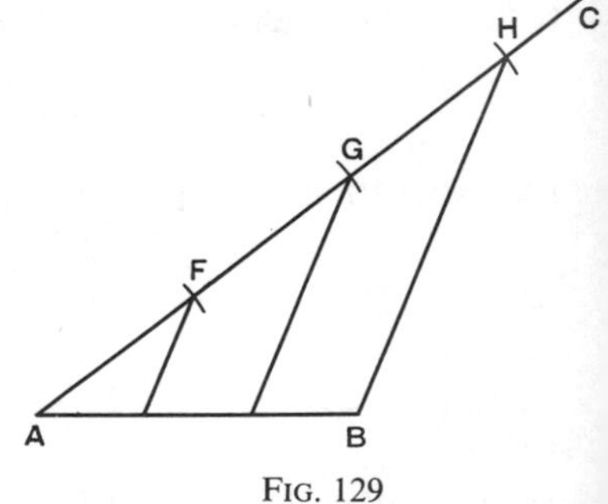

FIG. 129

4 In Fig. 130, AH = HB **and** AK = KC. **Prove that**
(i) HK **is parallel to** BC, (ii) HK = $\frac{1}{2}$BC.

Draw CP parallel to BA to meet HK produced at P.
(i) Explain why △CKP ≡ △AKH.
(ii) Prove that CP = BH. What then follows from the fact that CP is equal and parallel to BH?

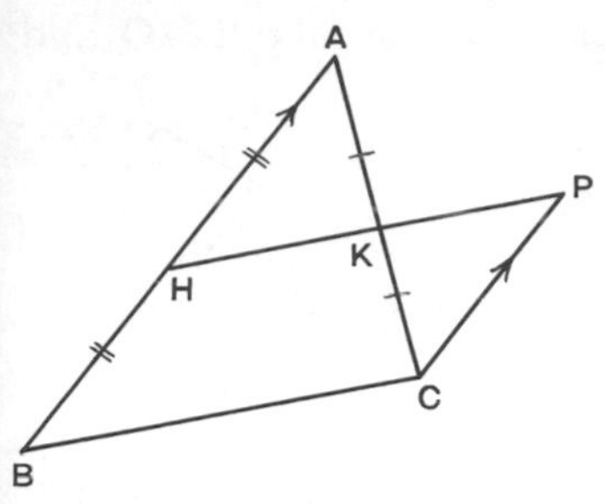

FIG. 130

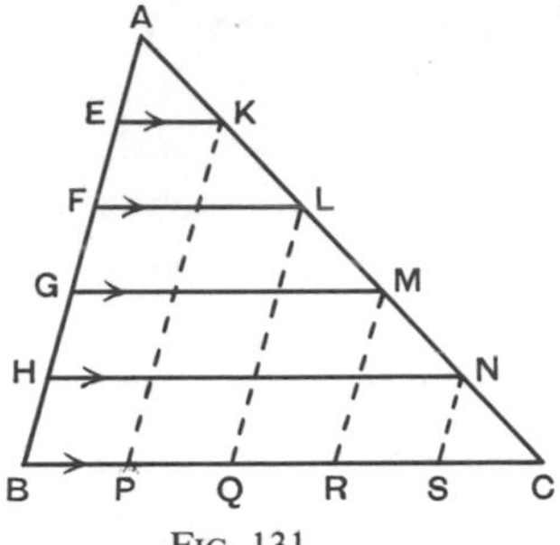

FIG. 131

5 In Fig. 131, the side AB of $\triangle$ABC is divided into 5 equal parts by the lines EK, FL, GM, HN drawn parallel to BC to cut AC at K, L, M, N.
Prove that EK $= \frac{1}{5}$BC, FL $= \frac{2}{5}$BC,
GM $= \frac{3}{5}$BC, HN $= \frac{4}{5}$BC.

Draw KP, LQ, MR, NS parallel to AB to cut BC at P, Q, R, S.

(i) What can you say about the points K, L, M, N?

(ii) What can you say about the points P, Q R, S?

6 If, in Fig. 131, AH $= \frac{7}{10}$AB, what can you say about the length of HN?

Similar Triangles The results in No. 5 can be obtained more shortly by using the properties of similar triangles.

Since in Fig. 131, EK is parallel to BC, $\triangle$AEK is equiangular to $\triangle$ABC and therefore similar to it.

$$\therefore \frac{AK}{AC} = \frac{AE}{AB} = \frac{1}{5} \quad \text{and} \quad \frac{EK}{BC} = \frac{AE}{AB} = \frac{1}{5};$$

$$\therefore AK = \tfrac{1}{5}AC \quad \text{and} \quad EK = \tfrac{1}{5}BC.$$

The reader should now obtain the other results in No. 5 and the answer to No. 6 in the same way.

EXERCISE 59

[Properties of similar triangles may be used]

1 ABC is a triangle; a line parallel to BC cuts AB, AC at P, Q; BC = 6 cm, CA = 7·2 cm, AB = 8 cm. Find AQ and PQ if (i) AP = PB, (ii) AP = 2PB.

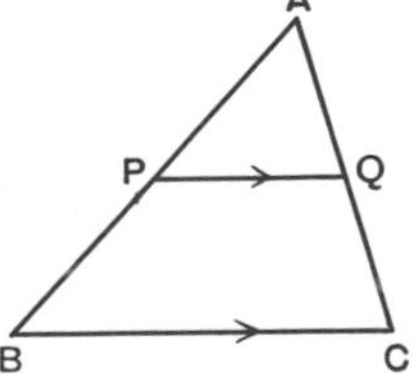

FIG. 132

[**2**] X, Y, Z are the mid-points of the sides BC, CA, AB of $\triangle$ABC. If BC = 5 cm, CA = 6 cm, AB = 7 cm, find the lengths of YZ, ZX, XY.

3 In Fig. 132, if AP = 2 cm, PB = 3 cm, BC = 4 cm, and AC = 4·5 cm, find the lengths of AQ and PQ. [Look at Fig. 131.]

[4] In Fig. 132, if $AP = \frac{3}{7}AB$, what can you say about AQ and about PQ? Give reasons.

5 In Fig. 133, find the values of a and b.

[6] In Fig. 133, find x if $y = 12$.

7 In Fig. 133, find z if $y = 12$.

[8] If, in Fig. 132, QR is drawn parallel to AB to meet BC at R, and if BP = 10 cm, AQ = 12 cm, BR = 15 cm and RC = 20 cm, find QC and AP.

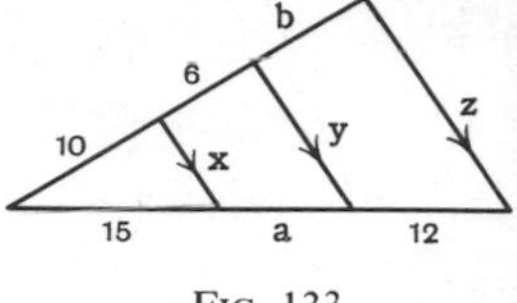

FIG. 133

9 In Fig. 134, if AC = CB = 4 cm, find CR. [Find CK and KR.]

10 In Fig. 134, if AC = 3 cm and CB = 6 cm, find the length of CR.

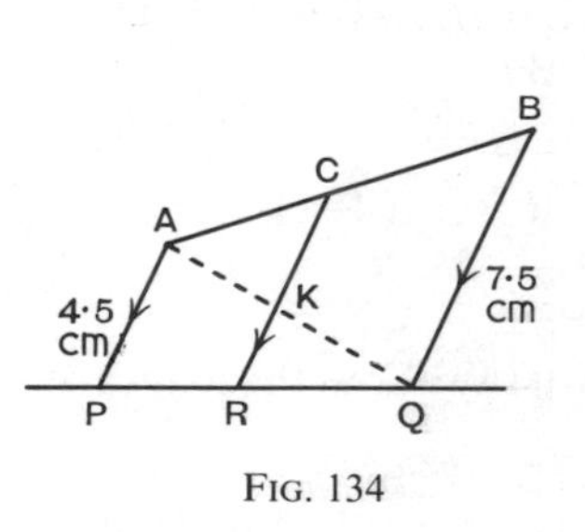

FIG. 134

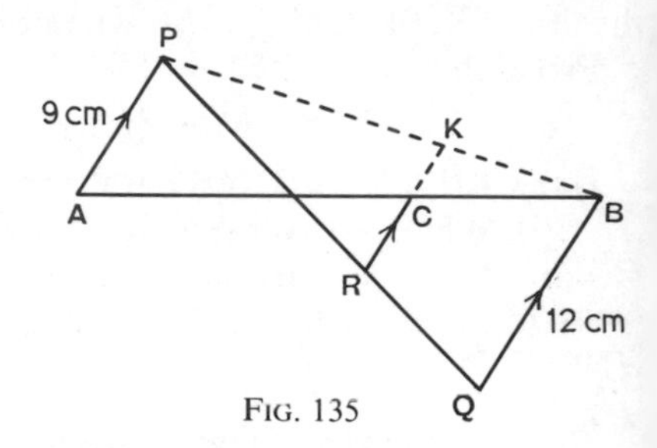

FIG. 135

11 In Fig. 135, if AC = CB = 10 cm, find CR. [Find CK and KR.]

12 In Fig. 135, if AC = 8 cm, CB = 4 cm, find the length of CR.

13 The tops A, C of two vertical poles AB, CD which stand on level ground are joined by a rod. If AB = 5 m and CD = 12 m, find the height of the mid-point of AC above the ground.

THEOREM

If three or more parallel straight lines make equal intercepts on a given transversal, they make equal intercepts on any other transversal.

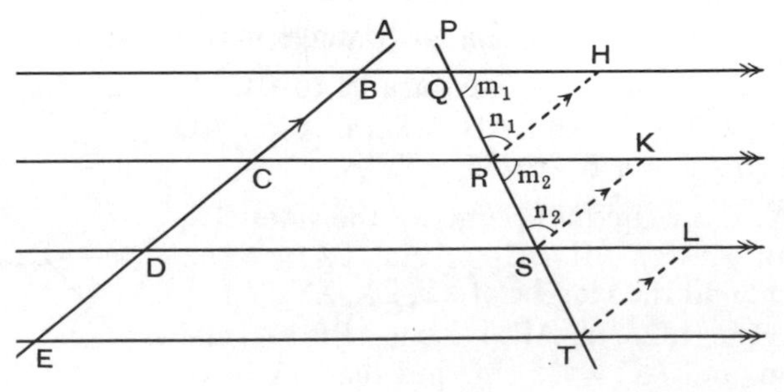

FIG. 136

Given the parallel lines BQ, CR, DS, ET, . . . cutting a transversal at B, C, D, E, . . . so that BC = CD = DE = . . . and cutting another transversal at Q, R, S, T, ...

To prove that

$$QR = RS = ST = \dots .$$

Construction Through R, S, T, . . . draw lines parallel to BCDE to meet BQ, CR, DS, . . . produced, if necessary, at H, K, L, ...

Proof Since BH$\parallel$CR *given*, and RH$\parallel$CB *constr.*,

CRHB is a parallelogram,

$\therefore$ RH = CB *opp. sides* $\parallel$*gram.*

Similarly it may be proved that SK = DC,
but BC = CD *given*, $\therefore$ RH = SK.

$\therefore$ in $\triangle$s RHQ, SKR, with the notation in the figure,

$m_1 = m_2$ *corr.* $\angle$*s*, QH$\parallel$RH,

$n_1 = n_2$ *corr.* $\angle$*s*, RH$\parallel$SK,

RH = SK *proved*,

$\therefore$ $\triangle$RHQ $\equiv$ $\triangle$SKR AAS.

$\therefore$ QR = RS, and similarly RS = ST, etc.

Abbreviation for reference: Intercept theorem

Construction

Divide a given straight line into any given number of equal parts.

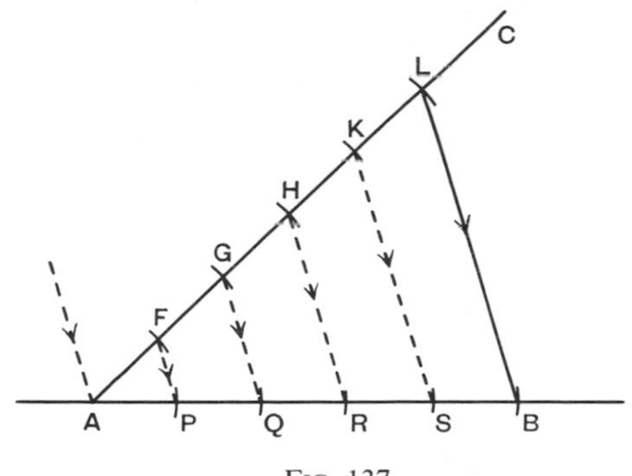

FIG. 137

Given a line AB.

To construct points dividing AB into any given number of equal parts, say 5.

Construction From A draw any line AC making any convenient angle with AB, and from AC cut off equal lengths AF, FG, GH, . . . the number of such lengths being the required number of equal parts, in this case 5.

Let the equal lengths be AF, FG, GH, HK, KL.

Join LB, and through F, G, H, K draw lines parallel to LB to meet AB at P, Q, R, S.

Then AP, PQ, QR, RS, SB are the required parts.

Proof Since the parallel lines FP, GQ, HR, KS, LB, together with a parallel through A, make equal intercepts on AC, they also make equal intercepts on AB. ∴ AP = PQ = QR = RS = SB.

THEOREM

The straight line joining the mid-points of two sides of a triangle is parallel to the third side and equal to one-half of it.

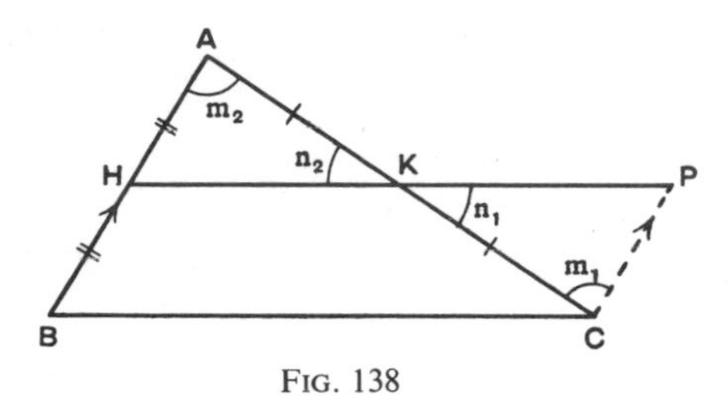

FIG. 138

Given the mid-points H, K of the sides AB, AC of △ABC.

To prove that (i) HK∥BC,
(ii) HK = $\frac{1}{2}$BC.

Construction Through C draw CP parallel to BA to meet HK produced at P.

Proof (i) With the notation in the figure,

in △s CKP, AKH,

$m_1 = m_2$ *alt.* ∠*s*, CP∥BA,

$n_1 = n_2$ *vert. opp.* ∠*s*,

CK = AK *given*,

∴ △s $\begin{matrix}\text{CKP}\\\text{AKH}\end{matrix}$ are congruent ASA,

∴ CP = AH and PK = HK,

but AH = BH *given*, ∴ CP = BH.

Also CP is drawn parallel to BH,

$\therefore$ the lines CP, BH are equal and parallel,

$\therefore$ BCPH is a parallelogram,

$\therefore$ HP $\parallel$ BC, *i.e.* HK $\parallel$ BC.

(ii) Also

BC = HP *opp. sides* $\parallel$*gram*,

but

HK = KP *proved*, $\therefore$ HK $= \frac{1}{2}$HP,

$\therefore$ HK $= \frac{1}{2}$BC.

Abbreviation for reference: Mid-point theorem

EXERCISE 60 (Class Discussion)

1 If X is the mid-point of the side AB of $\triangle$ABC and if XY is drawn parallel to BC to meet AC at Y, prove that Y is the mid-point of AC.

Draw a figure and mark the data on it.

Draw CP parallel to BA to meet XY produced at P.

(i) Explain why CP = BX = XA.

(ii) Prove $\triangle$CYP $\equiv$ $\triangle$AYX.

2 In Fig. 139, AC = CB and the parallel lines AP, CR, BQ meet another transversal at P, R, Q. Prove that

$$CR = \tfrac{1}{2}(AP+BQ).$$

Join AQ and let it cut CR at K.

(i) Explain why AK = KQ.

(ii) What do you know about the length of CK?

(iii) What do you know about the point R and the length of KR?

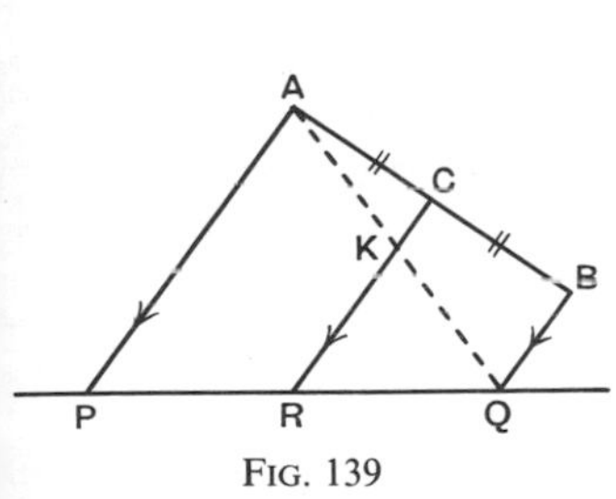

FIG. 139

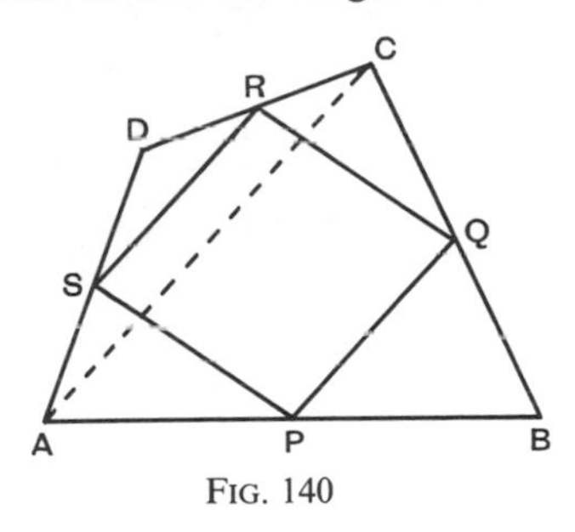

FIG. 140

3 In Fig. 140, P, Q, R, S are the mid-points of the sides of any quadrilateral ABCD. Prove that PQRS is a parallelogram.

Join AC.

(i) Explain why PQ is parallel to AC.

(ii) What do you know about the length of PQ?

(iii) What do you know about SR?

4 In Fig. 141, F **is the mid-point of the hypotenuse** AB **of the right-angled triangle** ABC. **Prove that** CF $= \frac{1}{2}$AB.

Draw FN parallel to BC to meet AC at N.

(i) What do you know about the length of AN?

(ii) Explain why $\triangle$FNC $\equiv$ $\triangle$FNA.

Complete the proof.

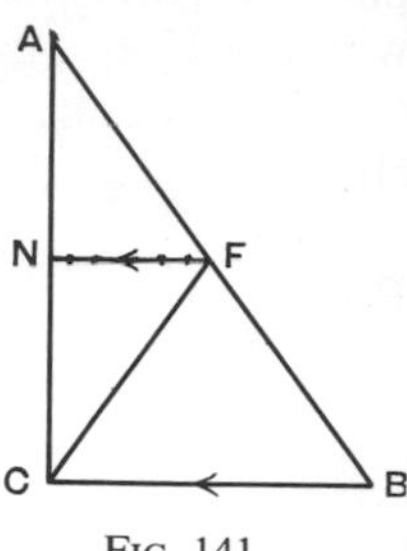

FIG. 141

5 What can you say about the circle whose diameter is the hypotenuse of a right-angled triangle?

6 In Fig. 142, E, F, G are the mid-points of AC, AB, AH.

(i) State two facts about EG.

(ii) Prove BGCH is a $\|$gram.

Draw Fig. 142. Join BC and let it cut AGH at D.

(iii) Prove BD = DC.

(iv) Prove DG $= \frac{1}{2}$GA $= \frac{1}{3}$DA.

(v) Prove EG $= \frac{1}{2}$GB $= \frac{1}{3}$EB.

(vi) Prove FG $= \frac{1}{2}$GC $= \frac{1}{3}$FC.

(vii) If AD = 3 cm, BE = 4·5 cm, CF = 6 cm, find the lengths of the sides of $\triangle$GHC.

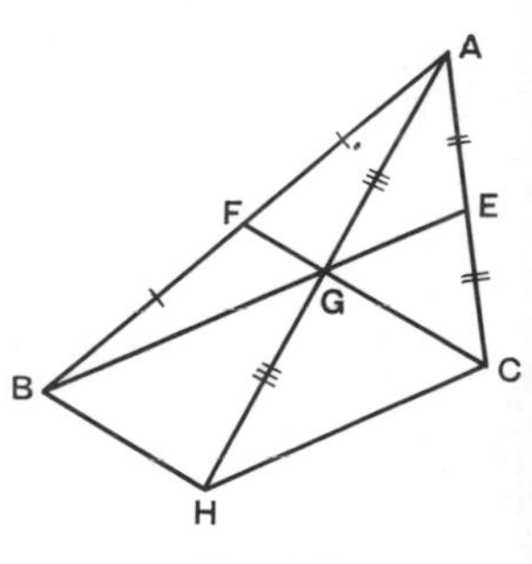

FIG. 142

The lines AD, BE, CF which join the vertices of $\triangle$ABC to the mid-points of the opposite sides are called the **medians** of $\triangle$ABC. It follows from (iii) that the medians of a triangle are concurrent.

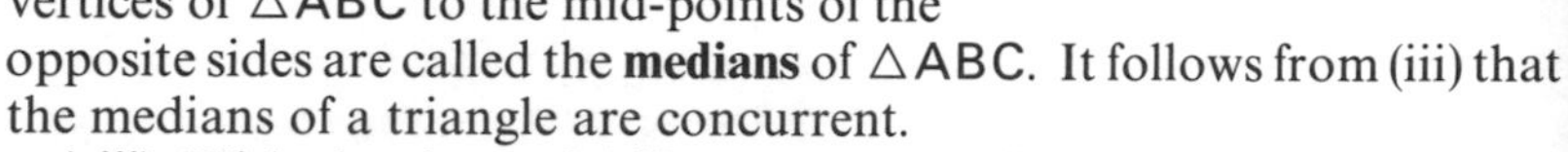

(viii) With the data of (vii), construct $\triangle$GHC and then construct $\triangle$ABC.

EXERCISE 61

1 In Fig. 143, P, Q, R are the mid-points of BC, CA, AB; prove that PQAR is a parallelogram.

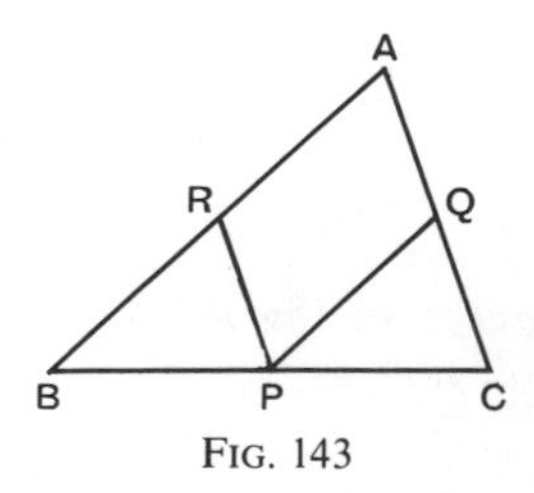

FIG. 143

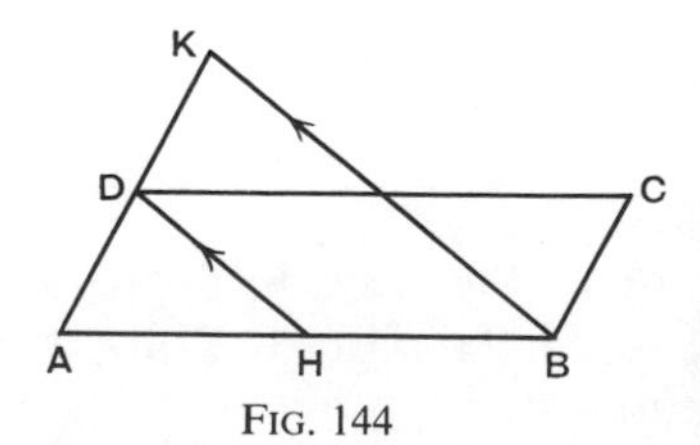

FIG. 144

2 In Fig. 144, ABCD is a parallelogram and H is the mid-point of AB. Prove that DK = BC.

3 P, Q are the mid-points of the sides AB, DC respectively of the parallelogram ABCD; PD, BQ cut AC at H, K respectively. (i) Explain why PBQD is a parallelogram. (ii) Prove that AH = HK = KC.

[**4**] In Fig. 147, if AK = 2KB, prove that AQ = 2QC. [Bisect AK at H.]

5 In Fig. 145, H, K, X, Y are the mid-points of PB, PC, QB, QC. Prove that HK = XY.

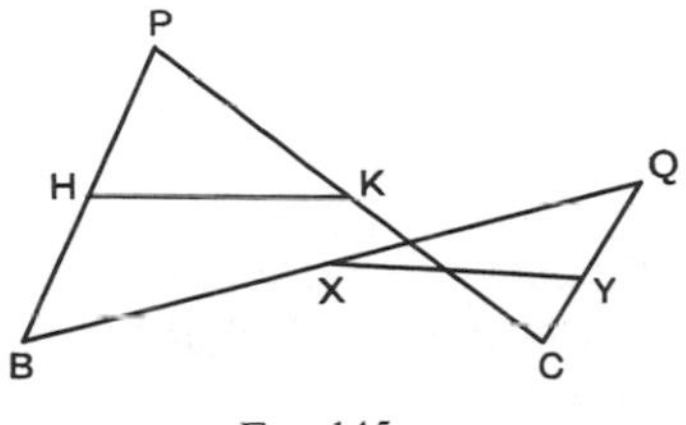

FIG. 145

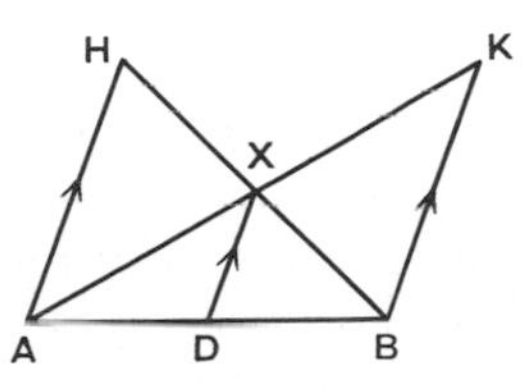

FIG. 146

6 In Fig. 146, AD = DB. Prove that AH = BK.

[**7**] Four parallel lines cut one transversal at A, B, C, D and cut another transversal at P, Q, R, S. If AB = CD, prove that PQ = RS. [Through Q, S draw QH, SK parallel to DA to meet AP, CR, produced if necessary, at H, K.]

8 In Fig. 147, AH = HK = KB; prove that (i) HP = $\frac{1}{3}$BC; (ii) KQ = $\frac{2}{3}$BC. [Draw PX and QY parallel to AB.]

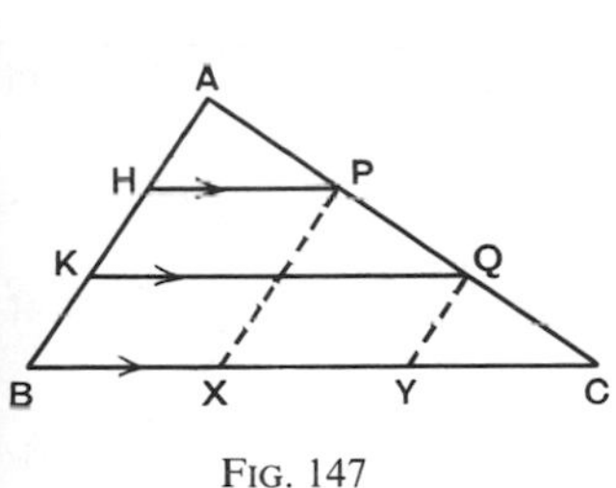

FIG. 147

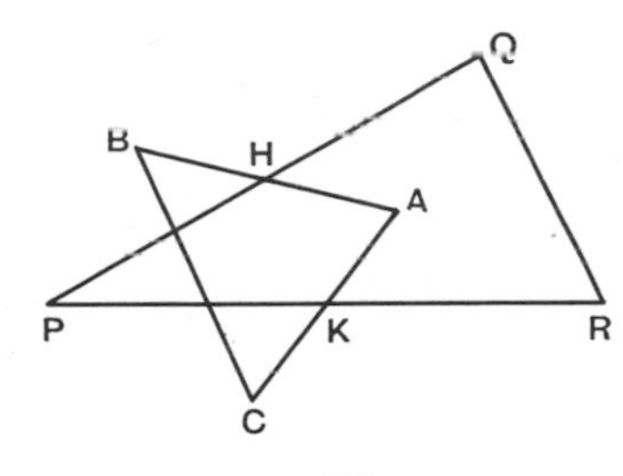

FIG. 148

9 In Fig. 148, H is the mid-point of AB and of PQ; K is the mid-point of AC and of PR. Prove that QR = BC.

10 Construct △ABC, given AC = CB = 4 cm, ∠ACB = 90°.

(i) Make a construction for dividing AB into 3 equal parts.

(ii) Construct a point P on AB so that AP:PB = 2:3.

[Start by dividing AB into 5 equal parts.]

11 Draw a line AB of length 8 cm. Construct on AB produced a point Q so that AQ:QB = 9:2.

[**12**] P is the mid-point of the side BC of $\triangle$ABC; Q is the mid-point of AP; BQ produced meets AC at R. Prove that AC = 3AR. [Draw PK parallel to BR to meet AC at K.]

[**13**] In $\triangle$ABC, $\angle$B = 1 right angle; BCX is an equilateral triangle. Prove that the line from X parallel to AB bisects AC. [There are two cases.]

14 In Fig. 135, p. 114, if AC = CB, prove that for any lengths of AP and BQ, CR = $\frac{1}{2}$(BQ $\sim$ AP).

15 A and B are given points; P is a variable point on a given circle, centre B. Prove that the mid-point of AP lies on a fixed circle whose centre is the mid-point of AB.

16 If the diagonals of a quadrilateral cut at right angles, prove that the mid-points of the four sides are the vertices of a rectangle.

[**17**] If the diagonals of a quadrilateral are equal, prove that the mid-points of the four sides are the vertices of a rhombus.

18 Q, R are the mid-points of the sides AB, BC of $\triangle$ABC; the perpendiculars AD, BE from A, B to BC, AC intersect at H; P is the mid-point of AH. Prove that $\angle$PQR is a right angle.

19 In Fig. 149, O is the centre of the circle; AOBD is a straight line such that AB = BD. Prove that PR = RD. [Join AP.]

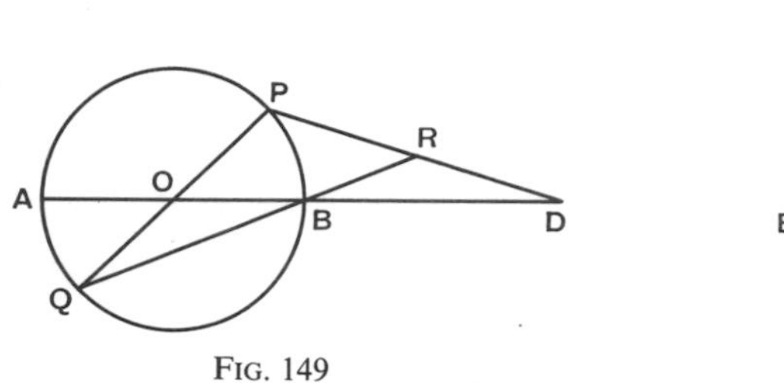

FIG. 149

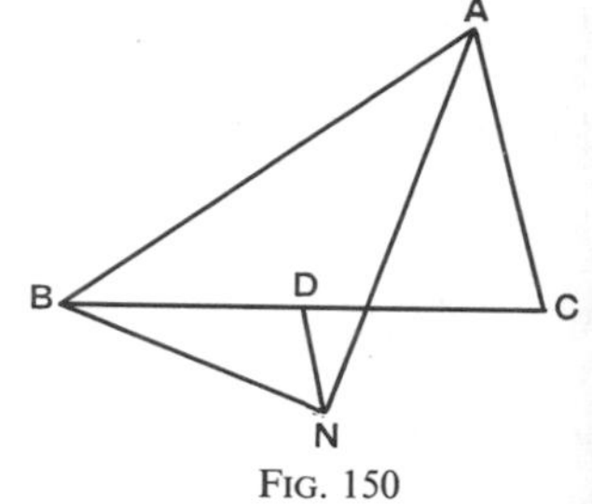

FIG. 150

20 In Fig. 150, BN is the perpendicular from B to the bisector of $\angle$BAC; D is the mid-point of BC. Prove that DN = $\frac{1}{2}$(AB − AC). [Produce BN and AC to cut at K.]

[**21**] Q, R are the mid-points of the sides AC, AB of $\triangle$ABC; AD is the perpendicular from A to BC. Prove that $\angle$RDQ = $\angle$BAC. [Look at No. 4, p. 118.]

22 BP, CQ are the bisectors of $\angle$B, $\angle$C of $\triangle$ABC; AH, AK are the perpendiculars from A to BP, CQ. Prove that KH is parallel to BC. [Produce AH, AK to cut BC at X, Y.]

CHAPTER 13

THE SINE AND COSINE

The Sine The two *right-angled* triangles PAM, QBN in Fig. 151 are **similar** because $\angle A = \angle B$.

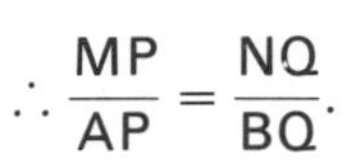

$$\therefore \frac{MP}{AP} = \frac{NQ}{BQ}.$$

The value of each of these ratios depends on the size of $\angle A$ or $\angle B$; it does not depend on the length of AP or BQ.

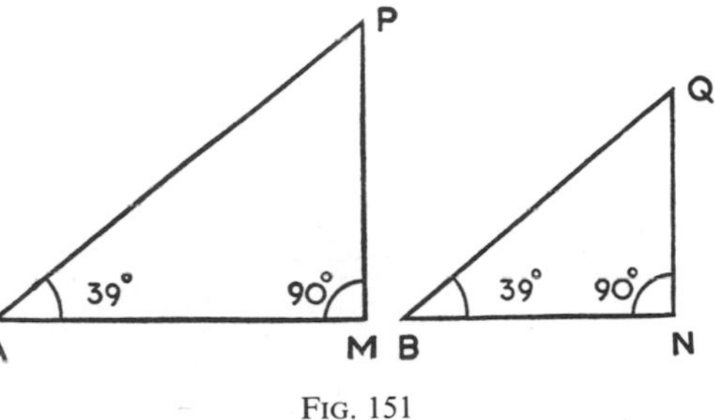

Fig. 151

By measurement, we have

$$MP = 2{\cdot}65 \text{ cm}, \quad AP = 4{\cdot}2 \text{ cm}, \quad NQ = 2{\cdot}0 \text{ cm}, \quad BQ = 3{\cdot}2 \text{ cm};$$

$$\therefore \frac{MP}{AP} = \frac{2{\cdot}65}{4{\cdot}2} = 0{\cdot}63; \quad \frac{NQ}{BQ} = \frac{2{\cdot}0}{3{\cdot}2} = 0{\cdot}63, \quad \text{to 2 figures.}$$

The ratio $\frac{MP}{AP}$ is called the **sine** of $\angle PAM$ and is written sin PAM. From Fig. 151, sin 39° = 0·63 approximately.

The Cosine With the data of Fig. 151, it follows that

$$\frac{AM}{AP} = \frac{BN}{BQ}.$$

The value of each ratio depends only on the size of $\angle A$ or $\angle B$. From Fig. 151, by measurement AM = 3·28 cm, BN = 2·5 cm;

$$\therefore \frac{AM}{AP} = \frac{3{\cdot}28}{4{\cdot}2} = 0{\cdot}78; \quad \frac{BN}{BQ} = \frac{2{\cdot}5}{3{\cdot}2} = 0{\cdot}78, \quad \text{to 2 figures.}$$

The ratio $\frac{AM}{AP}$ is called the **cosine** of $\angle PAM$ and is written cos PAM.

From Fig. 151, cos 39° = 0·78 approximately.

Definitions If a perpendicular PN is drawn from *any* point P in either arm of angle YAZ to the other arm, the ratio $\frac{NP}{AP}$ is called the **sine** of $\angle$YAZ, and the ratio $\frac{AN}{AP}$ is called the **cosine** of $\angle$YAZ.

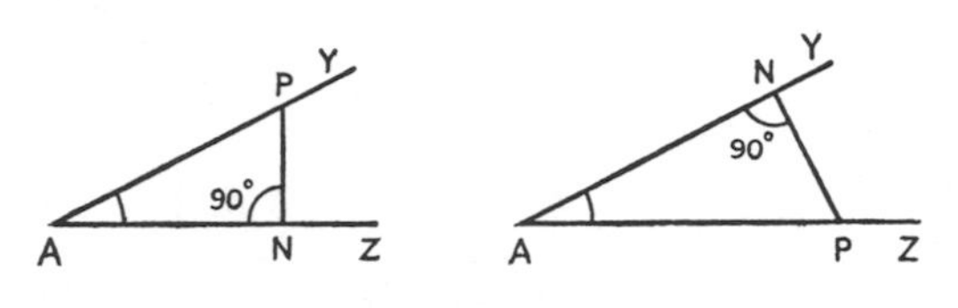

FIG. 152

In each part of Fig. 152, sin YAZ $= \frac{NP}{AP}$; cos YAZ $= \frac{AN}{AP}$.

These ratios may be described in words, as follows:

$$\textbf{sine} = \frac{\text{opposite}}{\text{hypotenuse}} = \frac{\textbf{opp.}}{\textbf{hyp.}}; \quad \textbf{cosine} = \frac{\text{adjacent}}{\text{hypotenuse}} = \frac{\textbf{adj.}}{\textbf{hyp.}}$$

Note The letters O.H.M.S. may be used to remind the beginner of the definition of the sine:

Opposite over Hypotenuse Means Sine.

Complementary Angles If two angles add up to 90°, they are called *complementary* angles. Thus in Fig. 152, $\angle$A and $\angle$P are complementary; either is called the *complement* of the other.

$$\text{Now the sine of } \angle P = \frac{\text{opp.}}{\text{hyp.}} = \frac{AN}{AP},$$

$$\text{and the cosine of } \angle A = \frac{\text{adj.}}{\text{hyp.}} = \frac{AN}{AP};$$

$\therefore$ the cosine of $\angle$A is equal to the sine of the complementary $\angle$P and this is what the word 'cosine' means:

Cosine of angle = sine of complementary angle.

For example, cos 20° = sin 70°, cos 80° = sin 10°.

Example for Class Discussion

Find, by drawing, approximate values of sin 20°, cos 20°; sin 40°, cos 40°; sin 70°, cos 70°.

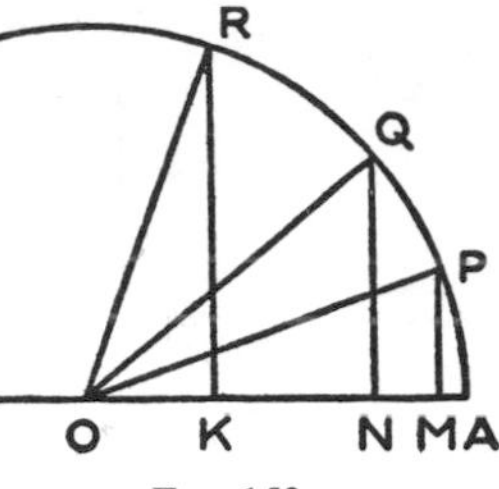

FIG. 153

Figure 153 shows part of a circle, centre O, radius 10 cm.
$\angle AOP = 20°$, $\angle AOQ = 40°$, $\angle AOR = 70°$; PM, QN, RK are drawn perpendicular to OA.

Copy and complete the following, either by measuring the printed figure or, preferably, by drawing and measuring a full-size figure.

It saves time to use squared paper. By measurement,

$$MP = \ldots; OM = \ldots; \text{also } OP = \ldots$$

$$\therefore \sin 20° = \frac{MP}{OP} = \ldots; \cos 20° = \frac{OM}{OP} = \ldots$$

Find in the same way, sin 40°, cos 40°; sin 70°, cos 70°.

Use of Tables

Figure 153 shows that as the angle $\theta°$ increases from 0° to 90°, the value of $\sin \theta°$ *increases* steadily from 0 to 1; but the value of $\cos \theta°$ *decreases* steadily from 1 to 0; and a glance at the tables confirms this.

The sine-table is used in exactly the same way as the tangent-table.

But, *in using the cosine-table*, figures in the difference-columns must be **subtracted,** because the cosine gets smaller as the angle gets larger.

Example 1 Find cos 53° 20′.

	cos. 53° 18′	0·5976
Subtract	diff. for 2′	5
	cos 53° 20′	0·5971

cos 53° 20′ = **0·5971.**

Greek Letters The following Greek letters are often used for denoting angles: α = alpha; β = bēta; γ = gamma; δ = delta; θ = thēta; ϕ = phī; ω = ōmega.

EXERCISE 62

1 Find by drawing, using squared paper, the sine and cosine of (i) 25°, (ii) 35°, (iii) 65°. What are the values of the sine and cosine of 90° and 0°?

2 Use tables to write down the sines of the following: 17°; 43°; 64°; 88°; 23° 30′; 23° 36′; 23° 31′; 23° 35′; 38° 21′; 64° 11′; 49° 2′; 8° 50′; 6° 47′.

3 Use tables to write down the cosines of the following: 14°; 28°; 56°; 89°; 66° 36′; 66° 42′; 66° 39′; 66° 41′; 62° 40′; 41° 4′; 28° 44′; 10° 16′; 51° 35′.

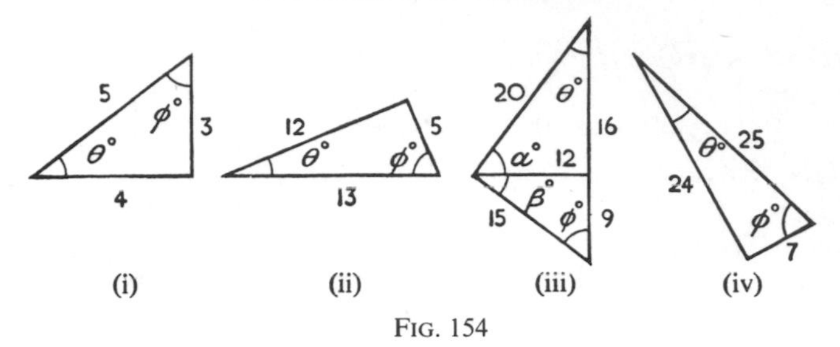

FIG. 154

4 It can be proved that, with the data of Fig. 154, the given triangles are right-angled. Write down the sine and cosine of each marked angle.

Example 2 A ladder 5 m long leans against a wall and is inclined at 53° to the ground. How far from the wall is the foot of the ladder? How high up the wall does the ladder reach?

AB is the ladder; AB = 5 m, ∠OAB = 53°, ∠AOB = 90°.

$$\frac{AO}{AB} = \cos 53° = 0{\cdot}6018,$$

$$\therefore AO = 5 \times 0{\cdot}6018 \text{ m} = \mathbf{3{\cdot}01\ m}, \text{ to 3 figures.}$$

$$\frac{OB}{AB} = \sin 53° = 0{\cdot}7986,$$

$$\therefore OB = 5 \times 0{\cdot}7986 \text{ m} = \mathbf{3{\cdot}99\ m}, \text{ to 3 figures.}$$

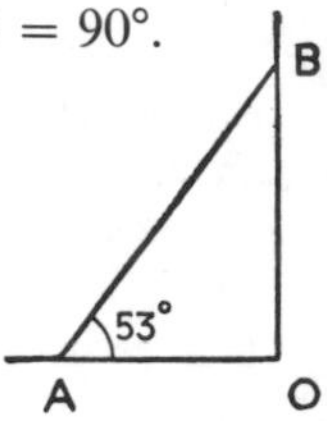

FIG. 155

If the figure does not contain a *right-angled* triangle, it may be necessary to make a construction.

Example 3 ABC is an isosceles triangle, in which AB = AC = 4 cm, ∠BAC = 54°; find BC.

Draw the perpendicular AD from A to BC.

$$\triangle ADB \equiv \triangle ADC \qquad \text{RHS}$$

$$\therefore \angle BAD = \tfrac{1}{2}\angle BAC = 27°,$$

$$\text{and } BD = DC.$$

$$\frac{BD}{BA} = \sin 27° = 0{\cdot}4540,$$

$$\therefore BD = 4 \times 0{\cdot}4540 \text{ cm},$$

$$\therefore BC = 2BD = 8 \times 0{\cdot}4540 \text{ cm} = \mathbf{3{\cdot}63\ cm}, \text{ to 3 figures.}$$

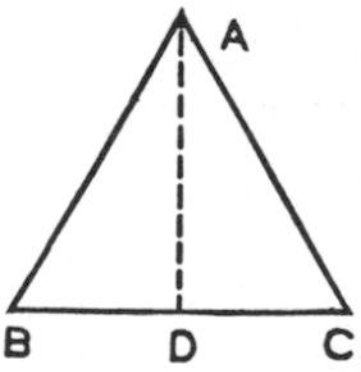

FIG. 156

EXERCISE 63

1 In Fig. 157 the triangles are right-angled, and the given side is in each case the hypotenuse. Find the other sides.

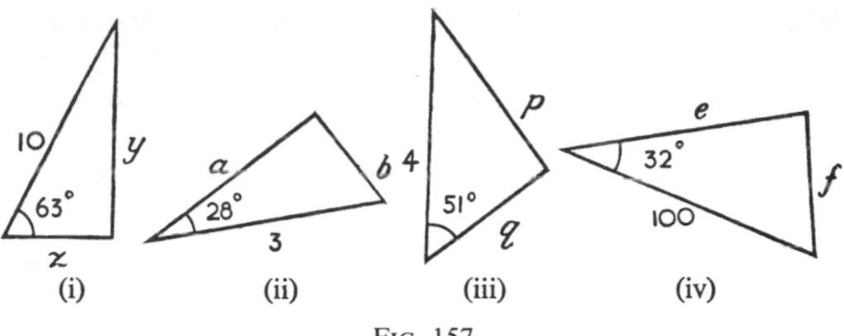

FIG. 157

2 A hill slopes upwards at an angle of 18° with the horizontal. What height does a man rise when he walks 100 m up the slope?

3 B is 2000 m N. 34° E. from A. How much is B (i) East, (ii) North of A?

[**4**] The string of a kite is 100 m long, and makes an angle of 62° with the horizontal. What is the height of the kite?

5 The legs of a pair of dividers are each 12 cm long, and are opened to an angle of 31°. Find the distance between their points.

[**6**] Repeat No. 5, if the angle is 170°.

7 In $\triangle ABC$, $AB = AC = 5$ cm, $\angle ABC = 62°$; find BC.

[**8**] In $\triangle ABC$, $AB = AC = 4$ cm, $\angle BAC = 114°$; find BC.

[**9**] The vertical angle of a cone is 23°, and the length of a slant edge is 2·5 cm. Find the diameter of the base.

10 Each side of a rhombus is 6 cm, and one of its angles is 48°. Find the lengths of the diagonals.

11 Find the cosine or tangent of each marked angle in Fig. 158; note that the triangles are *not* right-angled.

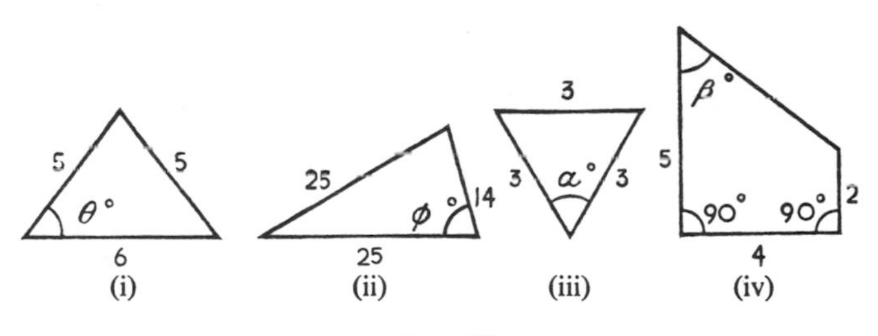

FIG. 158

[**12**] A regular pentagon is inscribed in a circle of radius 5 cm. Find the length of its side.

13 What are the values of cos 36° and sin 54°? Draw a neat but not accurate figure to explain why they are equal.

Notation The angle whose sine is x is often written $\sin^{-1}(x)$ or simply $\sin^{-1} x$, and pronounced 'sine minus one x'.

Similarly, $\cos^{-1} x$ means the angle whose cosine equals x, and $\tan^{-1} x$ means the angle whose tangent equals x.

Example 4 Find the angle $\sin^{-1}(0{\cdot}865)$.

(i) *By drawing.*

Figure 159 shows part of a circle, centre O, radius 10 cm; OA, OC are two perpendicular radii.

From OC, cut off ON equal to 8·65 cm; draw NP perpendicular to OC to meet the circle at P; draw PM perpendicular to OA.

Then $\sin \text{AOP} = \dfrac{\text{MP}}{\text{OP}} = \dfrac{\text{ON}}{\text{OP}} = \dfrac{8{\cdot}65}{10} = 0{\cdot}865.$

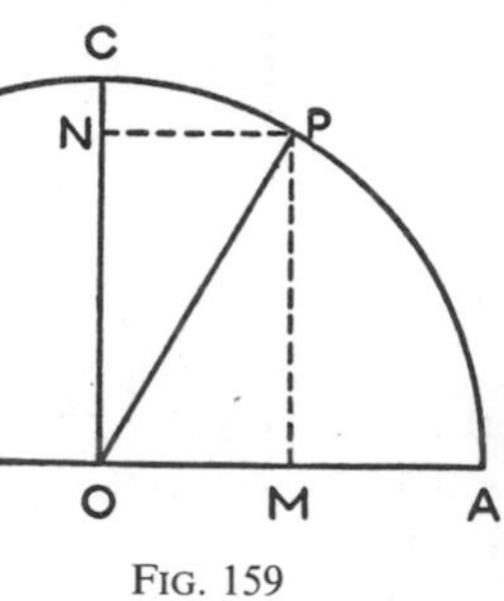

Fig. 159

By measurement, we find $\angle\text{AOP} \simeq 60°$.

(ii) *From tables.*

	0·8643	sin 59° 48′
	7	diff. for 5′ *add*
$\sin^{-1}(0{\cdot}8650) =$ **59° 53′**.	0·8650	sin 59° 53′

Example 5 Find from tables the angle $\cos^{-1}(0{\cdot}2901)$.

	0·2890	cos 73° 12′
	11	diff. for 4′ *subtract*
$\cos^{-1}(0{\cdot}2901) =$ **73° 8′**.	0·2901	cos 73° 8′

Note the subtraction in Example 5; the bigger the cosine, the smaller the angle.

EXERCISE 64

1 Find by drawing, using squared paper,
$\sin^{-1}(0{\cdot}4)$; $\sin^{-1}(\frac{3}{4})$; $\cos^{-1}(0{\cdot}31)$; $\cos^{-1}(0{\cdot}81)$.

[2] Find by drawing, using squared paper,
$\sin^{-1}(0{\cdot}7)$; $\sin^{-1}(0{\cdot}92)$; $\cos^{-1}(0{\cdot}63)$; $\cos^{-1}(0{\cdot}91)$.

3 Use tables to write down the angles whose sines are:
0·3907; 0·9613; 0·7694; 0·4493; 0·4509; 0·4498;
0·4504; 0·2345; 0·3199; 0·9648; 0·5986; 0·6645.

4 Use tables to write down the angles whose cosines are:
0·5592; 0·7880; 0·8712; 0·1805; 0·1788; 0·1794;
0·1802; 0·7585; 0·8631; 0·9834; 0·2513; 0·6312.

[5] Use tables to evaluate:
$\sin^{-1}(0{\cdot}5265)$; $\cos^{-1}(0{\cdot}3100)$; $\tan^{-1}(0{\cdot}6308)$;
$\cos^{-1}(0{\cdot}5203)$; $\sin^{-1}(0{\cdot}0114)$; $\tan^{-1}(3{\cdot}0090)$.

6 It can be proved with the data of Fig. 160 that all the triangles are right-angled. Use tables to evaluate the marked angles.

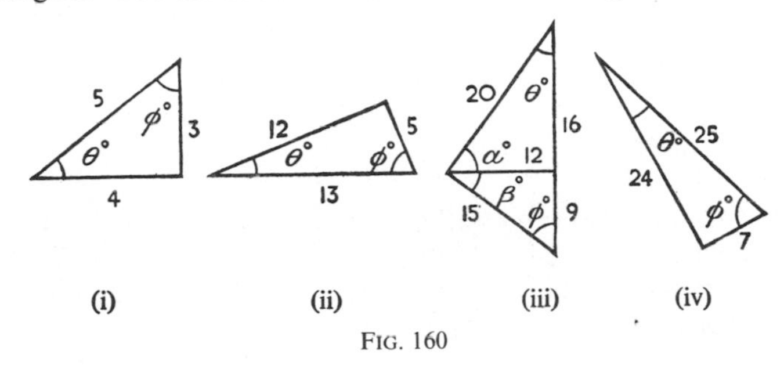

Fig. 160

7 A ladder 4 m long leans against the wall, and the lower end is 1 m from the wall. What angle does the ladder make with the wall?

[8] What is the angle of slope of a road if a man has risen 10 m vertically after walking 100 m up the road?

[9] The pole of a bell-tent is 2·5 m high, and the length of the slant side is 3·5 m. What angle does the side make with the ground?

10 The legs of a pair of dividers are each 12 cm long and are opened so that the points are 5 cm apart. What is the angle between the legs?

11 The tops of two vertical poles of heights 7, 5 m are joined by a taut wire 4 m long. What is the angle of slope of the wire?

[12] In Fig. 161, calculate ∠BAC.

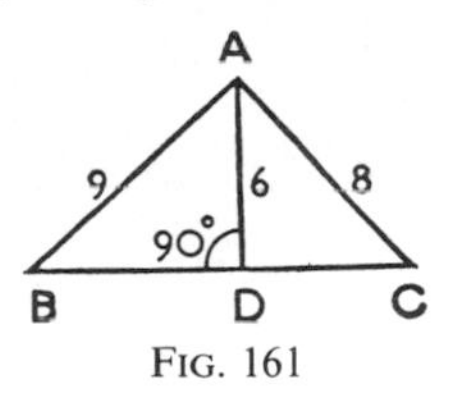

Fig. 161

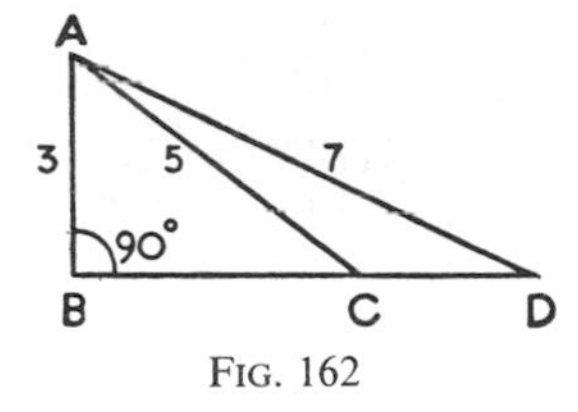

Fig. 162

13 In Fig. 162, calculate ∠CAD.

14 A man starts at O and walks 1 km N. 21° W. to A, then 2 km N. 43° E. to B. How far (i) North, (ii) East is B from O?

[15] A man walks 100 m up a slope of 22° and then 50 m up a slope of 18°. How far is he (i) vertically, (ii) horizontally, from his starting-point?

16 Find $\theta°$ if (i) $\sin\theta° = 2\sin\phi°$ and $\phi° = 23°$;
(ii) $\cos\theta° = 2\cos\phi°$ and $\phi° = 72°$.

MISCELLANEOUS EXAMPLES

EXERCISE 65

1 Find the height of a kite when the string is 100 m long and is inclined at 34° to the horizontal.

[**2**] In $\triangle$ABC, $\angle$A = 90°, $\angle$B = 25° 16′, AC = 10 cm; find AB.

3 In $\triangle$ABC, BC = CA = 6 cm, AB = 5 cm; find $\angle$C.

[**4**] A chord AB, 6 cm long, subtends an angle of 140° at the centre O of a circle. [This means that $\angle$AOB = 140°.] Find the radius of the circle, to the nearest millimetre.

5 Find the value of $\theta° + \phi°$ if $\tan \theta° = \frac{4}{3}$ and $\tan \phi° = \frac{3}{4}$.

6 A man starts at O and walks 1 km N. 27° E. to A, then turns to his right through 90° and walks $\frac{1}{2}$ km to B. Find the bearing of B from O.

[**7**] A man walks 1000 m on a bearing of 025° and then 800 m on a bearing of 035°. How far north is he of his starting-point?

8 Two men, one north and the other south of a tower, measure the angles of elevation of the top of its spire as 28° and 37°; the top of the spire is 40 m high. How far apart are the men?

9 In Fig. 163, AB = 4 cm. Calculate the lengths of AD, BD, BC.

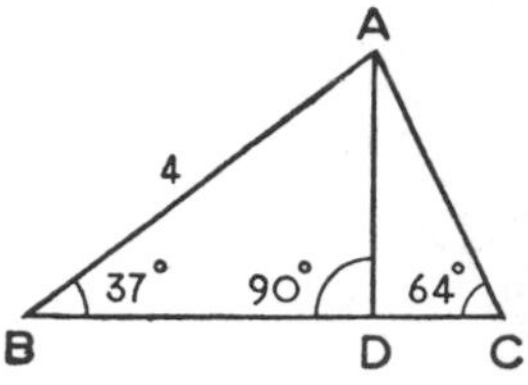

[**10**] A man sitting at a window with his eye 7 m above the ground can just see the sun over the top of a roof 15 m high, which is 30 m from him horizontally. Find the angle of elevation of the sun.

FIG. 163

11 A mining gallery descends for 100 m at an angle of 13° to the horizontal and then for 200 m at an angle of 7° to the horizontal. How far is the point reached below the level of the starting-point?

12 The diagonals of a rectangle are 12 cm long and contain an angle of 17° 30′. Find the length and breadth of the rectangle.

13 In Fig. 164, AB = 3 cm, BC = 7 cm, AP bisects $\angle$BAC. Find PC.

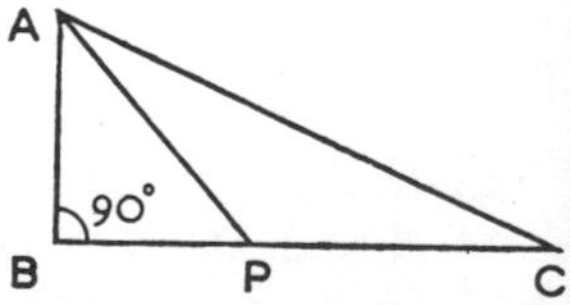

14 If, in Fig. 164, AC = 8 cm, PC = AB = 5 cm, find (i) $\angle$BAC, (ii) BC, (iii) $\angle$PAC.

[**15**] A regular heptagon (7 sides) is inscribed in a circle of radius 10 cm. Calculate its perimeter.

FIG. 164

16 In $\triangle$ABC, $\angle$B = 37° 15′, $\angle$C = 59° 40′, BC = 8 cm; the perpendicular bisector of BC cuts BA, CA produced at P, Q. Find the length of PQ.

17 Fig. 165 represents a rectangle ABCD held in a vertical plane; AB makes an angle 28° with the horizontal; AB = 6 cm, AD = 3 cm. Find (i) the vertical depths below A of B and D and C;
(ii) the distance between vertical lines through B and D;
(iii) the angles which AC and BD make with the vertical.

18 In Fig. 166, PQRS is a rectangle; PQ = 8 cm, PH = 10 cm, PK = 9 cm, ∠SPK = 39°. Find (i) ∠KPH, (ii) the lengths of SP, RH, RK.

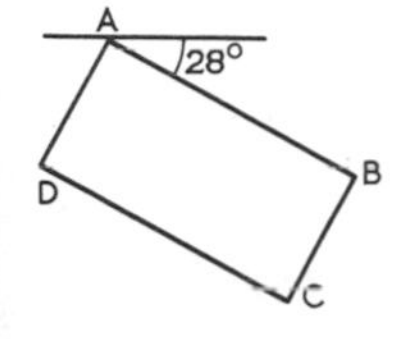

FIG. 165

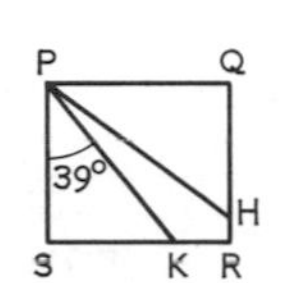

FIG. 166

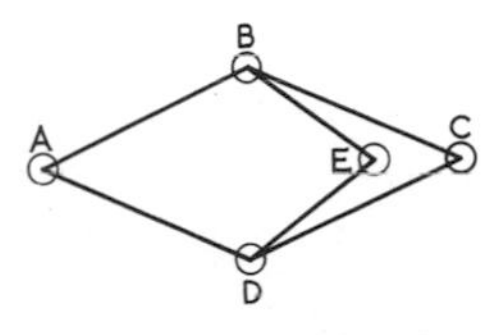

FIG. 167

19 Four thin rods AB, BC, CD, DA, each 10 cm long, and two thin rods BE, DE, each 7 cm long, are smoothly jointed together as shown in Fig. 167. (i) If ∠BAD = 32°, find ∠BED. (ii) Find the greatest possible size of ∠BAD.

20 In Fig. 168, not drawn to scale, AB, BC, CD represent straight paths in a vertical plane making angles $\sin^{-1}\frac{1}{4}$, $\sin^{-1}\frac{1}{3}$, $\sin^{-1}\frac{7}{25}$ respectively with the horizontal, AB = 800 m, BC = 450 m, CD = 1000 m. Find (i) the height of D above A, (ii) the distance between vertical lines through A and D, (iii) the angle which a straight tunnel driven from A to D makes with the horizontal.

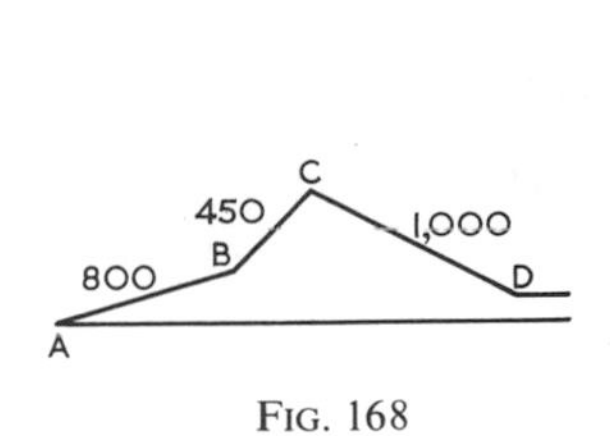

FIG. 168

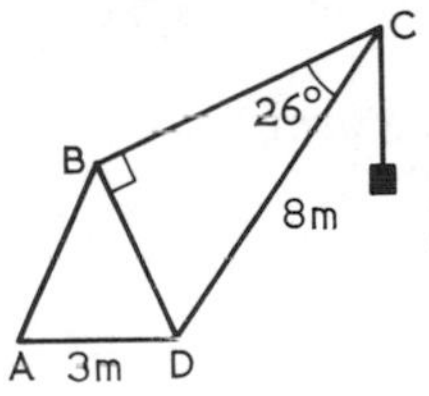

FIG. 169

21 Fig. 169, not drawn to scale, represents a hoist. AD is horizontal, AB = BD, AD = 3 m, CD = 8 m, ∠DBC = 90°, ∠DCB = 26°. Find (i) the length of BD, (ii) ∠BAD, (iii) the height of C above the level of AD.

CHAPTER 14
FACTORS

Factors by Grouping Terms

When factorising an expression, **first see if there is a common factor of each term.** If so, write it down **first,** and find the other factor by short division.

Example 1 Factorise $p(a+b)+q(a+b)$.
$(a+b)$ is a factor of each term. Short division of the expression by $(a+b)$ gives $(p+q)$.

$$\therefore\ p(a+b)+q(a+b) = \boldsymbol{(a+b)(p+q)}.$$

The short division should be done mentally.
Sometimes a common factor can be found by grouping terms.

Example 2 Factorise $ax-ay+bx-by$.

$$ax-ay+bx-by = a(x-y)+b(x-y).$$

Here $(x-y)$ is a factor of each term; $\therefore$ by short division, to be performed mentally,

$$ax-ay+bx-by = \boldsymbol{(x-y)(a+b)}.$$

When dividing by a common factor, write it down **first,** and treat it as the divisor in a division sum.

When you have obtained the factors, multiply them together **mentally,** to make sure that their product equals the given expression.

The fact that an expression can be arranged in two groups does not mean, necessarily, that it can be factorised.

Thus, $\quad ac+ax+bc+by = a(c+x)+b(c+y)$.

But *there is no factor common to these two terms,* so we cannot use the 'short division' method.

Actually, *this expression has no factors.*

An expression, such as $a(c+x)+b(c+y)$, which is written as the *sum* of two terms, **is not in factors.**

Example 3 Factorise $a^2+bc+ab+ac$.

$$a^2+bc+ab+ac = (a^2+bc)+a(b+c).$$

This is not in factors because it is the *sum* of two terms; also there is no factor common to the two terms, so we cannot use the 'short division' method.

But $$a^2+bc+ab+ac = a^2+ab+bc+ac$$
$$= a(a+b)+c(b+a).$$

Here, $(a+b)$ is a factor of each term, since $a+b = b+a$;

$\therefore$ by short division, $a^2+bc+ab+ac = (\boldsymbol{a}+\boldsymbol{b})(\boldsymbol{a}+\boldsymbol{c})$.

Note *Group together terms which have a common factor.*

Example 4 Factorise $ad+bc-ac-bd$.

$$ad+bc-ac-bd = ad-ac+bc-bd = a(d-c)+b(c-d)$$
$$= a(d-c)-b(d-c), \text{ since } c-d = -(d-c),$$
$$= (\boldsymbol{d}-\boldsymbol{c})(\boldsymbol{a}-\boldsymbol{b}).$$

The *form* of the answer may be altered by grouping differently:

$$ad+bc-ac-bd = bc-bd+ad-ac = b(c-d)+a(d-c)$$
$$= b(c-d)-a(c-d) = (\boldsymbol{c}-\boldsymbol{d})(\boldsymbol{b}-\boldsymbol{a}).$$

EXERCISE 66 (Oral)

In Nos. 1–15, if the answer is yes, state also the other factor.

1 Is $a+1$ a factor of (i) $2a+2$; (ii) $3+3a$; (iii) $a+2$?
[2] Is $b+1$ a factor of (i) $4+4b$; (ii) $-b-1$; (iii) $(1+b)^2$?
3 Is $c-1$ a factor of (i) $2c-1$; (ii) $1-c$; (iii) $c+1$?
[4] Is $x-3$ a factor of (i) $x+3$; (ii) $3-x$; (iii) x^2-3?
5 Is $x+y$ a factor of (i) $2y+2x$; (ii) $x-y$; (iii) $xz+yz$?
[6] Is $x-y$ a factor of (i) $ax-by$; (ii) $y-x$; (iii) x^2-y^2?
7 Is $a-b$ a factor of (i) $ac-bc$; (ii) $b-a$; (iii) b^2-a^2?
8 Is $c+d$ a factor of (i) c^2+d^2; (ii) c^2-d^2; (iii) $-d-c$?
9 Is $a+b$ a factor of $x(a+b)+y(b+a)$?
10 Is $c+d$ a factor of $x(c+d)-y(c-d)$?
11 Is $x+y$ a factor of $x(a+b)+y(a+c)$?
[12] Is $x-y$ a factor of $a(x-y)+b(y-x)$?
[13] Is $c-d$ a factor of $c(x+y)-d(x-y)$?
14 Is $a+b$ a factor of $x(a+b)+a+b$?
15 Is $x+y$ a factor of $a(x+y)-x+y$?

Have the following expressions factors? If so, find them and check by multiplication. *If there are no factors, say so.*

16 $a(c+d)-b(c+d)$
17 $p(x+y)-q(x-y)$
[18] $a(x+y)+b(x+z)$
[19] $a(c+d)-b(d+c)$
20 $c(a-b)+d(b-a)$
21 $x(2y+2)+z(y+1)$
[22] $a(b+1)+b(a+1)$
[23] $x-a+b(a-x)$
24 $a(x+y)+x+y$
25 $x(p+q)-p-q$
[26] $a(1+x)-b(1-x)$
[27] $c(y-z)-y+z$

EXERCISE 67

Factorise, *when possible*, the following expressions. *If there are no factors, say so.*

1 $ax-ay+bx-by$
2 $a^2+ab+ac+bc$
3 $ac+ad-bc-bd$
4 $cy-cz-dy-dz$
5 $x^2+xy+3x+3y$
6 $a^2+ac-5a-5c$
7 $ax+ay+bx+bz$
8 $a^2c^2+a^2d^2+b^2d^2+b^2c^2$
[9] $pr+ps-qr-qs$
[10] $cm-cn-km+kn$
[11] $dp-dq-4p+4q$
[12] $x^2-cx-dx+cd$
13 $5cx+5dy-5cy-5dx$
14 $6ab-3bx+2ay-xy$
15 $4x^2-2xy-6xz+3yz$
16 $ab-12xy+3bx-4ay$
17 $a^2-ab-2a-2b$
18 $2ab+2ac+b+c$
[19] $6cd+2cn-9md-3mn$
[20] $10cp-15pq-4ac+6aq$
[21] $x(a+b+c)+y(a+b+c)$
[22] $p(x+y)+q(x-y)$
23 $a(r-s)+b(s-r)$
24 $a^2(b+c)-bc(c+b)$
25 $ax-3+a-3x$
26 $xy+y^2-x-y$
27 x^3+x^2+x+1
28 $ac-bc-a+b$
[29] $ap+pq-a-q$
[30] $xy^2-1+x-y^2$
[31] $a(b-c)-d(c-b)$
[32] $x^2-(a+2b)x+2ab$
33 $ca-cd-bd+ba$
34 $2px-py+qy-2qx$
[35] $ac-a^2+ad-cd$
[36] $x^2-2x-xy-2y$
37 $x^2-x+y-xy$
38 $4-4x+cx-c$
[39] $l(a-b)-m(a+b)$
[40] $r(a-x)+z(x-a)$
41 $1+c^2+cd+c^3d$
42 $2a^4-2a^3x+x-a$
***43** $x^2-y^2+ax+ay$
***44** $z^2-1+cz-c$
***45** $a^2-b^2-ac+bc$
***46** $1-m^2-t-tm$
***47** $rt-2st-r^2+4s^2$
***48** $3cy+dy-9c^2+d^2$

Quadratic Functions The product of two first-degree functions of x is a quadratic function of x.

Thus,
$$(2x-5)(3x+4) = 2x(3x+4)-5(3x+4)$$
$$= 6x^2+8x-15x-20 = 6x^2-7x-20.$$

To factorise a quadratic function, we express it so that this process can be *worked backwards*. This requires that the term in x should be replaced by two equivalent terms chosen in such a way that the grouping method can be used.

Example 5 Factorise $8x^2+10x+3$.

Replace $+10x$ by two equivalent terms whose product is equal to $8x^2 \times 3, = 24x^2$, i.e. *the product of the other two terms.*

Since the product, $+24x^2$, is *positive*, the terms have the *same* sign; since the sum, $+10x$, is *positive*, each term is *positive*.

$$24x^2 = 4x \times 6x \quad \text{and} \quad 4x+6x = 10x;$$

$$\begin{aligned} \therefore\ 8x^2+10x+3 &= 8x^2+4x+6x+3 \\ &= 4x(2x+1)+3(2x+1) \\ &= \mathbf{(2x+1)(4x+3)}. \end{aligned}$$

Always check the factors by multiplying mentally.

Example 6 Factorise $y^2-22y+96$.

Replace $-22y$ by two equivalent terms whose product is $y^2 \times 96, = 96y^2$.

Since the product, $+96y^2$, is *positive*, the terms have the *same* sign; since the sum, $-22y$, is *negative*, each term is *negative*.

$$96y^2 = y \times 96y = 2y \times 48y = 3y \times 32y = 4y \times 24y = 6y \times 16y;$$

and $6y+16y = 22y$; we therefore replace $-22y$ by $-6y-16y$.

$$\begin{aligned} y^2-22y+96 &= y^2-6y-16y+96 \\ &= y(y-6)-16(y-6) \\ &= \mathbf{(y-6)(y-16)}. \end{aligned}$$

Example 7 Factorise $6x^2-11xy-10y^2$.

Replace $-11xy$ by two equivalent terms whose product is

$$6x^2 \times (-10y^2), \qquad = -60x^2y^2.$$

Since the product is *negative*, the terms have *opposite* signs; since the sum is *negative*, the *numerically larger* term is *negative*.

$$-60x^2y^2 = 2xy \times (-30xy) = 3xy \times (-20xy) = 4xy \times (-15xy).$$

$$\begin{aligned} 6x^2-11xy-10y^2 &= 6x^2+4xy-15xy-10y^2 \\ &= 2x(3x+2y)-5y(3x+2y) \\ &= \mathbf{(3x+2y)(2x-5y)}. \end{aligned}$$

Example 8 Factorise $2+t-10t^2$.

Replace $+t$ by two equivalent terms whose product is

$$2 \times (-10t^2), \qquad = -20t^2.$$

The terms have *opposite* signs and the *numerically larger* term is positive:

$$-20t^2 = (-4t) \times (+5t) \quad \text{and} \quad -4t+5t = t;$$

$$\begin{aligned} \therefore\ 2+t-10t^2 &= 2-4t+5t-10t^2 \\ &= 2(1-2t)+5t(1-2t) \\ &= \mathbf{(1-2t)(2+5t)}. \end{aligned}$$

If the coefficient of the term of degree 2 is negative and if the constant term is positive, work in *ascending* powers as in Example 8. *Do not turn the expression round.*

The above examples show that we need only consider the *coefficient* of the product and that if this coefficient is **positive,** the numerical factors required have a known **sum;** and if this coefficient is **negative,** the numerical factors required have a known **difference.** If this coefficient has a large number of factors, it saves time to express it in *prime factors.*

Example 9 Factorise $12a^2-16a-35$.

Replace $-16a$ by two equivalent terms whose product is $12a^2\times(-35)$. Look for two numbers whose product is 12×35 and which *differ* by 16.

$$12\times35 = 2^2\times3\times5\times7 = 14\times30$$

and

$$14a-30a = -16a;$$

$$\begin{aligned}\therefore\ 12a^2-16a-35 &= 12a^2+14a-30a-35\\ &= 2a(6a+7)-5(6a+7)\\ &= \mathbf{(6a+7)(2a-5)}.\end{aligned}$$

EXERCISE 68

Write down the coefficient of x in Nos. 1–9:

1 $(2x+3)(x+4)$ **2** $(2x-3)(x-4)$ **3** $(2x+3)(x-4)$
4 $(2x-3)(x+4)$ **5** $(1-2x)(2+5x)$ **6** $(7-3x)(2-3x)$
[**7**] $(3x-7)(2x+3)$ [**8**] $(x+1)(7x-5)$ [**9**] $(1-4x)(2-7x)$

Find by *inspection* two numbers to fit the conditions, Nos. 10–21:

10 Product 15, sum 8. **11** Product 15, sum 16.
12 Product 45, difference 4. **13** Product 48, difference 13.
[**14**] Product 12, sum 8. [**15**] Product 108, sum 24.
[**16**] Product 60, difference 11. [**17**] Product 54, difference 3.
18 Product 24, sum -11. **19** Product -18, sum 7.
20 Product -45, sum 12. **21** Product -96, sum -10.

Factorise the following; *check your answers mentally*:

22 $3x^2+5x+2$ **23** $2y^2+7y+3$ **24** $3z^2+8z+4$
25 a^2+6a+8 **26** $b^2+10b+9$ **27** $c^2-7c+12$
28 $d^2-5d-14$ **29** $x^2+7xy-30y^2$ **30** $r^2-8rs-84s^2$
31 $2z^2+5z-3$ **32** $2p^2-11p+12$ **33** $3t^2+13t+4$
[**34**] $2a^2+11a+5$ [**35**] $3b^2-10b+8$ [**36**] $4c^2+11c-3$
[**37**] $d^2+7d+10$ [**38**] $t^2+6t-16$ [**39**] $n^2+7n-18$
[**40**] $x^2-3x-28$ [**41**] $y^2+15y+14$ [**42**] $z^2+3z-70$
43 $15+2a-a^2$ **44** $4-3b-b^2$ **45** $12+11c-c^2$
46 $4x^2+13xy+3y^2$ **47** $12a^2-11ab+2b^2$ **48** $2y^2-22yz+48z^2$

49 $2c^2-22cd+60d^2$ **[50]** $9r^2-39rs-30s^2$ **[51]** $p^2-13pq+40q^2$
[52] $8m^2-10mn-3n^2$ **[53]** $2-k-6k^2$ **[54]** $18-3x-x^2$
[55] $6+5y-6y^2$ **[56]** $3a^2+60a-63$ **[57]** $2b^2-16b-96$
58 $12n^2+33n-9$ **59** $27t^2+18t-24$ **60** $60+3y-3y^2$
61 $4z^2+16z+15$ **62** $24-14x-20x^2$ **63** $6a^2-13ab+6b^2$

Factors by Inspection

After a little practice, *simple* quadratic functions can often be factorised at sight, without using the grouping method. **But always check the answer by (mental) multiplication.**

Example 10 Factorise (i) $x^2+11x+24$; (ii) $x^2-2x-35$.

(i) Find two numbers whose product is $+24$ and sum $+11$. These are $+8$, $+3$.

$$x^2+11x+24 = (\boldsymbol{x+8})(\boldsymbol{x+3}).$$

(ii) Find two numbers whose product is -35 and sum -2. These are -7, $+5$.

$$x^2-2x-35 = (\boldsymbol{x-7})(\boldsymbol{x+5}).$$

Check by mental multiplication in each case.

Example 11 Factorise $2y^2+5y-3$.

Write down pairs of factors whose products introduce $2y^2$ and -3, and select that pair which also introduces $+5y$. Possible pairs are

$$(2y+3)(y-1);\quad (2y-3)(y+1);\quad (2y-1)(y+3);\quad (2y+1)(y-3).$$

$$2y^2+5y-3 = (\boldsymbol{2y-1})(\boldsymbol{y+3}).$$

Using the grouping method, whenever you are not able to obtain the factors by inspection, *quickly.*

If there is a common factor, **write it down first** and find the other factor by short division.

EXERCISE 69

Factorise the following; *check your answers mentally*:

1 a^2+5a+6 **2** b^2+6b+9 **3** $c^2-7c+12$
4 $d^2-3d-10$ **5** k^2+2k-8 **6** n^2-n-6
7 p^2+p-30 **8** $r^2+12r+36$ **9** $t^2-5t-50$
10 $x^2+14x+45$ **11** $y^2+7y-60$ **12** $z^2-9z-70$
13 $m^2+16m+64$ **14** n^2-64 **15** $r^2-16r+64$
[16] $a^2+8a+12$ **[17]** $b^2-8b+16$ **[18]** c^2+4c-5
[19] $d^2+9d+18$ **[20]** $k^2-3k-28$ **[21]** $n^2-11n+30$
[22] $p^2-14p+49$ **[23]** $q^2+6q-27$ **[24]** $r^2-18r+72$

[**25**] t^2-36 [**26**] x^2+x-90 [**27**] $y^2-2y-63$

[**28**] $z^2-7z-120$ [**29**] $m^2-20m+100$ [**30**] s^2-100

31 $x^2-12xy+32y^2$

32 $a^2+3ab-10b^2$

33 $c^2-4cd-32d^2$

34 $1+3a-10a^2$

35 $1-2b-24b^2$

36 $1-12c+35c^2$

37 $1-14d+49d^2$

38 $1-49e^2$

[**39**] $1+14g+49g^2$

[**40**] $1-k-20k^2$

41 $24-5n-n^2$

42 $28-11p+p^2$

[**43**] $12-a-a^2$

[**44**] $35-2c-c^2$

45 $2x^2+6xy-20y^2$

46 $3-27z^2$

47 $a^2b^2-7abc+10c^2$

48 $1-3xy-18x^2y^2$

[**49**] $24r^2+2rs-s^2$

[**50**] $2p^2-7p-15$

[**51**] $5q^2-16q+3$

[**52**] $4r^2+5r-6$

[**53**] $10x^2-13x+4$

[**54**] $12y^2-11y-5$

55 $10z^2-21z+9$

56 $2a^2+2a-12$

57 $5b^2-5b-150$

58 $4c^2-100$

[**59**] $2d^2-5d+2$

[**60**] $3e^2+5e-2$

[**61**] $4b^2-12b+9$

[**62**] $9c^2+c-10$

[**63**] $9m^2-64n^2$

[**64**] $25x^2+40xy+16y^2$

65 $42c^2-cd-30d^2$

66 $10y^2-43y+12$

67 If $x+2$ is a factor of $x^2+ax+10$, what is the other factor? Hence find a.

[**68**] If $x+3$ is a factor of $x^2+bx-12$, what is the other factor? Hence find b.

69 Find c if $x-6$ is a factor of $x^2+cx+30$.

70 Find c if $x+4$ is a factor of x^2+7x+c.

[**71**] Find a if $x+2$ is a factor of x^2-5x+a.

72 Find b if $2x-3$ is a factor of $6x^2+bx-12$.

CHAPTER 15
AREAS AND VOLUMES

Area of a Rectangle

Example for Oral Discussion Take a piece of squared paper ruled in centimetres and millimetres. (i) How many square millimetres are there in each square centimetre? Express as a decimal of a square centimetre the area of a square millimetre. (ii) Draw on the squared paper a rectangle 0·9 cm by 0·7 cm. How many square millimetres does it contain? Express the area of the rectangle in square centimetres.

Repeat this argument for a rectangle 1·7 cm by 1·3 cm. This example illustrates the fact that:

the number of units of area in a **rectangle** *is the product of the numbers of the units in its length and breadth, whether these numbers are integers or fractions.*

Thus the relations between the length l cm, the breadth b cm and the area A cm^2 of a rectangle are as follows:

$$A = l \times b; \qquad l = A \div b; \qquad b = A \div l.$$

Similarly, if the length is l metres and the breadth b metres, the area A square metres is given by the formula $A = l \times b$. The sign for square metres is m^2.

It was explained in Vol. I (Chapter 18, p. 192) that the table of areas is:

$$\mathbf{100\,(=10^2)\,mm^2 = 1\,cm^2}$$
$$\mathbf{10\,000\,(=100^2)\,cm^2 = 1\,m^2}$$
$$\mathbf{10\,000\,m^2 = 1\,hectare}$$

Example 1 Find the area of a rectangle 2·6 m long, 1·2 m broad.
The area = (2·6 × 1·2) m^2 = **3·12 m^2**.

EXERCISE 70

Find the areas of the following rectangles, in terms of the unit indicated in brackets:

1 2 m by 1·25 m (m^2)
[**2**] 1·5 m by 60 cm (m^2)
[**3**] 12 m by 8·75 m (m^2)
4 6·4 cm by 3·8 cm (cm^2)
[**5**] 3·2 m by 50 cm (m^2)
[**6**] 2·5 m square (m^2)

Find the lengths of the following rectangles:

7 Area 48 m^2, breadth 4 m
[**8**] Area 10 m^2, breadth 1·25 m
9 Area 63 m^2, breadth 6·75 m
[**10**] Area 18·9 cm^2, breadth 3·5 cm

Find the areas of Figs. 170 and 171 in which all the corners are right-angled and the dimensions are shown in centimetres.

11 [**12**]

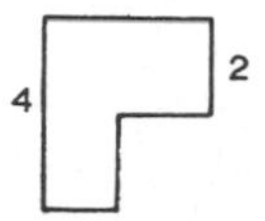

FIG. 170

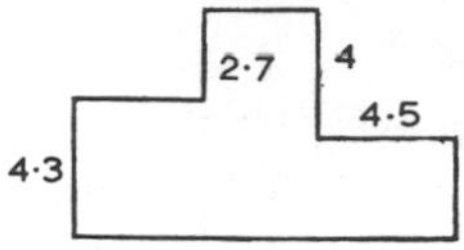

FIG. 171

***13** A rectangular sheet of cardboard, 25 cm by 16 cm, weighs 18·3 g; an oval is cut out of it and is found to weigh 10·2 g.

Find the area of the oval, to the nearest cm^2.

***14** The length and breadth of a sheet of paper are measured as 6·2 cm, 4·8 cm to the nearest millimetre. Within what limits, correct to 0·1 cm^2, does its area lie?

***15** How many pieces of cardboard, each 5 cm square, can be cut from a sheet 48 cm long, 36 cm wide? What area remains over?

Find the total area of the four walls of a room:

16 5·5 m long, 4 m wide, 2·5 m high.

17 5 m long, 4 m wide, 2·7 m high.

18 Find the area of cardboard used for making a closed box, 10 cm by 8 cm by 6 cm.

[**19**] Repeat No. 18 for a closed box 4·5 cm by 3·5 cm by 3 cm.

20 Find the total area of the external surface of an *open* tank, 2·5 m long, 1·5 m wide, 1 m high external measurements.

21 Find the total area of the walls of a room 4 m high, if the breadth is 8 m and the floor area is 68 m^2.

22 A cistern 2 m long, 1·5 m wide contains water to a depth of 36 cm. Find the area of the wet surface.

Example 2 A room 5 m long, 4 m wide has a carpet in the middle, leaving a margin 50 cm wide all round which is covered with linoleum at 70p per m^2. Find the cost of the linoleum.

In Fig. 172, not drawn to scale, PQRS represents the carpet and ABCD the floor. The margin is $\frac{1}{2}$ m wide;

$\therefore$ PQ $= \{5-(\frac{1}{2}\times 2)\}$ m $= 4$ m;

PS $= \{4-(\frac{1}{2}\times 2)\}$ m $= 3$ m;

$\therefore$ area of carpet $= (4\times 3)$ m^2 and

area of floor $= (5\times 4)$ m^2;

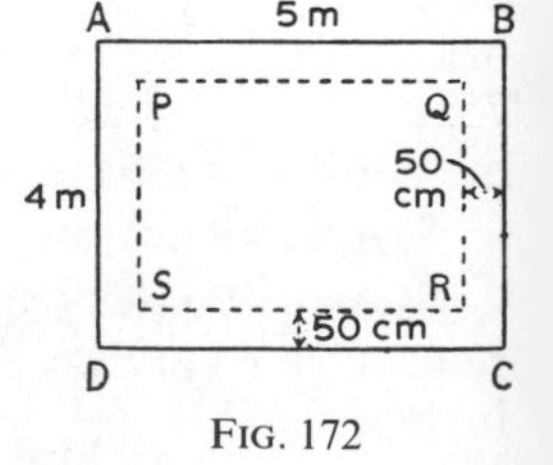

FIG. 172

$\therefore$ area of linoleum $= (20-12)\ \text{m}^2$
$= 8\ \text{m}^2.$

$\therefore$ cost of linoleum $= 70\text{p} \times 8$
$= 560\text{p} =$ **£5·60.**

Note It is better to use the subtraction method than to divide up the border into rectangles.

EXERCISE 71

1 Find the area of a path running all round a lawn 20 m long, 13 m wide, if the path is (i) 2 m wide, (ii) 1·5 m wide.

[**2**] Find the area of a frame 6 cm broad round the edge of a picture whose overall measurements are 90 cm wide, 75 cm high.

3 A sheet of tin measures 47 cm by 36 cm. If a strip 2·5 cm wide is cut off all round, find the area of the part cut off.

[**4**] A photograph, 12 cm wide and 20 cm high, is mounted on a card so that there is a margin 3 cm wide at top and bottom and 1·5 cm wide along the sides. Find the area of the part of the card which is not covered.

5 A room 5 m long, 4 m wide, has a carpet in the middle of the floor, leaving a margin all round which is stained. What area is stained if the margin is 30 cm wide?

6 Find by the subtraction method the shaded area in Fig. 173, the units being centimetres:

(i) if $a = 2{\cdot}5$, $b = 1{\cdot}5$, $c = 2$, $d = 1$;
(ii) if $a = 12{\cdot}5$, $b = 5$, $c = 9{\cdot}5$, $d = 3$.

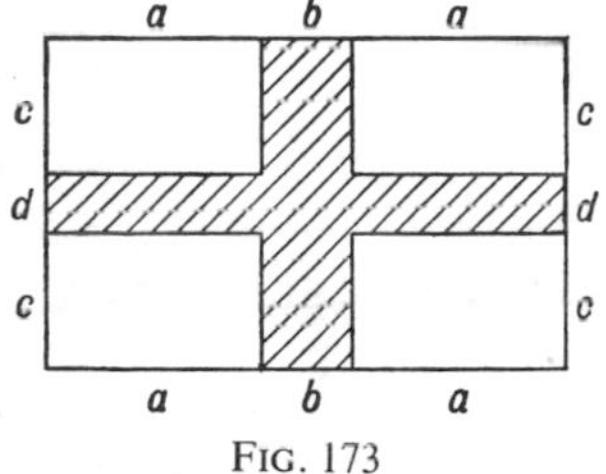

FIG. 173

7 Linoleum at 75p per m^2 is put down in a room 7 m by 6 m so as to leave a margin 50 cm all round. Find the cost.

[**8**] A room is 5 m by 4 m. Find the cost of staining a border 75 cm wide all round the edge of the floor at 16p per m^2.

9 Fig. 174 shows a rectangular brick wall pierced with four equal rectangular windows. Find in m^2 the area of the surface of brickwork, if $a = 80$, $b = 45$, $c = 95$, $d = 75$, $e = 630$, $f = 450$, the units being centimetres.

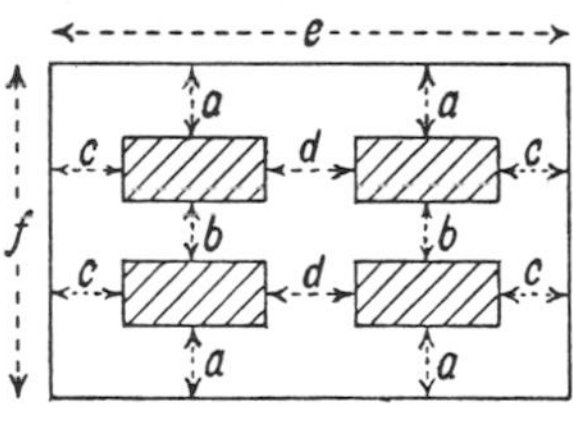

FIG. 174

Volume of a Cuboid

Example for Oral Discussion (i) How many cubes, edge 1 mm, are required for building up a cubic-centimetre block? What is the volume in cubic centimetres of a cube whose edge is 1 mm?

(ii) If a cuboid, 2·6 cm long, 2·1 cm wide, 1·5 cm high, is built up of small cubes, edge 1 mm, how many such cubes will there be in each layer, and how many layers will there be? What is the total number of these cubic millimetres in the volume of the cuboid? Divide your answer by 1000, since 1000 mm^3 = 1 cm^3, and obtain the volume of the cuboid in cubic centimetres. Compare the result with the value of $2{\cdot}6 \times 2{\cdot}1 \times 1{\cdot}5$.

This and similar examples illustrate the fact that:

the number of units of volume in a **cuboid** *is obtained by multiplying together the numbers of units in the length, breadth and height of the cuboid, whether these numbers are integers or fractions.*

Thus the relations between the length l cm, the breadth b cm, the height h cm and the volume V cm^3 of a cuboid are:

$$\boldsymbol{V = l \times b \times h; \qquad l = \frac{V}{b \times h}; \qquad b = \frac{V}{l \times h}; \qquad h = \frac{V}{l \times b}.}$$

It was explained in Vol. I (Chapter 18, p. 192) that the table of volumes is:

$$\mathbf{1000\ (= 10^3)\ mm^3 = 1\ cm^3}$$
$$\mathbf{1\,000\,000\ (= 100^3)\ cm^3 = 1\ m^3}$$
$$\mathbf{1000\ cm^3 = 1\ litre\ (l) = 1000\ millilitres\ (ml)}$$

Since 1 cm^3 of water at 4°C. weighs 1 g, it follows that the weight of 1 litre of water is 1000 g, or 1 kg.

Example 3 A rectangular tank, with a horizontal base 2 m long and 1·5 m wide, internal measurements, contains 2250 litres of water. Find the depth of the water.

$$d = \frac{V}{l \times b}.$$

Now $$V = 2250\ l = 2\,250\,000\ \text{cm}^3$$
$$= 2{\cdot}25\ \text{m}^3.$$
$$l \times b = 2 \times 1{\cdot}5 = 3\ \text{m}^2$$
$$\therefore d = \frac{2{\cdot}25}{3}\ \text{m} = \mathbf{0{\cdot}75\ m.}$$

EXERCISE 72

Find the volumes of the following rectangular blocks:

1 4·25 m by 4 m by 2·75 m

[**2**] 45 cm by 24 cm by 5 mm

3 Find the volume of air-space in a room 5·5 m long, 4 m wide and 2·8 m high.

[4] How many cm^3 of wood are there in a table-top 4·5 m long, 1·8 m wide, 3 cm thick?

5 How many rectangular blocks, each 5 cm by 4 cm by 3 cm, can be packed in a box 2·25 m by 1·96 m by 1·05 m, internal measurements?

[6] How many bricks, each 25 cm by 15 cm by 8 cm, are required for a wall 32 m long, 3 m high, 40 cm thick?

7 Find the capacity in litres of a tank, 1·2 m by 80 cm by 50 cm, internal measurements.

[8] Find the capacity in litres of a tin 24 cm long, 7·5 cm wide, 12 cm high.

9 A lock in a canal is 40 m long, 7 m wide. When the sluices are opened, the depth of water in the lock decreases from 5 m to 3·5 m. How many cubic metres of water run out?

10 A beam 3 m long, 15 cm wide, 10 cm deep, is made of wood which weighs 0·8 g per cm^3; find the weight of the beam.

[11] Find the weight of a wooden plank 4 m long, 20 cm wide, 6 cm thick, if the wood weighs 800 kg per m^3.

[12] Find, to the nearest gramme, the weight of a rectangular steel plate 7·2 cm by 6·5 cm by 8 mm, if the steel weighs 8 g per cm^3.

13 Find the weight of petrol which a tin, 15 cm long, 12 cm wide, 40 cm high, can hold, if 1 cm^3 of petrol weighs 0·7 g.

14 The volume of a rectangular block, 4·5 cm long, 2·5 cm wide, is 36 cm^3; what is its height?

[15] A tank 10·5 m long, 8 m wide, contains 33·6 m^3 of water. Find the depth of the water.

16 A rectangular tank 1·5 m long, 88 cm wide, contains water to a depth of 65 cm. The water is transferred to an empty tank 2 m long, 1 m wide; find the depth of the water.

17 A tank 2 m long, 1·5 m wide, 1 m high, contains water to a depth of 50 cm. A metal block 1·2 m by 1 m by 36 cm is put into the tank and totally submerged. Find the amount the water-level rises.

18 How many litres of water will a tank 1 m long, 60 cm wide, 35 cm deep, hold?

19 A tank 1·5 m long, 70 cm wide, contains 630 litres of water; find the depth of the water.

[20] A swimming-bath is 100 m long, 25 m wide. How many litres must be pumped into it to raise the water-level 4 cm?

21 Find in kg the weight of $1\frac{1}{2}$ litres of water.

22 How many litres of water weigh 2·75 kg?

[23] Find in kg the weight of water which a cistern 1·5 m long, 1·2 m wide, 64 cm high, will hold.

24 560 kg of water are drawn out of a tank 2·5 m long, 1·4 m wide. What distance does the water-level fall?

Example 4 Find the weight of an empty open rectangular pan made of aluminium 0·5 cm thick, if the base is 20 cm long, 15 cm wide, measured internally, and if the pan can hold $1\frac{1}{2}$ litres of water, given that 1 cm^3 of aluminium weighs 2·5 g.

$$1 \text{ litre} = 1000 \text{ cm}^3, \qquad \therefore \text{ internal volume of pan} = 1500 \text{ cm}^3.$$

$$\therefore \text{ internal height of pan} = \{1500 \div (20 \times 15)\} \text{ cm} = 5 \text{ cm}.$$

Since the aluminium is 0·5 cm thick,

$$\text{external length} = (20+1) \text{ cm} = 21 \text{ cm},$$

$$\text{external breadth} = (15+1) \text{ cm} = 16 \text{ cm},$$

$$\text{(no lid), external height} = (5+0{\cdot}5) \text{ cm} = 5{\cdot}5 \text{ cm}.$$

Therefore, by the subtraction method,

$$\text{volume of aluminium} = (21 \times 16 \times 5{\cdot}5 - 20 \times 15 \times 5) \text{ cm}^3$$

$$= (1848 - 1500) \text{ cm}^3 = 348 \text{ cm}^3.$$

But 1 cm^3 of aluminium weighs 2·5 g;

$$\therefore \text{ weight of empty pan} = (2{\cdot}5 \times 348) \text{ g} = \mathbf{870 \text{ g}}.$$

EXERCISE 73

Find the volume of wood required for the *closed* boxes, Nos. 1, 2:

1 External dimensions: 20 cm by 18 cm by 16 cm; wood 1 cm thick.

2 Internal dimensions: 20 cm by 12·5 cm by 9·5 cm; wood 1·25 cm thick.

Find the volume of wood required for the *open* boxes, Nos. 3, 4:

3 External dimensions: 17·5 cm long, 14 cm wide, 10 cm high; wood 7·5 mm thick.

4 Internal dimensions: 24 cm long, 11·5 cm wide, 9·5 cm high; wood 5 mm thick.

5 Find the weight of a closed box, made of wood 1 cm thick, measuring internally 30 cm by 20 cm by 12 cm, if the wood weighs 0·8 g per cm^3.

[6] An open rectangular tank is made of concrete, the sides and base being 30 cm thick. Externally, the tank is 4 m long, 3 m broad, 1·5 m high. Find its weight if the concrete weighs 2·5 g per cm^3.

7 The external dimensions of a closed rectangular cistern are 1 m by 84 cm by 70 cm, and the thickness of the material is 2 cm. How many litres, to the nearest whole number, will the cistern hold?

Volume of Solid of Uniform Cross-section

The *abbreviated* statement

Volume = area of cross-section × length

is true for *any shape of cross-section, provided only that it is uniform.*

Example 5 The dimensions of the cross-section of a steel girder are shown in centimetres in Fig. 175. If the steel weighs 7·8 g per cm^3, find the weight of the girder per metre length in kilogrammes, correct to 3 significant figures.

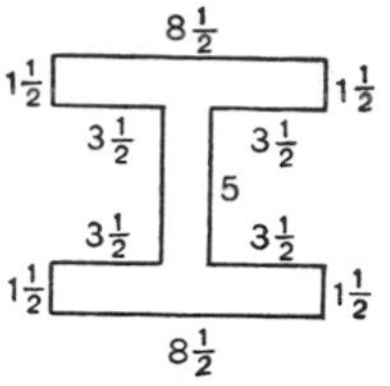

First find the area of the cross-section.

$$\text{Area of two cross-pieces} = (8\tfrac{1}{2} \times 1\tfrac{1}{2} \times 2)\ cm^2 = 25\tfrac{1}{2}\ cm^2;$$

$$\text{width of connecting portion} = (8\tfrac{1}{2} - 3\tfrac{1}{2} - 3\tfrac{1}{2})\ cm = 1\tfrac{1}{2}\ cm.$$

$$\therefore \text{area of connecting portion} = (5 \times 1\tfrac{1}{2})\ cm^2 = 7\tfrac{1}{2}\ cm^2;$$

FIG. 175

$$\therefore \text{area of cross-section} = (25\tfrac{1}{2} + 7\tfrac{1}{2})\ cm^2 = 33\ cm^2.$$

$\therefore$ volume of portion of girder 1 m long (100 cm long),

$$= \text{area of cross-section} \times \text{length}$$
$$= (33 \times 100)\ cm^3 = 3300\ cm^3.$$

But 1 cm^3 of steel weighs 7·8 g.

$$\therefore \text{weight of girder per metre length} = (7{\cdot}8 \times 3300)\ g$$
$$= 25740\ g = 25{\cdot}74\ kg$$

= **25·7 kg,** to 3 sig. figs.

Example 6 A water-can of uniform cross-section holds 10 litres. The area of its base is 250 cm^2, internal measurement; find its internal height.

$$10\ \text{litres} = 10\,000\ cm^3.$$

$$\text{Internal height} = \text{volume} \div (\text{area of base}) = \frac{10\,000}{250}\ cm$$

= **40 cm.**

EXERCISE 74

1 Find the volume of a rail 1·25 m long, uniform cross-section 12·8 cm^2.

2 Find the length of a girder, volume 4400 cm^3, uniform cross-section 12·5 cm^2.

Find the uniform cross-section of a solid, given:

3 Volume 10·5 m^3, length 28 m

4 Volume 92·8 cm^3, length 6·4 m.

5 A vessel of uniform cross-section of area 840 cm^2 contains 5·46 litres of water. What is the depth of the water?

6 What is the base area of a vessel of uniform cross-section of internal height 37·5 cm, if its capacity is $1\frac{1}{2}$ litres?

7 The dimensions of the L-shaped cross-section of a bar, 80 cm long, are shown in centimetres in Fig. 176. Find (i) the volume of the bar, (ii) the weight if the material weighs 7·6 g per cm^3.

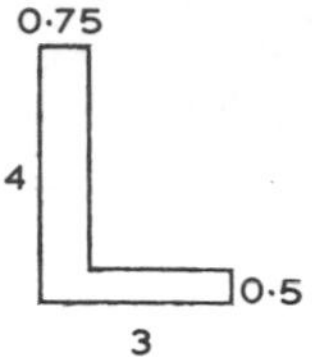

FIG. 176

[**8**] The area of the cross-section of a steel rail is 15 cm^2; find the weight of the rail per metre-run, if 1 cm^3 of steel weighs 7·8 g.

9 2 cm of rain over an area of 150 m^2 is collected in a tank 3 m by 2·4 m. Find the rise of water in the tank.

[**10**] An empty rectangular tank is 1·5 m long, 1·2 m wide, 1 m deep. Rain-water runs into it from roofs of total horizontal area 100 m^2. What depth of rainfall will fill the tank?

11 The dimensions of the cross-section of a girder 2·5 m long are shown in cm in Fig. 177. Find (i) the volume of the girder, (ii) the weight if the material weighs 7·8 g per cm^3.

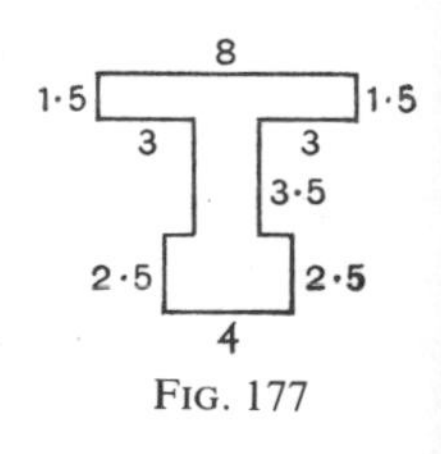

FIG. 177

12 A log of wood 3 m long has a uniform cross-section of area 1180 cm^2. Find its weight if the wood weighs 0·8 g per cm^3.

13 A can of uniform cross-section contains 4 litres of water. Find, to the nearest square centimetre, the area of the cross-section if the water is 28 cm deep.

***14** A rectangular gutter is 12 cm wide, 5 cm deep. If water flows along it at 1·2 m per second, find the number of litres which pass a given point in 8 min, if the gutter remains full.

***15** A metal rail 7 m long of uniform cross-section weighs 390 kg. If the metal weighs 7·8 g per cm^3, find the area of the cross-section of the rail.

***16** The cross-section of a pipe is 25 cm^2, and water is pouring out of it at the rate of 2 m per second. If the pipe remains full, find the number of litres discharged by the pipe in 5 min.

***17** A coil of wire 375 m long weighs 2·88 kg. Find the area of the cross-section of the wire if it weighs 9·6 g per cm^3.

***18** A block of copper, 5 cm by 4·5 cm by 15 cm, weighs 3 kg. It is drawn out into wire of cross-sectional area 0·15 cm^2. Find the weight of a length of 10 m of the wire.

REVISION EXERCISE R 2 (Ch. 1–15)

1 A case full of tea weighs 50 kg; when empty it weighs 2 kg; what does it weigh when one-quarter full?

2 50 m^2 of lint are cut up into bandages 5 cm wide; find the total length.

3 A man spent $\frac{7}{15}$ of his income at home and $\frac{1}{6}$ of his income on holidays; this left £660; find his income.

4 On a map of scale 4 km to the centimetre, the distance between two churches is 4·5 cm; what is it on a map of scale 5 km to the centimetre?

5 Find to the nearest penny the value of 0·318 of £2·25.

6 Find the area of thin sheeting required for making an *open* cistern 1·8 m long, 1·2 m wide, 0·9 m deep.

7 $x-y = 11$, $x+y = 8$ **[8]** $x = 5y$, $2x-7y = 15$ **9** $y = \frac{1}{3}(5x+1)$, $x = \frac{2}{5}(y+3)$

10 $0{\cdot}5x+1{\cdot}2y = 1{\cdot}4$, $0{\cdot}6x-7y = 5{\cdot}9$ **[11]** $5x+3y = \frac{2}{3}(3x-5y)$, $13x-57y = 190$

12 $2x+y = 5-7x-3y = \frac{1}{4}y$ **[13]** $x+y = 0{\cdot}1$, $x = \frac{1}{4}y$

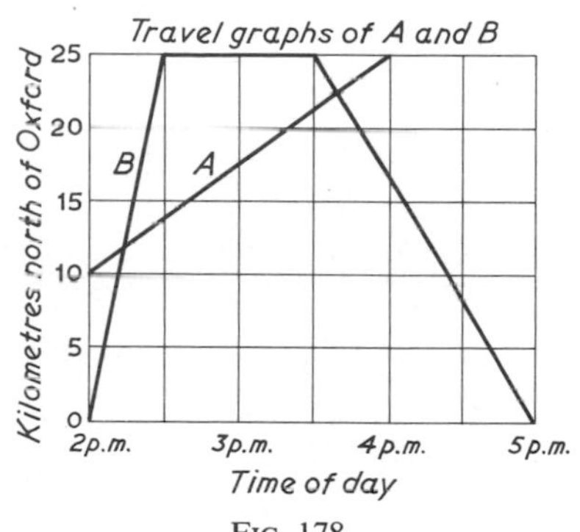

FIG. 178

14 Interpret the given graphs *in detail*. When and where do A and B pass one another?

15 Tests for the breaking strain of a wire rope gave these results:

Circumference in centimetres	1·5	2	2·5	3	3·5	4
Breaking strain in tonnes	4·0	7·5	12	18	26	37

Find from a graph (i) the breaking strain if the circumference is ·8 cm, 3·7 cm; (ii) the girth if the breaking strain is 9 t, 20 t.

16 It costs £960 to fence a square field of area 160 000 m^2. Find the cost of the fencing per metre length.

[**17**] Find the rates on an assessment of £85 at 89p in the £.

18 Find the three smallest numbers proportional to a, b, c if $a:b = 7:8$ and $b:c = 9:10$.

19 $2\frac{1}{2}$ litres of petrol weigh 1·7 kg; find the weight of petrol in grammes per cm^3.

20 Find from tables the sine and cosine of each of the angles: (i) 32° 40′; (ii) 70° 16′; (iii) 81° 52′.

21 Find from tables the angles whose sines and cosines have each of the values: (i) 0·9397; (ii) 0·4483; (iii) 0·1542.

22 What is $x°$ if (i) $\sin x° = \cos 18°$; (ii) $\cos x° = \sin 34°$?

[**23**] A man starts from O and walks 2·5 km north and then 3·5 km west. Calculate his bearing from O.

24 A man walks 3 km in a direction S. 73° W. How far is he (i) south, (ii) west of his starting-point?

25 The vertical angle of a cone is 118° and the diameter of its base is 3 cm. Calculate its height.

[**26**] The diagonals of a rhombus are 6 cm, 4 cm. Calculate the angles of the rhombus.

27 What is the angle of slope of a road up a hill if a man has risen 15 m vertically after walking 80 m up the road?

[**28**] In △ABC, ∠B = 72°, ∠C = 57°; AD is the perpendicular from A to BC; AD = 4 cm. Calculate BC.

29 In △ABC, AB = 4 cm, AC = 9 cm, ∠B = 90°; the bisector of ∠BAC cuts BC at P. Calculate ∠BAP and PC.

30 In △ABC, AB = AC = 4·5 cm, BC = 7·2 cm. Calculate ∠BAC.

[**31**] In △ABC, ∠A = 90°, AB = 4 cm, AC = 5 cm; P is a point on AB such that AP = 3 cm. Calculate ∠BCP.

32 A regular polygon with 11 sides is inscribed in a circle of radius 10 m. Find its perimeter.

33 For £9·60 I can hire for 1 hour either 18 men and 20 boys or 12 men and 40 boys. How much do I pay per hour for a man and how much for a boy?

[**34**] If A gives B 5p, B will have 3 times as much as A. If B gives A 7p, A will have 3 times as much as B. How much has each?

35 The formula for the effort P kg required to raise a load W kg by a machine is $P = a + bW$, where a, b are constants. For loads of 130 kg and 200 kg the necessary efforts are 37 kg and 51 kg. Find a, b, and the effort required to raise an 80-kg load.

[**36**] A fraction is such that if 1 is added to the numerator and 5 to the denominator, it reduces to $\frac{1}{3}$. If the numerator of the original fraction is doubled and if the denominator is increased by 13, it reduces to $\frac{1}{2}$. Find the fraction.

37 Find a number of 2 digits which exceeds four times the sum of its digits by 3 and which is increased by 18 when the digits are interchanged.

[**38**] The ratios of the sides of a quadrilateral are 3:4:6:7 and its perimeter is 12 cm. Find the length of each side.

39 A boy's marks for 3 tests are 22 out of 24, 29 out of 40, 12 out of 20. What percentage does he get on the whole?

40 Find the profit if an article is sold for £2·16 at a gain of 8%.

[**41**] A man buys screws at $13\frac{1}{2}$p per dozen and sells them at 8 for 12p; find his gain per cent.

42 Calculation shows that an angle is $37\frac{1}{2}°$; the size obtained by drawing and measurement is 36°; find the error per cent.

[**43**] Find the value after 3 years of a car bought for £400 which loses each year 20% of its value at the beginning of that year.

44 If 12% of a bill is deducted, £55 remains. What is the bill?

45 Simplify $(x^2-3x+2)(x^2-2x-3)\div(x^2-1)$.

46 Express $(5x^2+14x+13)^2-16(x^2+x-2)^2$ as the product of four factors.

47 Find the square root of

$$(a^2+2a-15)(a^2+3a-10)(a^2-5a+6).$$

[**48**] Divide $(3x^2-2x+4)^2-(2x^2+x-1)^2$ by x^2-3x+5.

[**49**] Is a^2+2 a factor of a^4+4?

50 Find the value of c if $x-2$ is a factor of $x^2+cx-10$.

51 Find the numerical value of c if $xy-3x+5y+c$ has factors.

[**52**] A photograph 15 cm by 9 cm is mounted on a card and framed. The frame is 2 cm wide all round and measures 27 cm by 20 cm externally. Find the area of the visible part of the card.

53 A can of uniform cross-section contains 6 litres of water. If the water is 20 cm deep, find the area of the cross-section.

54 An armour plate is 7 m long, 5 m wide, and weighs 31·92 tonnes. If 1 m³ of iron weighs 7600 kg, find its thickness.

55 M is the mid-point of the side QR of $\triangle$PQR. If MP = MQ, prove that $\angle$QPR = 90°.

56 The line PBCQ cuts a circle, centre O, at B, C. If $\angle$BOQ = $\angle$OPQ, prove that $\angle$POC = $\angle$OQP.

57 ABCD is a parallelogram; DP, DQ are the perpendiculars from D to AB, BC. If DP = DQ, prove AB = BC.

58 In Fig. 179, arrows indicate that lines are given parallel. Find the values of *a*, *b*. If the shortest of the three parallels is 10 units long, find the lengths of the other two.

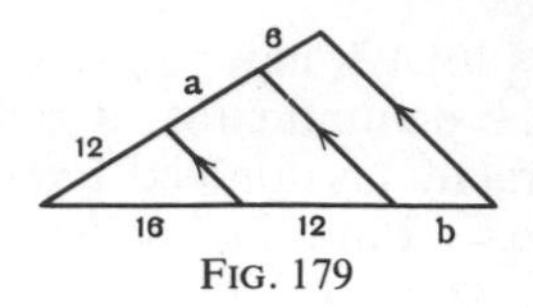

FIG. 179

59 A parallelogram ABCD is drawn on the surface of a wall so that the heights of A, B, C above the ground are 5·7 m, 2·1 m, 3·3 m respectively; AC cuts BD at N. Find the height of N and of D above the ground.

60 D is the mid-point of the side BC of $\triangle ABC$; CA is produced to E. If BR is the perpendicular from B to the bisector of $\angle BAE$, prove $DR = \frac{1}{2}(AB+AC)$. [Produce BR, CA to meet at N.]

CHAPTER 16

COORDINATES: THE LINEAR FUNCTION

Coordinates The position of a point in a plane can be described by making use of two given perpendicular lines $x'Ox$, $y'Oy$, graduated in the usual way.

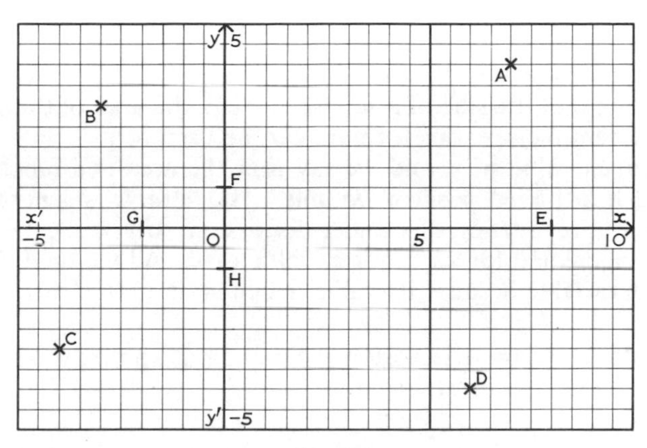

FIG. 180

In Fig. 180, the position of the point A is fixed by saying that, if we start from O and move 7 units x-wards and then 4 units y-wards, we arrive at A; 7 is called the *x-coordinate* of A, 4 is called the *y-coordinate* of A, and A is called the point **(7, 4)**.

The *x*-coordinate is always named first.

Similarly to arrive at B, we start from O and move (-3) units x-wards, that is 3 units along Ox', and then 3 units y-wards;

$$\therefore \text{ B is the point } (\mathbf{-3, 3}).$$

To arrive at H, we start from O and move no distance x-wards but only (-1) unit y-wards, that is 1 unit along Oy';

$$\therefore \text{ H is the point } (\mathbf{0, -1}).$$

The point which represents $x = 0$, $y = 0$, is the point O. It is called the **origin.** The line $x'Ox$ *across* the page through O, along which values of x are measured, is called the ***x*-axis,** and the line $y'Oy$ *up* the page through O, along which values of y are measured, is called the ***y*-axis.**

Oral Work

(i) Write down the coordinates of C, D, E, F, G in Fig. 180.

(ii) Draw on squared paper the axes $x'Ox$, $y'Oy$. Choose as scales, 2 cm represents 5 units on each axis. Mark the following points, M(−8, −10), N(5, −7), P(6·5, 8·5), Q(0, 3·5), R(−7·5, 0), T(−3·5, 1·5).

Functions Any expression containing x, whose value can be found when the value of x is given, is called a **function of x**. Thus

$$7x, x^2 - \tfrac{3}{4}x + 5, \frac{2x-1}{x+3}, \sqrt{x}, \text{etc.}$$

are all functions of x.

If the length of the side of a square is x cm, the area of the square is x^2 square centimetres; this formula can be used to calculate the area for any given value of x, and we say that the area is a function of x and we call x the **independent variable.** The letter y is generally used to represent the function of x under discussion. If $y = x^2$, the value of y can be calculated for any chosen value of x, and we then call y the **dependent variable** because its value depends on that of x.

Example 1 (Class Discussion) Draw a graph representing the squares of numbers from 0 to 5.

Denote the function x^2 by y.

Make a table showing the values of y for selected values of x.

x	0	1	2	3	4	5
$y = x^2$	0	1	4	9	16	25

Choose scales as follows:

On the x-axis, 1 unit to 1 cm;

on the y-axis, 5 units to 1 cm.

Graduate the axis *across* the page to show values of x and graduate the axis *up* the page to show values of y.

Plot the points (1, 1), (2, 4), (3, 9), (4, 16), (5, 25), *representing the values in the above table*; the last three points are rather far apart; we therefore take additional values of x to make the drawing of the graph easier and more accurate; add these to the table and plot the corresponding points.

x	3·5	4·5
y	12·25	20·25

N.B. Work in decimals, *not* in fractions.

Now draw a smooth curve through the plotted points, and compare the graph you have drawn with Fig. 181.

The graph in Fig. 181 represents the relation between a number x and its square x^2; it is called the **graph of the function x^2**. Figure 181 also represents the graph of y, where $y = x^2$, for values of x from 0 to 5.

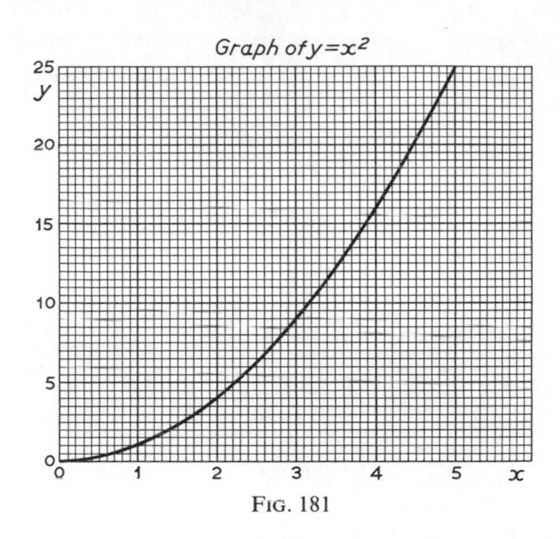

FIG. 181

Oral Work

Obtain from your graph (or from Fig. 181) the approximate values of y when $x = 3{\cdot}8, 2{\cdot}4, 4{\cdot}4, 2{\cdot}6$; and the approximate values of x when $y = 10, 23, 5, 14$.

The following convention must be obeyed when drawing the graph of a function:

Draw **across** *the page the axis for the values of the* independent *variable (the x-axis) and graduate it from left to right; draw the axis for the values of the* dependent *variable (the y-axis)* **up** *the page and graduate it upwards.*

Functions of the First Degree

The graph of y, where $y = 3x - 2$.

Consider the following table of values:

x	-2	-1	0	1	2	3	4
$3x-2$	-8	-5	-2	1	4	7	10

We have selected values of x which increase by 1, and we find that the corresponding values of $3x-2$ increases by equal amounts, namely 3. This means that *the graph has the same slope throughout and must therefore be a straight line.* The values in the table are plotted in Fig. 182 on the next page; they lie on the straight line AB.

The Linear Function

The same argument applies to every first-degree function of x; since its graph is a straight line, it is called a **linear function** of x.

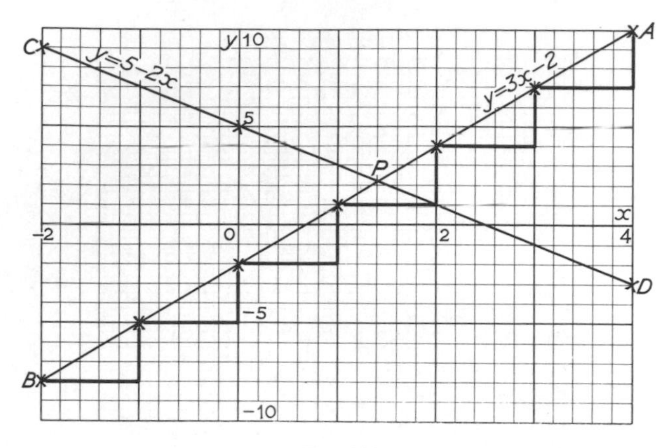

FIG. 182

Thus, any function of the form $bx+c$, where b, c are constants, is called a *linear function* of x, and its graph can be drawn by plotting two points only, and drawing the straight line which joins them; *it is, however, advisable to plot a third point as a check.*

Example 2 Find from the graph the value of x for which the functions $5-2x$ and $3x-2$ are equal.

x	-2	0	2	4
$5-2x$	9	5	1	-3

We have selected values of x which increase by 2, and we find that the corresponding values of $5-2x$ decrease by equal amounts, namely 4. This means that *the graph has* **the same slope throughout** (*here a downward slope*) *and must therefore be a* **straight line.**

The graph is the straight line CD in Fig. 182.

From the figure, we see that the two functions have the same value when $x = 1{\cdot}4$, and this value is $2{\cdot}2$.

The reader should check this result by calculation.

Graphs and Equations The graph of y, where $y = x^2$, in Fig. 181 was drawn by selecting values of x and then calculating the corresponding values of y which satisfy the equation $y = x^2$. The smooth curve drawn through the plotted points is a *locus-graph*; it is the locus of a variable point (x, y) whose position is subject to the condition $y = x^2$.

For this reason, $y = x^2$ is called the **equation of the curve** in Fig. 181.

We usually speak of 'the graph of $y = x^2$', instead of 'the curve whose equation is $y = x^2$'.

Similarly the straight line AB in Fig. 182 is the locus of a variable point (x, y) whose position is subject to the condition $y = 3x-2$, and so $y = 3x-2$ is called the **equation of the straight line** AB. In the same way it follows from Example 2 that $y = 5-2x$ is the *equation of the straight line* CD in Fig. 182.

Since AB and CD meet at the point P(1·4, 2·2), the solution of the simultaneous equations $y = 3x-2$, $y = 5-2x$, is $x = 1{\cdot}4$, $y = 2{\cdot}2$; but this result can be obtained more easily by solving the equations algebraically.

Gradient For the line AB in Fig. 182, if x increases by 1, y increases by 3. Thus, y increases three times as fast as x. For the line CD, if x increases by 1, y *decreases* by 2. Thus, y decreases twice as fast as x increases. The rate at which y increases compared with the rate at which x increases is called the **gradient** of the line. A line sloping upwards to the right has a positive gradient. One sloping downwards to the right has a negative gradient. Parallel lines all have the same gradient. We can move from a point P on a line to another point Q on the line by a step x-wards and then a step y-wards. The value of the ratio $\dfrac{y\text{-step}}{x\text{-step}}$ is the same all along the line and this is the gradient of the line.

The gradient of $y = 3x-2$ (see Fig. 182) is 3; that of $y = 5-2x$ is -2.

Gradient of Line joining (x_1, y_1) and (x_2, y_2)

Suppose that (x_1, y_1) and (x_2, y_2) are two points.

From the definition of gradient given above, it will be seen that the gradient of the line joining these points is

$$\frac{y_1-y_2}{x_1-x_2}.$$

Gradient of the Linear Function $bx+c$

Suppose that (x_1, y_1), (x_2, y_2) are two points on the line $y = bx+c$.

Then $\quad y_1 = bx_1+c$

and $\quad y_2 = bx_2+c.$

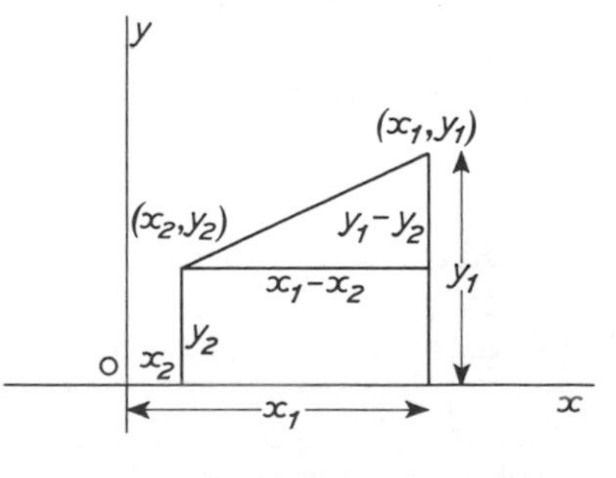

FIG. 183

$$\text{The gradient of the line} = \frac{y_1 - y_2}{x_1 - x_2}$$

$$= \frac{(bx_1 + c) - (bx_2 + c)}{x_1 - x_2}$$

$$= \frac{bx_1 + c - bx_2 - c}{x_1 - x_2}$$

$$= \frac{b(x_1 - x_2)}{x_1 - x_2} = b.$$

Hence, the **gradient of the function $(bx + c)$ is b.**

The linear function $(bx + c)$ The graph of $y = bx + c$ meets the y-axis where $x = 0$. When $x = 0$, $y = c$. $\therefore$ the line meets the y-axis at $(0, c)$.

Hence we see that $y = bx + c$ has gradient b and meets the y-axis at the point $(0, c)$.

Note on different uses of the word 'gradient' When we say that the gradient of a road is 1 in 220, we usually mean that the road rises 1 m vertically in a distance of 220 m measured along the road. The definition of gradient given above would mean that the road rises 1 m vertically in a distance of 220 m measured horizontally. In Fig. 184, the word

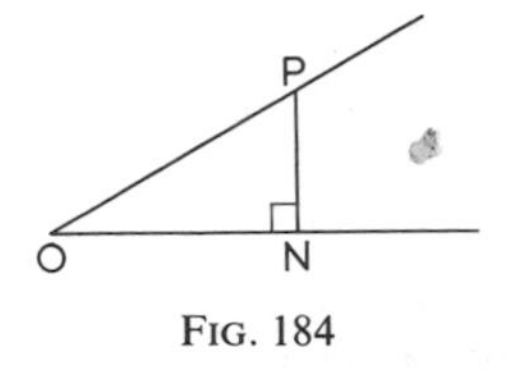

FIG. 184

'gradient' in ordinary speech would often mean $\frac{\text{PN}}{\text{OP}}$, which is the *sine* of $\angle$PON; whereas in mathematics the gradient is $\frac{\text{PN}}{\text{ON}}$, which is the *tangent* of $\angle$PON.

Gradient of a Travel Graph Look again at the travel graphs in Fig. 25, p. 27. The equation of the line OP is $y = \frac{1}{3}x$, and the corresponding speed is $\frac{1}{3}$ km per minute; the equation of OS is $y = 2x$, and the corresponding speed is 2 km per minute. Similarly for the other travel graphs. It will thus be seen that the **gradient of a travel graph represents the speed.**

Half Planes Any line in a plane may be said to divide it into two **half planes,** one on each side of the line. Every point (x, y) on the line AB in Fig. 182 (p. 152) has its coordinates x, y connected by the equation $y = 3x-2$. Now consider the point (2, 6), which lies *above* AB. The y-coordinate of that point is 6, and the value of $(3x-2)$ is $3\times2-2$, or 4. Now 6 is greater than 4, and so y is *greater than* $(3x-2)$ for this point. The same could be said for every point above AB. If we now take the point (2, 3), which lies *below* AB, it will be seen that y is *less* than $(3x-2)$ for this point, and similarly for every point below AB.

The symbols for 'is greater than' and 'is less than' are, respectively, $>$ and $<$.

The statement '10 is greater than 7' is written

$$10 > 7.$$

The statement '10 is less than 12' is written

$$10 < 12.$$

The statement 'x is either greater than 1 or equal to 1' is written

$$x \geqslant 1.$$

Particular care must be taken when we are dealing with negative numbers. Referring to the number scale in Fig. 2, p. 3, it will be at once obvious that any number on this scale is greater than a number below it. Thus,

$-1 > -2$ (although $1 < 2$),

$1 > -3$,

$-\frac{1}{3} < -\frac{1}{4}$.

Using these inequality signs, we can say that the upper half plane (above AB) in Fig. 182 is $y > 3x-2$, and the lower half plane is $y < 3x-2$. The line AB itself is called $y = 3x-2$. The region above or on AB is called $y \geqslant 3x-2$.

The usual method of indicating a half plane in a diagram is to leave *unshaded* the region named, and to shade instead a part of the *other* half plane. Thus, in Fig. 185, the half plane indicated is $y \geqslant 3x-2$.

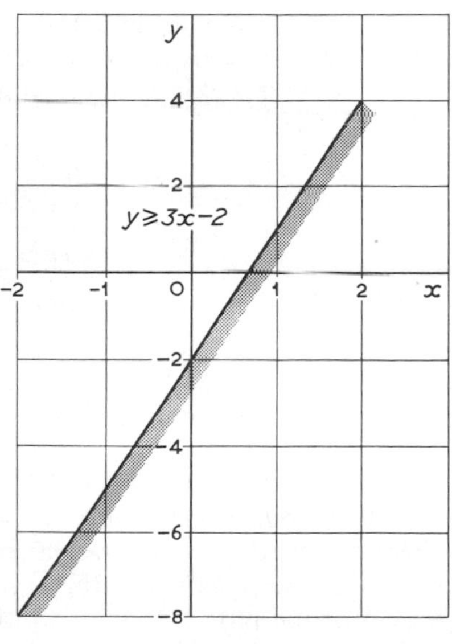

FIG. 185

Example 3 Show in a diagram the region of the plane denoted by:

$$x \geqslant -1, \quad y \geqslant 0, \quad y \leqslant 3-2x.$$

The region $x \geqslant -1$ lies to the right of or on the line $x = -1$ (which is a line parallel to the y-axis, 1 unit to the left of it), so we shade to the left of this line. See Fig. 186.

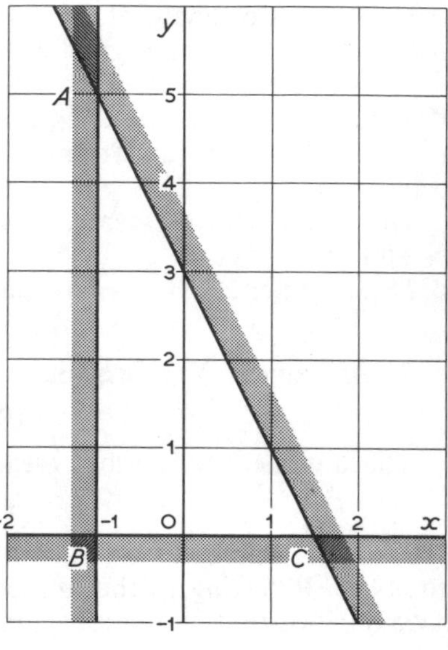

FIG. 186

The region $y \geqslant 0$ lies above or on the line $y = 0$, which is the x-axis, so we shade below this line.

The region $y \leqslant 3-2x$ lies below or on the line $y = 3-2x$. To draw this line, plot the points (0, 3), (1, 1), $(1\frac{1}{2}, 0)$ and join up with a ruler. In order to decide which side of the line is required, we can test with $x = 0$, $y = 0$. These values make y zero and $(3-2x)$ equal to 3. Now $0 < 3$. Therefore the origin O belongs to the region $y \leqslant 3-2x$. Hence the half plane required lies below the line $y = 3-2x$, and we therefore shade above this line.

The required region is the triangle ABC in Fig. 186.

EXERCISE 75

In Nos. 1–3, *the scales on each axis are* 1 *unit to* 1 *cm.*

1 P, Q are the points $(-2, -1)$, $(1, 1)$. Find by trigonometry the angle which PQ makes with the x-axis.

2 A, B are the points (3, 1), (4, 1); P is a point such that AP = 10 cm and $\angle$BAP = 27°. Find by trigonometry the coordinates of P. (Two answers.)

3 The lengths of the sides of a rectangle are 8 cm, 5 cm, and the longer sides are parallel to the x-axis. Find the coordinates of the vertices of the rectangle if the centre of the rectangle is (i) the point (3, 1), (ii) the point $(-1, -2)$. Find also in each case the equations of the sides of the rectangle.

4 Draw, *with as little calculation as possible*, the graphs of $y = 2x+3$ and $y = 9-3x$ from $x = -2$ to $x = 3$. Use the graphs to solve the simultaneous equations $2x-y+3 = 0$, $3x+y-9 = 0$, and compare by solving algebraically.

5 If y varies directly as x, then $y = kx$ where k is a constant. Draw the graph of $y = kx$ if (i) $k = 2$, (ii) $k = 3$, (iii) $k = -2$, (iv) $k = -3$. What features are common to all four graphs? What is the difference between the graphs for which k is positive and those for which k is negative?

6 Draw on the same axes, from $x = -3$ to $+3$, the graphs of (i) $y = 3x+2$, (ii) $y = -x+2$, (iii) $y = \frac{1}{2}x+2$. State the coordinates of the point where these lines cut the y-axis. What are the coordinates of the point where $y = bx+c$ cuts the y-axis?

7 Draw on the same axes, from $x = 1$ to 3, the graphs of (i) $y = 5-3x$, (ii) $y = -3x$, (iii) $y = 1-3x$. What are the gradients of these three lines.

8 Draw the graph of $y = 5x-7$. If points are marked on the graph for values of x which increase by 2, say $x = 1, 3, 5, 7$, etc., how do the corresponding values of y change?

[**9**] Repeat No. 8 for the graph of $y = 5-\frac{3}{4}x$.

10 Solve graphically $y = \frac{1}{3}(x-2)$, $y = \frac{1}{4}(3-x)$, and compare by solving algebraically.

[**11**] If $3x+2y = 6$, express y in terms of x and draw the line whose equation is $3x+2y = 6$. Draw in the same figure the line whose equation is $4x-4y+7 = 0$ and solve graphically $3x+2y-6 = 0$, $4x-4y+7 = 0$.

12 What are the gradients of the graphs of: (i) $y = 3x+7$, (ii) $y = 2-x$, (iii) $y = 3-4x$, (iv) $3y = x+2$, (v) $4y = 3-x$? State the equations of the lines through the origin parallel to these lines.

***13** Find the equation of the line joining (2, 1) and (3, 6). (Assume that the equation is $y = bx+c$.)

***14** Find the equation of the line through the point (2, -1) and with gradient 3. (Suppose that the equation is $y = 3x+c$.)

15 A cyclist starts at 10 a.m. to ride to a place 14 km away, riding at 20 km per h till he has a puncture; he waits 10 minutes and then walks on at 6 km per h, and reaches his destination at 11.15 a.m. Find graphically how far he had ridden when the puncture occurred.

EXERCISE 76

Indicate in a sketch each of the half-planes named in Nos. 1–6.

1 $y < -2$ **2** $x > -1$ [**3**] $x < 3$ **4** $y > 1\frac{1}{2}$

5 $y > x$ **6** $y < 3x$ **7** $2x+y > 1$ [**8**] $x-3y < 1$

Name by means of a pair of inequalities the area shaded in Figs. 187–189.

9

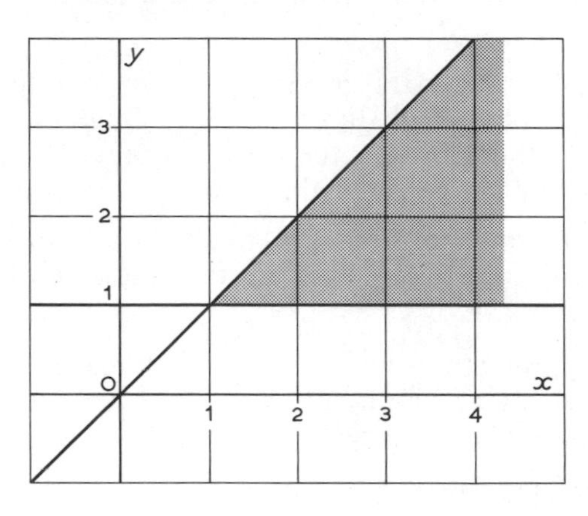

Fig. 187

10

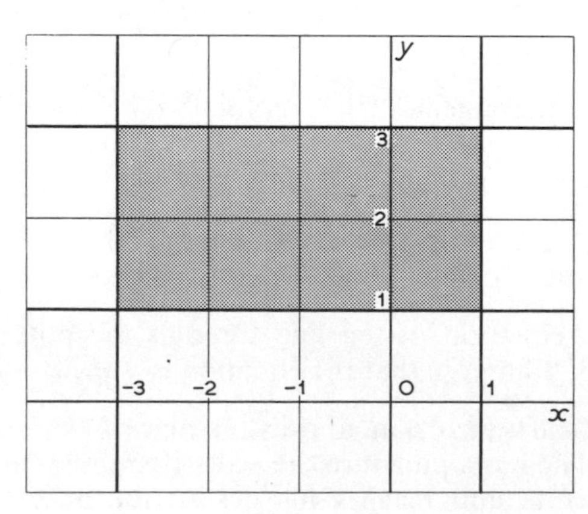

Fig. 188

11

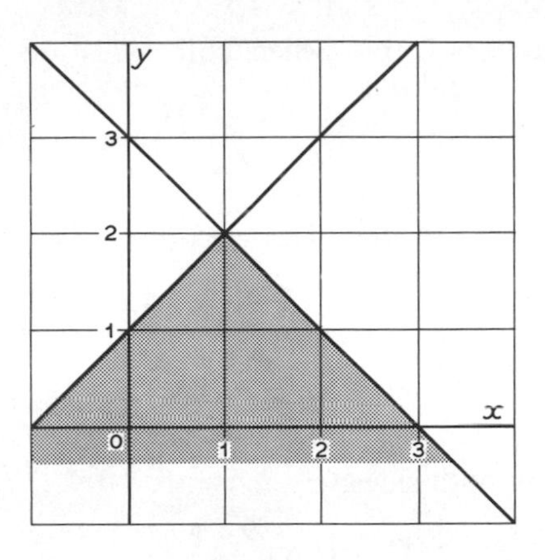

FIG. 189

Show in a diagram the regions indicated in Nos. 12–19, leaving the region named *unshaded*.

12 $x \leqslant 2, y \leqslant -1$

13 $-1 \leqslant y \leqslant 2$

14 $y \geqslant x, y \leqslant 2x$

15 $x \geqslant 0, y \geqslant 0, 2x+3y \leqslant 6$

[**16**] $x \geqslant 1, y \leqslant 2, y \leqslant \frac{1}{2}x+1$

[**17**] $y \leqslant x+3, x \leqslant 2, x \geqslant 0$

18 $y \geqslant \frac{1}{2}x, y \leqslant 4x, x+\frac{1}{2}y \geqslant 2$

19 $2y \leqslant x+4, 2x-y \leqslant 4, x+y \geqslant -5$

Linear Inequations The relation $4x-2 > 6$ is an example of an **inequation**, and it is **linear** because $4x-2$ is a linear function of x. In order to find what values x may have if

$$4x-2 > 6,$$

we add 2 to both sides, making

$$4x > 8.$$

Divide both sides by 4,

$$x > 2.$$

$\therefore$ x can have any value greater than 2.

This example shows that the method of solving a linear inequation is the same as that of solving a linear equation. We can add the same quantity to both sides (whether positive or negative), and we can multiply or divide both sides by the same number. But there is one very important difference; namely, that if both sides of an inequation are multiplied or divided by a negative number, then **the direction of the inequality sign must be reversed.**

For example, $3 > 2$; but when both sides are multiplied by -1, we get $-3 < -2$, and *not* $-3 > -2$.

Similarly, if $-x > 3$, then

$$x < -3.$$

Example 4 Solve the inequation $15-5x > 4x+33$.
Given that

$$15-5x > 4x+33,$$

subtract 15 from both sides;

$$-5x > 4x+18.$$

Subtract $4x$ from both sides,

$$-9x > 18.$$

Multiply both sides by -1 and reverse the inequality sign,

$$9x < -18.$$

Divide both sides by 9,

$$\boldsymbol{x < -2.}$$

EXERCISE 77

Solve the following inequations:

1 $2x > 3$ **2** $2x < -3$ **3** $3x > -2$

[**4**] $-3x > 3$ **5** $-3x < 9$ [**6**] $x-2 < 0$

7 $2x > 3x-1$ [**8**] $\frac{1}{3}x+12 < x$ **9** $\frac{4}{5}x-1 < x$

10 $4-3x > 1-2x$ **11** $8-2x < 10+x$ **12** $3x < 4(x+1)$

13 $4x-3 > 1+4x$ **14** $16-10x > 2x$ [**15**] $13 > 3(x-2)$

[**16**] $5(x-4) < 7(x-3)$ **17** $x^2+5 > x(x+1)$

18 $12x < 10-(1-x)$ **19** $2x-4\frac{2}{3} < 1\frac{1}{2}x-\frac{2}{3}x$

***20** $6-\frac{3}{4}(x-6) < \frac{1}{2}(x+1)$

***21** $\dfrac{18}{x}-3 > 4-\dfrac{3}{x}$ ***22** $\dfrac{3x}{2}-\dfrac{x-1}{3} < \dfrac{3x-2}{2}$

23 $6x-3 < 2+6x$ **24** $5(x-5) < 4$

CHAPTER 17
USE OF TABLES

Approximate values of the squares, square-roots and reciprocals of numbers may be found by using tables; but in some cases a higher degree of accuracy than is given by the available tables is required. Throughout this chapter, *four-figure* tables will be used. Decimal points are usually omitted from the tables and must be inserted by making a rough estimate of the answer.

Table of Squares The arrangement of the tables is best explained by taking an example.

By direct multiplication, $237^2 = 237 \times 237 = 56169$.

This result can be obtained, *correct to 4 figures*, from 4-figure tables. In the table of squares, look for the number 23 in the left-hand column; the figures in the '23' row, together with the headings at the top of the page, are:

MAIN COLUMNS

	0	1	2	3	4	5	6	7	8	9
23	5290	5336	5382	5429	5476	5523	5570	5617	5664	5712

There is also at the end of this row a column of 'mean differences' which is used in the same way as the mean difference columns in trigonometric tables.

In the '23' row, under the 7 in the main columns, we find the figures 5617; this means that *the first* 4 *significant figures* in 237^2 are 5617.

Since $200^2 = 40\,000$ and $300^2 = 90\,000$, the square of 237 lies between 40 000 and 90 000,

$$\therefore\ 237^2 = \mathbf{56\,170} \text{ correct to 4 figures.}$$

This statement can be checked by comparing it with the exact value 56 169 obtained by direct multiplication.

Similarly the first 4 significant figures in 23^2, being the same as in 230^2, are 5290; and so $23^2 = \mathbf{529}$ *exactly*.

The figures obtained in working out the square of 23·7 are exactly the same as those obtained in working out the square of 237, the only difference being the position of the decimal point.

Since $20^2 = 400$ and $30^2 = 900$, the square of 23·7 lies between 400 and 900; therefore from the tables

$$23{\cdot}7^2 = \mathbf{561{\cdot}7}, \text{ correct to 4 figures.}$$

Similarly $2{\cdot}37^2 = 5{\cdot}617$, $0{\cdot}237^2 = 0{\cdot}05617$, etc., to 4 figures.

Example 1 Use tables to find approximately $2{\cdot}374^2$.

In the mean-difference column for the 23 row, under the heading 4, we find 19; this is the required compensation.

$2{\cdot}37^2$	5·617
diff. for 4	19
$2{\cdot}374^2$	5·636

$$2{\cdot}374^2 = \mathbf{5{\cdot}636}, \text{ to 4 figures.}$$

Direct multiplication shows that $2{\cdot}374^2 = 5{\cdot}635876$, so that the result given by the tables is actually correct to 4 figures. In general, when mean-difference columns are used, the 4th figure is not absolutely reliable, but the error will not usually be more than a unit in the fourth figure; this fact may be indicated by writing the result in the form, $2{\cdot}374^2 = 5{\cdot}63(6)$.

EXERCISE 78

Find from 4-figure tables the squares, to four figures, of:

1 47	**2** 4·7	**3** 72	**4** 720	**5** 8·3	**6** 290
7 3100	**8** 2·3	**9** 0·24	**10** 660	**11** 4·13	**12** 72·5
13 619	**14** 1·76	**15** 2870	**16** 31·6	**17** 3·17	**18** 0·346

19 0·895	**20** 695	**21** 0·327	**22** 5040	**23** 0·0101
24 0·703	**25** 0·094	**26** 0·0633	**27** 70·8	**28** 3180
29 3·826	**30** 38·12	**31** 387·4	**32** 3868	**33** 6·453
34 0·6427	**35** 0·06472	**36** 64·84	**37** 9·031	**38** 90·76
39 9058	**40** 0·09011	**41** 80·04	**42** 5308	**43** 400·7
44 0·7047	**45** 1·798	**46** 0·1656	**47** 14·88	**48** 0·02067
49 3·162	**50** 3·163	**51** 0·3108	**52** 3187	**53** 0·2009
54 30·08	**55** 100·7	**56** 1009	**57** 8487	**58** 920·8
59 0·9907	**60** 90·01	**61** 4·5753	**62** 16·367	**63** 227·84

The **square roots** of 900, 9, 0·09, 0·0009, ...

are 30, 3, 0·3, 0·03, ...

If the decimal point in a number is moved an *even* number of places, the decimal point in the square root of the number is moved *half as many places* and the significant figures are unchanged.

But if the decimal point is moved an *odd* number of places, the significant figures in the square root of the number are changed:

$\sqrt{5}$, the square root of 5, is a number between 2 and 3; but $\sqrt{50}$, the square root of 50, is a number between 7 and 8.

The first significant figure in the square root of a number can be written down at sight by moving the decimal point an *even* number of places so as to give a number between 1 and 100.

For example, the first figure in $\sqrt{6927}$ and in $\sqrt{0{\cdot}6927}$ is the same as the first figure in $\sqrt{69{\cdot}27}$; this is 8 because

$$8^2 = 64 \quad \text{and} \quad 9^2 = 81.$$

Similarly the first figure in $\sqrt{692{\cdot}7}$ and $\sqrt{0{\cdot}06927}$ is the same as the first figure in $\sqrt{6{\cdot}927}$; this is 2 because

$$2^2 = 4 \quad \text{and} \quad 3^2 = 9.$$

The simplest way of finding the first figure and the position of the decimal point in the square root of a number is to mark off *pairs* of digits of the number *starting from the decimal point*, as follows:

Square root	8	*·			
Number	69	27·			
Square root	0·	8	*		
Number	0·	69	27		
Square root	0·	0	8	*	
Number	0·	00	69	27	
Square root	2	*·	*		
Number	6	92·	7		
Square root	0·	2	*	*	
Number	0·	06	92	7	
Square root	0·	0	2	*	*
Number	0·	00	06	92	7

This shows why a table of square roots must be arranged on *pairs* of pages. (Some tables have a double *row* against each number.)

On one page or row, the first figure in the 69 row is 8; on the corresponding page or row, the first figure in the 69 row is 2.

In order to decide on which page or row the square root of a given number will be found, it is necessary to start by finding the *first significant figure* in the square root.

Example 2 Use tables to find to 4 figures, (i) $\sqrt{537}$; (ii) $\sqrt{0{\cdot}537}$.

(i) The pattern on the right shows that

2	*·
5	37·

$$\sqrt{537} = 2{*}.{*}$$

Look on the page or row where the square root of 537 starts with 2.

$$\therefore \sqrt{537} = \mathbf{23{\cdot}17}, \text{ to 4 figures.}$$

(ii) The pattern on the right shows that

0·	7	*
0·	53	7

$$\sqrt{0{\cdot}537} = 0{\cdot}7{*}$$

Look on the page or row where the square root of 537 starts with 7.

$$\therefore \sqrt{0{\cdot}537} = \mathbf{0{\cdot}7328}, \text{ to 4 figures.}$$

If a number is given to 4 significant figures, its square root to 4 figures is found by using the column of 'mean differences'.

Example 3 Use tables to find $\sqrt{0{\cdot}06184}$ to 4 figures.
Look on the page where the square root starts with 2.

0·	2	*	*
0·	06	18	4

$\sqrt{0{\cdot}0618}$	0·2486
Add diff. for 4	1
$\sqrt{0{\cdot}06184}$	0·2487

$$\therefore \sqrt{0{\cdot}6184} = \mathbf{0{\cdot}2487}, \text{ to 4 figures.}$$

Example 4 Use tables to find $\sqrt{0{\cdot}1138}$, to 4 figures.

Look on the page where the square root starts with 3.

0·	3	*
0·	11	38

When mean differences are used, the 4th figure is not reliable. Here, the result is more likely to be correct to 4 figures, if 1138 is treated as $1140-2$, instead of $1130+8$.

$\sqrt{0{\cdot}113}$	0·3362
Add diff. for 8	12
$\sqrt{0{\cdot}1138}$	**0·3374**

$\sqrt{0{\cdot}114}$	0·3376
Subtract diff. for 2	3
$\sqrt{0{\cdot}1138}$	**0·3373**

The rule for square root gives

$$\sqrt{0{\cdot}1138} = 0{\cdot}33734$$

to 5 figures, so that the second result is actually correct to 4 figures.

If however answers are required only to 3 figures, it is sufficient to use the addition method for differences.

EXERCISE 79

Find to 4 figures, as given by 4-figure tables, the square roots of:

1 2·6	**2** 26	**3** 268	**4** 0·268	**5** 73
6 730	**7** 7340	**8** 0·0734	**9** 616	**10** 59·7
11 0·923	**12** 0·0864	**13** 50·7	**14** 3040	**15** 0·00529
16 0·000609	**17** 17·43	**18** 1·743	**19** 356·4	**20** 0·3564
21 707·4	**22** 60·86	**23** 92050	**24** 0·9407	**25** 0·3863
26 4·007	**27** 5718	**28** 0·05074	**29** 77060	**30** 57387
31 0·20738	**32** 815·67			

Pythagoras' Theorem Examine the following diagrams:

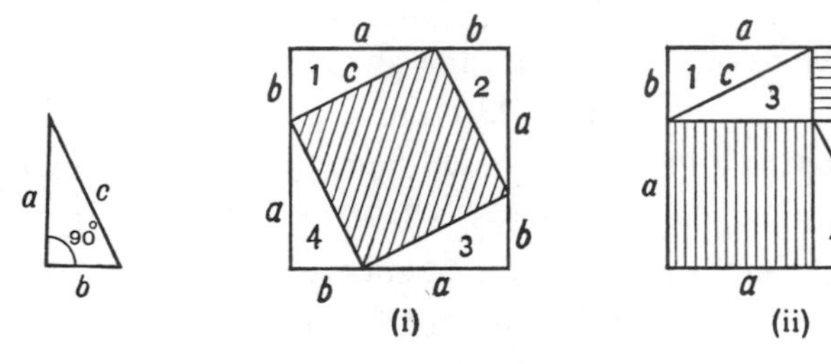

FIG. 190

Cut out 4 congruent right-angled triangles and arrange them in the form of Fig. 190 (i); the complete figure is a square of side $(a+b)$ units and the portion not covered by the triangles is a square of side c units, area c^2 units of area. Now rearrange the triangles in the form of Fig. 190 (ii); the complete figure is again a square of side $(a+b)$ units and the portion not covered by the triangles is made up of two squares of sides a, b, units, areas a^2, b^2, units of area.

Hence

$$c^2 = a^2 + b^2.$$

This result is called **Pythagoras' Theorem** and may be stated:

The area of the square on the hypotenuse of a right-angled triangle is equal to the sum of the areas of the squares on the other two sides.

This theorem can be proved by trigonometry as follows:

Let ABC be a triangle right-angled at C, and CF an altitude. With the notation of Fig. 191,

from $\triangle$ABC, $\cos B = \dfrac{a}{c}$;

from $\triangle$CFB, $\cos B = \dfrac{q}{a}$;

$$\therefore \frac{a}{c} = \frac{q}{a}; \qquad \therefore a^2 = cq.$$

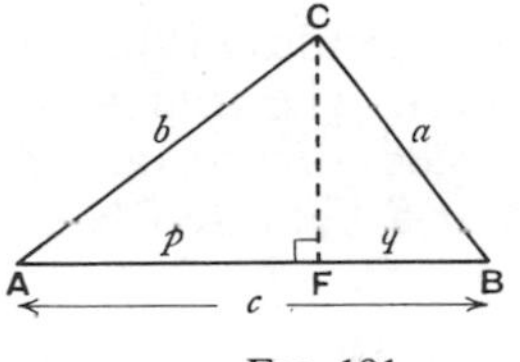

FIG. 191

Similarly from $\triangle$ABC, $\triangle$CFA,

$$\cos A = \frac{b}{c} \quad \text{and} \quad \cos A = \frac{p}{b}; \qquad \therefore \frac{b}{c} = \frac{p}{b}; \qquad \therefore b^2 = cp.$$

$\therefore$ by addition,

$$a^2 + b^2 = cq + cp = c(q+p);$$

but $q + p = c$,

$$\therefore \boldsymbol{a^2 + b^2 = c^2}.$$

Example 5 ABC is a triangle right-angled at C, see Fig. 191.
(i) If BC = 3 cm, AC = 4 cm, find AB.
(ii) If AB = 13 cm, BC = 5 cm, find AC.

(i) With the notation of Fig. 191,

$$c^2 = 3^2 + 4^2 = 9 + 16 = 25; \qquad \therefore c = 5.$$

$$\therefore \text{AB} = \mathbf{5\ cm}.$$

(ii) With the notation of Fig. 191,

$$13^2 = 5^2 + b^2,$$

$$\therefore b^2 = 13^2 - 5^2 = 169 - 25 = 144; \qquad \therefore b = 12.$$

$$\therefore \text{AC} = \mathbf{12\ cm}.$$

Example 6 If the lengths of the sides of a rectangle are 4·7 cm, 3·2 cm, find to 3 significant figures the length of a diagonal.

If the length of a diagonal is c cm,

$$c^2 = 4{\cdot}7^2 + 3{\cdot}2^2$$
$$= 22{\cdot}09 + 10{\cdot}24 \text{ (table of squares)}$$
$$= 32{\cdot}33;$$
$$\therefore c = \sqrt{32{\cdot}33} = 5{\cdot}686 \text{ (table of square roots)};$$

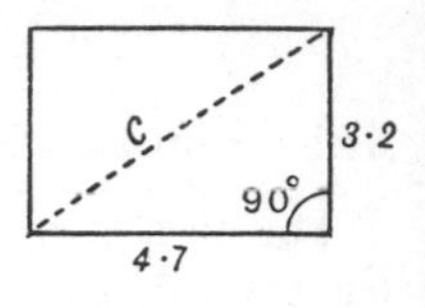

FIG. 192

$\therefore$ the length of the diagonal is **5·69 cm,** to 3 significant figures.

In **solid geometry,** it is often necessary to look for right-angled triangles in different planes. Figure 193 represents a rectangular room; AB, BC are sides of the floor and CN is the line in which two of the walls meet. The triangle ABC is right-angled at B because AB and BC are sides of a rectangle. The triangles NCB, NCA are each right-angled at C because CN is a vertical line and CB, CA are horizontal lines.

Example 7 A hall is 17 m long, 14 m wide, 9 m high. Find the distance from a corner A of the floor to the opposite corner N of the ceiling.

Suppose the diagonal AC of the floor is x m and the diagonal AN of the hall is y m.

From the right-angled triangle ABC,

$$x^2 = 17^2 + 14^2,$$

and from the right-angled triangle ACN,

$$y^2 = x^2 + 9^2;$$

$$\therefore y^2 = 17^2 + 14^2 + 9^2 = 289 + 196 + 81 = 566;$$
$$\therefore y = \sqrt{566} = 23{\cdot}79 \text{ (table of square roots)};$$

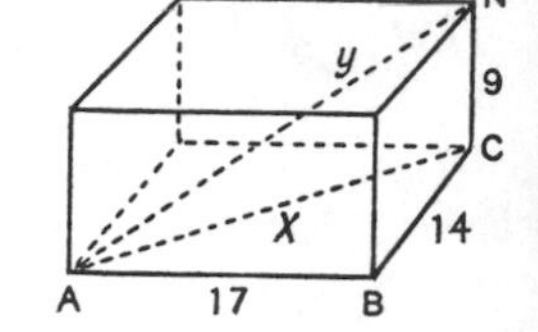

FIG. 193

$\therefore$ AN is **23·8 m,** to 3 significant figures.

Notice that it is unnecessary to find the value of x.

EXERCISE 80

[*Give answers correct to 3 significant figures, unless otherwise stated*]

Find the length of a side of a square field of area:

1 8000 m^2 [**2**] 13750 m^2 [**3**] $\frac{1}{20}$ km^2 **4** 4840 m^2

In Nos. 5–8, the length of the hypotenuse of a right-angled triangle is c cm, and the lengths of the other sides are a cm, b cm.

[**5**] If $a = 5{\cdot}2$, $b = 7{\cdot}8$, find c. **6** If $a = 4{\cdot}73$, $b = 6{\cdot}14$, find c.

7 If $b = 5{\cdot}67$, $c = 8{\cdot}06$, find a. [**8**] If $a = 23{\cdot}26$, $c = 37{\cdot}08$, find b.

[**9**] The sides of a rectangle are 7·85 cm, 6·42 cm long, find the length of a diagonal.

10 The length of a diagonal of a rectangular courtyard is 23·7 m, and the length of one side is 18·8 m; find the perimeter.

[**11**] The area of a square field is 12 100 m^2. Find the length of a diagonal.

12 A ladder 4·35 m leans against a wall, and the foot of the ladder is 1·72 m from the wall. How high up the wall does the ladder reach?

13 ABC is an equilateral triangle in which BC = 6 cm; find the length of the perpendicular from A to BC.

[**14**] The equal sides AB, AC of an isosceles triangle ABC are 8·54 cm long, and BC = 6·08 cm. Find the distance of A from BC.

15 The slant edge of a cone is 8 cm, and its base-diameter is 10 cm. Find the height of the cone.

16 Find the distance between two points whose coordinates are (1, 2) and (5, 8).

17 A room is 4·5 m long, 3·9 m wide, 2·55 m high. Find the distance from a corner of the floor to the opposite corner of the ceiling.

[**18**] Find the length of the diagonal of a match-box, 6 cm by 3 cm by 2 cm.

[**19**] What is the length of the longest straight thin rod, measuring a whole number of centimetres, that can be put into a rectangular box whose internal measurements are 1·8 m by 1·5 m by 1·2 m?

20 A pyramid of height 7·5 cm stands on a square base of side 4 cm; all the slant edges are equal, find their lengths.

Table of Reciprocals Any two numbers whose product is 1 are called **reciprocals** of one another. Thus 7 and $\frac{1}{7}$, $\frac{8}{11}$ and $\frac{11}{8}$, $2\frac{1}{3}$ and $\frac{3}{7}$, are pairs of reciprocals.

The reciprocal of 4 is $\frac{1}{4}$, that is 0·25; and the reciprocals of 40, 400, etc. are 0·025, 0·0025, etc.; also the reciprocals of 0·4, 0·04, etc. are 2·5, 25, etc. Thus, as a number gets larger, its reciprocal gets smaller, and as a number gets smaller, its reciprocal gets larger.

Example 8 Find the reciprocal of 0·365 to 4 figures.

The reciprocal of 0·365 is $\frac{1}{0{\cdot}365}$; $\quad \frac{1}{0{\cdot}3|65} \simeq \frac{10}{3} \simeq 3.$

By long division, $\frac{1}{0{\cdot}365} = 2{\cdot}740$, to 4 figures.

```
      27397
365)1000
     730
     2700
     2555
      1450
      1095
       3550
       3285
        2650
```

It is quicker to use a table of reciprocals which is arranged on the same plan as tables of squares, etc.

In a table of reciprocals in the '36' row under the heading 5 in the main columns, we find the figures 2740; the position of the decimal point must be found by making a rough approximation as above;

$\therefore$ from the tables, $\frac{1}{0{\cdot}365} =$ **2·740,** to 4 figures.

Example 9 Find the reciprocal of 18 460 to 4 figures.

The reciprocal of 18 460 is *less* than the reciprocal of 18 400 and therefore the mean difference must be **subtracted.**

$$\frac{1}{18460} \simeq \frac{1}{20000}$$
$$\simeq 0{\cdot}000\,05;$$
$$\therefore \frac{1}{18460} = \mathbf{0{\cdot}000\,0541(7)}.$$

Number	Reciprocal
184	5435
Subtract diff. for 6	18
1846	5417

Example 10 Use tables to find (i) $\frac{1}{\sqrt{3}}$, (ii) $\frac{2}{6{\cdot}17^2}+\frac{3}{8{\cdot}02^2}$.

(i) $\frac{1}{\sqrt{3}} \simeq \frac{1}{1{\cdot}732} \simeq \mathbf{0{\cdot}577(3)}.$

Number	Reciprocal
173	5780
Sub. diff. for 2	7
1732	5773

Check: $\frac{1}{\sqrt{3}} = \frac{1\times\sqrt{3}}{\sqrt{3}\times\sqrt{3}}$

$$= \frac{\sqrt{3}}{3} \simeq \frac{1{\cdot}732}{3} \simeq 0{\cdot}577(3).$$

(ii) $\frac{2}{6{\cdot}17^2}+\frac{3}{8{\cdot}02^2} \simeq \frac{2}{38{\cdot}07}+\frac{3}{64{\cdot}32}$

$$= \frac{1}{19{\cdot}03(5)}+\frac{1}{21{\cdot}44}$$
$$\simeq 0{\cdot}05254+0{\cdot}04664$$
$$\simeq \mathbf{0{\cdot}0992}.$$

Number	Reciprocal
190	5263
Sub. diff. for 3(5)	9
1903(5)	5254

Number	Reciprocal
214	4673
Sub. diff. for 4	9
2144	4664

EXERCISE 81

Find to 4-figures, as given by 4-figure tables, the values of:

1 $\frac{1}{2{\cdot}8}$ **2** $\frac{1}{28}$ **3** $\frac{1}{0{\cdot}28}$ **4** $\frac{1}{2800}$ **5** $\frac{1}{2{\cdot}9}$

6 $\frac{1}{290}$ **7** $\frac{1}{77}$ **8** $\frac{1}{707}$ **9** $\frac{1}{43{\cdot}6}$ **10** $\frac{1}{0{\cdot}527}$

11 $\frac{1}{0{\cdot}0816}$ **12** $\frac{1}{6090}$ **13** $\frac{1}{0{\cdot}409}$ **14** $\frac{1}{10{\cdot}1}$ **15** $\frac{1}{0{\cdot}106}$

16 $\frac{1}{0{\cdot}0204}$ **17** $\frac{1}{2{\cdot}645}$ **18** $\frac{1}{2{\cdot}648}$ **19** $\frac{1}{0{\cdot}4916}$ **20** $\frac{1}{54{\cdot}73}$

21 $\frac{1}{0{\cdot}2036}$ **22** $\frac{1}{20080}$ **23** $\frac{1}{0{\cdot}02017}$ **24** $\frac{1}{10{\cdot}63}$ **25** $\frac{1}{0{\cdot}1107}$

Find, to 3 significant figures, the values of:

26 $\dfrac{1}{\sqrt{5}}$ [**27**] $\dfrac{3}{17{\cdot}87}$ **28** $\dfrac{11}{400{\cdot}8}$ **29** $\dfrac{1}{\sqrt{16{\cdot}74}}$

30 $\sqrt{\dfrac{3}{40{\cdot}96}}$ **31** $\dfrac{2}{\sqrt{7}}$ **32** $\dfrac{1}{(0{\cdot}4071)^2}$ [**33**] $\dfrac{5}{\sqrt{0{\cdot}1107}}$

34 $\dfrac{0{\cdot}03}{\sqrt{0{\cdot}02}}$ [**35**] $\dfrac{25}{(12{\cdot}7)^2}$ [**36**] $\dfrac{1}{\sqrt{4{\cdot}93}}+\dfrac{1}{\sqrt{6{\cdot}48}}$

37 $\dfrac{4}{(3{\cdot}817)^2}+\dfrac{5}{\sqrt{130{\cdot}4}}$

38 If $\dfrac{1}{u}+\dfrac{1}{v}=\dfrac{1}{f}$, find f if $u = 8{\cdot}63$, $v = 6{\cdot}07$.

[**39**] If $\dfrac{1}{u}-\dfrac{1}{v}-\dfrac{1}{f}$, find f if $u = 27{\cdot}4$, $v = 40{\cdot}8$.

40 Find $\sqrt[4]{0{\cdot}2546}$, to 3 significant figures.

41 Find $\frac{1}{3{\cdot}78}+\frac{1}{6{\cdot}78}$ to 3 significant figures.

42 Find to one significant figure the error per cent in taking $\sqrt{3}$ as $1\frac{3}{4}$.

43 The two shorter sides of a right-angled triangle are $\sqrt{2}$ cm, $\sqrt{3}$ cm; find the perimeter of the triangle, correct to 3 figures.

44 The area of a rectangular field is 24 200 m^2, and its length is 3 times its breadth; find its perimeter in metres.

45 The lengths of the sides of a rectangle are in the ratio 2:3, and the length of a diagonal is 5 m. Find the perimeter in metres.

46 The diagonal of a cube is 8 cm long. Find (i) the length of an edge, (ii) the total area of the surface of the cube.

CHAPTER 18

LOCI

If we look at the tip of the seconds-hand of a watch we see that it occupies a series of different positions in the course of each minute, and if we combine together all these different positions we obtain the circumference of a circle which the tip of the seconds-hand traces out each minute. This *aggregate* of all possible positions of the tip is called its **locus.**

When it is stated that the locus of a small object which moves about subject to some given law is a certain curve, two complementary ideas are involved:

(1) The position of every point on the curve satisfies the given law.
(2) Every position of the object which satisfies the given law lies on the curve.

It often happens that the conditions of the problem prevent the object from describing the whole of a curve; in this case it must be stated what part of the curve forms the locus.

Oral Example A, B are two given points in a given plane and P is a point in the plane such that $\angle APB = 60°$. Find experimentally the aggregate of all the possible positions of P.

Stick two pins into the paper at A and B, perpendicular to the paper, and slide your set-square between the pins so that the arms of the angle 60° of the set-square pass through A and B. Prick in the paper a number of possible positions of the vertex of the angle. It appears from the figure that the locus consists of two arcs of two distinct circles AP_1B and AP_2B, not the whole circumference of one circle; for the present this must be treated as an experimental result.

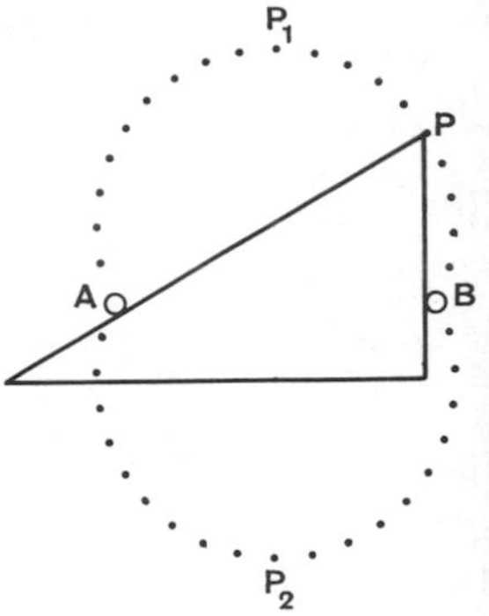

FIG. 194

In this example, if the restriction that P lies in a given plane is removed, the aggregate of all its possible positions forms a *surface* which can be obtained by revolving the arc AP_1B about the line AB as axis through 4 right angles.

Definition The aggregate of all points whose positions are determined by a given law is called the **locus** of points subject to that law.

When there are an unlimited number of possible positions of a point P, all subject to some given law, we call any possible position of P a

variable point, and it is often said that the variable point P moves in accordance with the given law and traces out the resulting locus. This form of words is not strictly accurate, because a point marks a position in a plane or in space and cannot move; its use is justified by regarding the phrase, '*variable point*', as meaning some small object, such as the tip of a pencil, which is free to move.

EXERCISE 82 (Class Discussion)

What are the loci described in Nos. 1–11?

1 A mark on the platform of a merry-go-round.

2 The tip of the pendulum of a clock.

3 The top of your head if you slide downstairs on a tea-tray.

[**4**] The tip of your nose in a swing.

5 The centre of the wheel of an engine running along (i) a straight railway line, (ii) a circular railway line.

[**6**] A mark on a see-saw.

7 A mark on the top of a trap-door in the floor when the trap-door is opened to its full extent.

[**8**] The tip of a man's nose on a moving staircase if he stands still from the time he steps on till the time he steps off.

[**9**] The centre of a marble which rolls inside a spherical bowl.

10 The highest point of the shade of an electric light hanging from the ceiling when the light swings about.

11 The right-angled corner of your set-square if you rotate it round the hypotenuse as axis.

12 The distance of a variable point P from a given point A is 3 cm. What is the locus of P (i) if P lies in a plane through A, (ii) without this restriction?

13 The distance of a variable point P from a given line AB, produced if necessary, is 5 cm. What is the *complete* locus of P (i) if P lies in a plane through AB, (ii) without this restriction?

14 ABC is a piece of wire bent so that the straight portions AB and BC, whose lengths are 3 cm, 4 cm, are at right angles. The end A is fixed; what is the locus of C?

[**15**] A coin, diameter 2 cm, is held flat on a table, and a similar coin, also flat on the table, is made to roll round it. What is the locus of the centre of the moving coin?

16 Four rods form a parallelogram ABCD; AB is fixed, but AD and BC can turn round A and B respectively. P is a marked point on the rod CD such that CP = 3 cm, PD = 2 cm; BC = 4 cm. Find the locus of P. [Draw PQ parallel to DA to meet AB at Q.]

[17] A, B are fixed points; P is a variable point on a given circle, centre A, radius 5 cm; ABPQ is a parallelogram. Find the locus of Q. [Produce BA to C so that BA = AC; join CQ.]

18 AB is a string 2 m long with a weight attached to B; the end A is free to slide on the rim of a fixed vertical circular ring, centre O, radius 1 m, and AB itself remains vertical. What is the locus of B?

19 A ladder 4 m long rests with one end against a vertical wall and the other end on a horizontal floor. If the ladder slips down, remaining in a plane perpendicular to the wall, find the locus of its mid-point. [Use Exercise 60, No. 4, p. 118.]

Locus-Theorems The *complete* proof of a locus-theorem involves proving two distinct statements, either of which is the *converse* of the other:

(1) It must be proved that each point whose position satisfies the given law lies on the stated locus.

(2) It must be proved that the position of each point on the stated locus satisfies the given law.

At this stage we consider two locus-theorems which have important applications. Other locus-theorems, including the locus on p. 170, will be discussed later.

THEOREM

(1) A point which is equidistant from two given points lies on the perpendicular bisector of the straight line joining the given points.

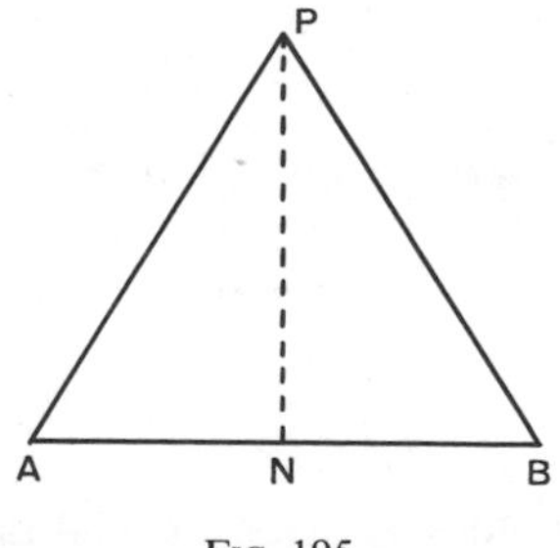

Fig. 195

Given two points A, B and a point P such that PA = PB.

To prove that P lies on the perpendicular bisector of AB.

Construction Let N be the mid-point of AB. Join PN.

Proof In $\triangle$s ANP, BNP,

$$AN = BN \quad \textit{constr.},$$
$$PA = PB \quad \textit{given},$$
$$PN = PN.$$

$\therefore \triangle s \begin{matrix} ANP \\ BNP \end{matrix}$ arc congruent SSS.

$\therefore \angle ANP = \angle BNP$,

but these are adjacent $\angle$s on a straight line, therefore each is a right angle.

$\therefore$ PN is perpendicular to AB and bisects it.

$\therefore$ P lies on the perpendicular bisector of AB.

Abbreviation for reference: perp. bisector locus.

THEOREM

(2) Any point on the perpendicular bisector of the line joining two given points is equidistant from the given points.

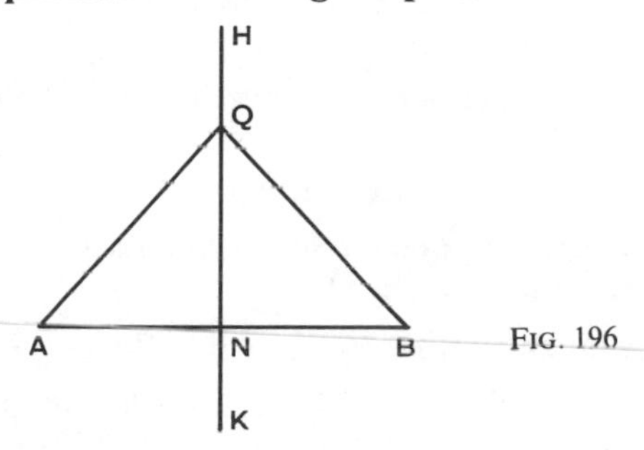

FIG. 196

Given two points A, B and any point Q on the perpendicular bisector HK of AB.

To prove that QA = QB.

Construction Let HK cut AB at N.

Proof In $\triangle$s ANQ, BNQ,

$$AN = BN \quad \textit{given},$$
$$QN = QN,$$
$$\angle ANQ = \angle BNQ \quad \textit{rt. } \angle\textit{s, given.}$$

$\therefore \triangle s \begin{matrix} ANQ \\ BNQ \end{matrix}$ are congruent SAS.

$\therefore QA = QB$.

The enunciations (1) and (2) together form the locus-theorem:

The locus of points which are equidistant from two given points is the perpendicular bisector of the straight line joining the given points.

Abbreviation for reference: perp. bisector locus.

EXERCISE 83 (Class Discussion)

1 If AB is a chord of a circle, centre O, then O lies on the perpendicular bisector of AB.

OA = OB *radii.*

Use locus-theorem (1), p. 172.

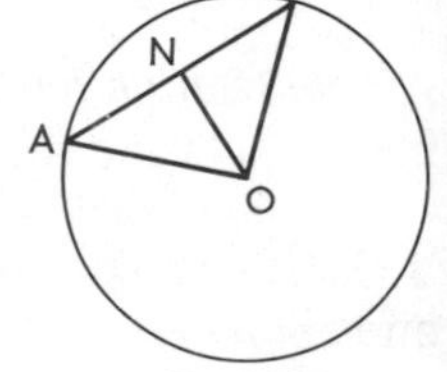

FIG. 197

2 If N is the mid-point of a chord AB of a circle, centre O, then ON is perpendicular to AB.

Use the *proof* of locus-theorem (1), p. 172.

3 If N is the foot of the perpendicular from the centre O of a circle to a chord AB, then AN = NB.

Explain why $\triangle$ONA ≡ $\triangle$ONB, **RHS.**

4 There is one and only one circle which passes through three given points A, B, C, not on a straight line.

Draw the perpendicular bisectors of AB, AC; let them meet at O.

OA = OB and OA = OC, *locus-theorem* (2);

∴ the circle, centre O, radius OA, passes through A, B, C.

Conversely, if OB = OA = OC, O lies on the perpendicular bisectors of AB and AC, *locus-theorem* (1);

∴ there is only one possible position for the centre O of the circle which passes through A, B, C.

The circle which passes through A, B, C *is called the circle* ABC.

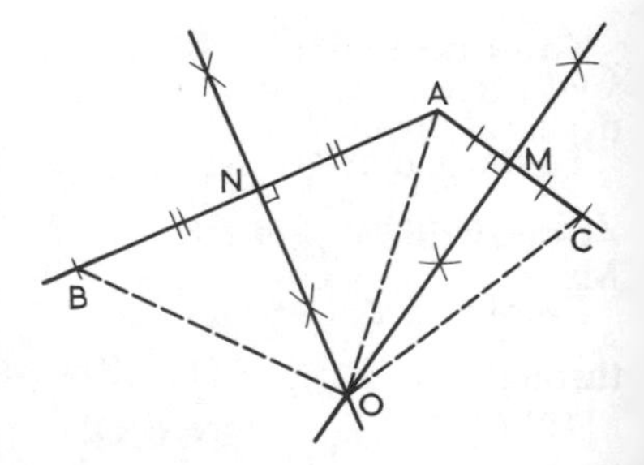

FIG. 198

Definitions Given a triangle ABC, the circle ABC is called the **circumcircle** of $\triangle$ABC, its centre is called the **circumcentre,** and its radius is called the **circumradius.**

It follows from No. 4 that the circumcentre of a triangle ABC is the point of intersection of the perpendicular bisectors of any two of the sides.

5 Draw $\triangle ABC$ in which $BC = 6$ cm, $\angle B = 50°$, $\angle C = 75°$. *Construct* the circumcircle of $\triangle ABC$ and measure the circumradius.

6 Prove the perpendicular bisectors of the three sides of a triangle are concurrent.

EXERCISE 84

1 *Construct* a rhombus ABCD in which $AC = 6$ cm, $BD = 8$ cm; *construct* a point P on AD so that $PB = PC$. Measure AP.

2 Draw a triangle ABC in which $BC = 7$ cm, $CA = 6$ cm, $AB = 5$ cm. Construct the circle which passes through A, B, C and measure its radius.

[**3**] *Construct* a triangle PQR in which $PQ = 6$ cm, $PR = 4$ cm, $\angle QPR = 120°$. *Construct* the circle which passes through P, Q, R and measure its radius.

4 Draw a triangle ABC in which $BC = 7$ cm, $\angle B = 66°$, $\angle C = 48°$. *Construct* the position of a point P inside $\triangle ABC$ such that $PA = PC$ and $PB = 4$ cm. Measure PA.

[**5**] Draw a triangle ABC in which $CA = 7$ cm, $CB = 4$ cm, $\angle ACB = 40°$. *Construct* the position of a point P inside $\triangle ABC$ such that $PB = PC$ and $PA = 5$ cm. Measure PB.

6 Construct, with ruler and compasses only, a quadrilateral ABCD in which $AB = 3$ cm, $BC = 4{\cdot}5$ cm, $CD = 6$ cm, $\angle ABC = 105°$, $\angle BCD = 90°$. Construct the position of a point P such that $PB = PC$ and $PA = PD$.

[**7**] Draw a triangle XYZ in which $YZ = 6$ cm, $ZX = 5$ cm, $XY = 4$ cm. *Construct* the position of a point P such that $PX = PZ$ and XP produced is perpendicular to YZ.

8 *Construct* a triangle ABC in which $AB = 4$ cm, $BC = 6$ cm, $\angle B = 60°$. Find a point P on CB produced such that $PC - PA = 3$ cm. Measure PC. [Find two points from which P is equidistant.]

9 Given two points A, B and a line CD, *construct* a circle to pass through A and B and have its centre on CD. Is this always possible?

[**10**] A, B, C, D are four points on the circumference of a circle. Prove that the perpendicular bisectors of AB, AC, AD, BC, BD, CD are concurrent.

11 The diagonals of the quadrilateral ABCD cut at K. Circles are drawn through A, K, B; B, K, C; C, K, D; D, K, A. Prove that their centres are the vertices of a parallelogram.

EXERCISE 85 (Class Discussion)

1 If in Fig. 199, AB = 8 cm and the radius of the circle is 5 cm, find the distance OH of the centre O from AB.

(i) What is the length of AH?

(ii) Use Pythagoras to find the length of OH.

2 If in Fig. 199, the distance OK of the centre O from CD is 6 cm and the radius is 7·5 cm, find the length of CD.

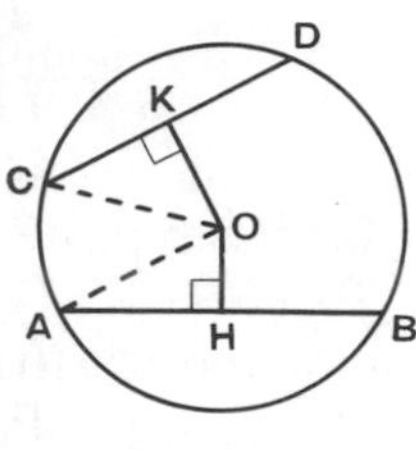

FIG. 199

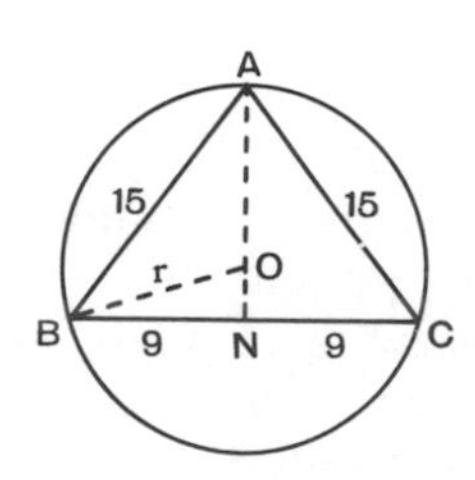

FIG. 200

3 Calculate the circumradius, r cm, of $\triangle$ABC, given that AB = AC = 15 cm, BC = 18 cm.

(i) Explain why the circumcentre O lies on the line joining A to the mid-point N of BC, see Fig. 200.

(ii) Find AN; then express the length of ON in terms of r.

(iii) Use $\triangle$ONB to find r.

EXERCISE 86

[*Give answers which are approximate, correct to 3 figures*]

1 A chord of length 10 cm is at a distance of 12 cm from the centre of the circle. Find the radius.

2 A chord of a circle of radius 6 cm is 8 cm long. Find the distance of the chord from the centre.

[**3**] A chord of a circle radius 7 cm is at a distance of 4 cm from the centre. Find the length of the chord.

4 PQ is a variable chord of a given circle of radius 7·5 cm. If PQ = 9 cm, find the locus of the mid-point of PQ.

5 N is a point on the diameter AB of a circle APBQ; PNQ is the chord through N perpendicular to AB. If AN = 8 cm, NB = 2 cm, find the length of PQ.

[**6**] In a circle of radius 5 cm, there are two parallel chords of lengths 6 cm, 4 cm. Find the distance between the chords. [Two answers.]

[**7**] Two concentric circles are of radii 7 cm, 4 cm; a line PQRS cuts one circle at P, S and the other at Q, R. If QR = 6 cm, find the length of PS.

8 A chord of a circle is 10 cm long and is 4 cm from the centre. Find the length of a chord 3 cm from the centre.

[**9**] The length of the common chord of two equal intersecting circles is 10 cm, and the distance between the two centres is 6 cm. Find the radius of each circle.

10 A hemispherical bowl, internal diameter 12 cm, is partly full of water. If the water-surface is 4 cm below the centre of the bowl, find the diameter of the circular water-surface.

11 If Fig. 201 represents a section of a circular cone, vertex V, by a plane through its axis VN, and the section of the sphere of radius 5 cm in which the cone is inscribed, and if the base-radius of the cone is 3 cm, find the height VN of the cone.

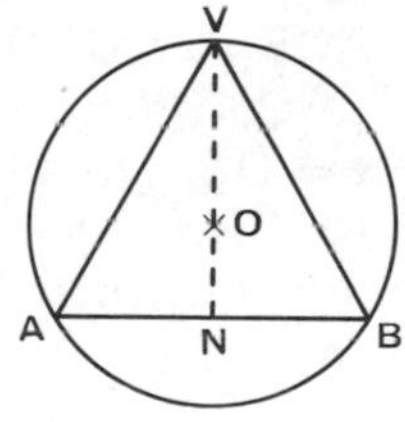

FIG. 201

[**12**] In Fig. 201, if VA = VB = 10 cm and if the diameter of the circle VAB is 12 cm, find the lengths of VN and AB. [Let ON = h cm, AN = d cm, where O is the centre.]

13 In Fig. 201, if VA = VB = 13 cm and AB = 10 cm, find the radius of the circle VAB. [Use the right-angled triangle ONA where O is the centre; let the radius = r cm.]

14 In Fig. 201, VAB is an equilateral triangle, side 6 cm, inscribed in a circle, centre O. Prove (i) $\angle$AON = 60°, (ii) $\triangle$AON is 'half an equilateral triangle'. Hence find OA.

***15** AB, CD are parallel chords of a circle, 3 cm apart, on the same side of the centre O; AB = 4 cm, CD = 10 cm, find OA. [Draw ONM perpendicular to AB to meet AB, CD in M, N; let ON = x cm, OA = OC = r cm.]

***16** Two spheres, radii 6 cm, 8 cm, have their centres 10 cm apart. Find the radius of the circle in which the spheres cut, and the distances of the plane of this circle from the centres of the spheres.

EXERCISE 87 (Class Discussion)

1 OH, OK are the perpendiculars from the centre O of a circle to two chords AB, CD. If AB = CD, prove that OH = OK.

(i) Explain why AH = CK.

(ii) Prove $\triangle$OHA $\equiv$ $\triangle$OKC.

2 OH, OK are the perpendiculars from the centre O of a circle to two chords AB, CD. If OH = OK, prove that AB = CD.

(i) Prove $\triangle$OHA $\equiv$ $\triangle$OKC.

(ii) Explain why AB = 2AH and complete the proof.

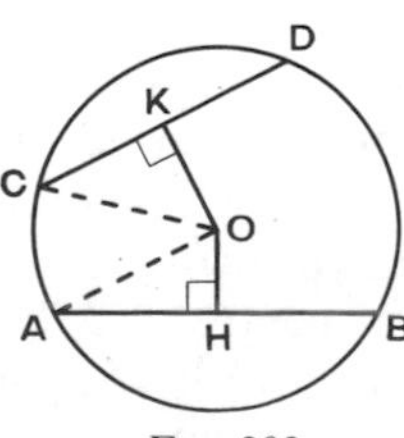

FIG. 202

3 (i) If the chord AB, length l cm, of a circle, radius R cm, is at a distance p cm from the centre, find l in terms of R, p.
(ii) In equal circles, prove that equal chords are at equal distances from the centres.
(iii) State and prove a converse of (ii).

EXERCISE 88

1 A straight line PQRS cuts two concentric circles at P, S and Q, R. Prove that PQ = RS. [Draw the perpendicular from the centre of the circles to PS; no other construction.]

[**2**] If AB and CD are equal chords of a circle, centre O, prove that ∠AOB = ∠COD.

[**3**] O is the centre of the circle PQRS. If PQ = RS, prove that PR = QS. [Join O to P, Q, R, S.]

4 Given a point K inside a given circle, show how to construct the chord AKB such that AK = KB.

5 Two circles, centres A, B, intersect at X, Y. Prove that AB bisects XY at right angles.

[**6**] Two circles intersect at X, Y; a line PQRS parallel to XY cuts one circle at P, S and the other circle at Q, R. Prove that PQ = RS. [Use No. 5.]

7 Two circles, centres A, B, intersect at C, D; PCQ is a line parallel to AB cutting the circles at P, Q. Prove that PQ = 2AB.

8 APB, CPD are intersecting chords of a circle, centre O. If OP bisects ∠APC, prove that AB = CD. [Draw OH, OK perpendicular to AB, CD.]

[**9**] AB and CD are two equal chords of a circle; M, N are their mid-points. Prove that MN makes equal angles with AB and CD.

10 PQ is a variable chord of given length of a given circle. Find the locus of the mid-point of PQ.

[**11**] Two chords of a circle bisect each other at K. Prove that K is the centre of the circle. [If possible, *join* K to the centre.]

12 ABCD is a given quadrilateral. Show how to construct two concentric circles, one of which passes through A, B and the other through C, D. What can you say about ABCD (i) if there is more than one solution, (ii) if two *distinct* circles cannot be drawn, (iii) if there is no solution?

[**13**] The diagonals of the rhombus ABCD meet at O; circles are drawn through A, O, B; B, O, C; C, O, D; D, O, A. Prove that their four centres are the vertices of a rectangle.

14 In Fig. 203, if PXQ is parallel to RYS, prove PQ = RS. [Draw perpendiculars from the centres to PQ, RS.]

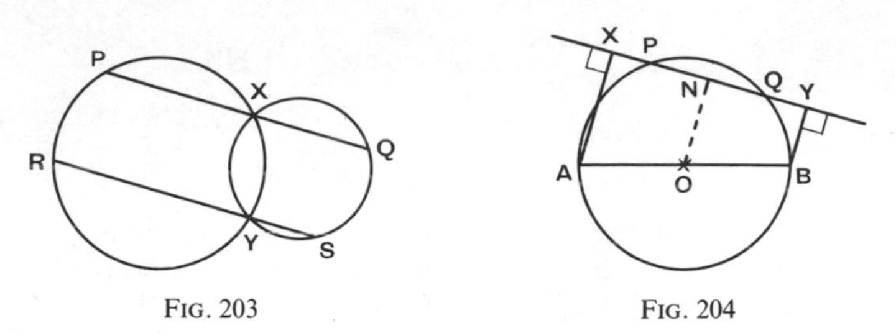

FIG. 203 FIG. 204

15 In Fig. 204, AX, BY are the perpendiculars to a chord PQ from the ends A, B of a diameter. Prove XP = QY. [Use the intercept theorem.] Is the same result true if P, Q lie on opposite sides of AB?

THEOREM

(1) A point which is equidistant from two intersecting straight lines lies on one of the lines which bisect the angles between the given lines.

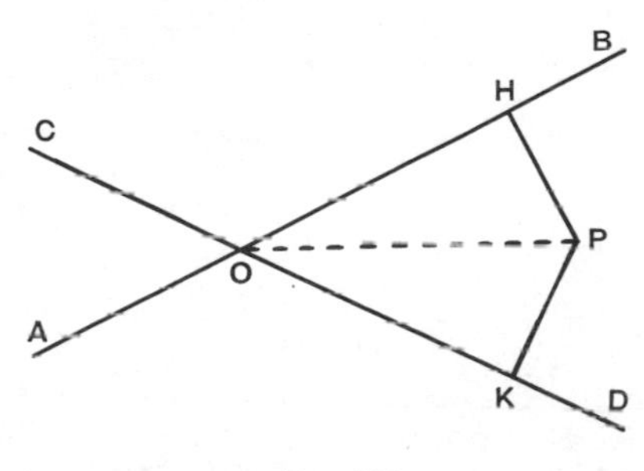

FIG. 205

Given two straight lines AOB, COD and a point P such that the perpendiculars PH, PK from P to AB, CD are equal.

To prove that P lies on the bisector of one of the angles between AOB, COD.

Construction Join OP.

Proof Suppose that P lies within the angle BOD. In $\triangle$s PHO, PKO,

$\angle$PHO = $\angle$PKO *rt. $\angle$s, given,*
PH = PK *given,*
PO = PO.

$\therefore \triangle$s PHO, PKO are congruent RHS.

$\therefore \angle POH = \angle POK$.

$\therefore$ P lies on the bisector of $\angle$BOD.

In the same way, it may be proved that, if P lies within any one of the angles BOC, COA, AOD, it lies on the bisector of that angle.

Abbreviation for reference: $\angle$bisector locus.

THEOREM

(2) A point which lies on the bisector of a given angle is equidistant from the arms of that angle.

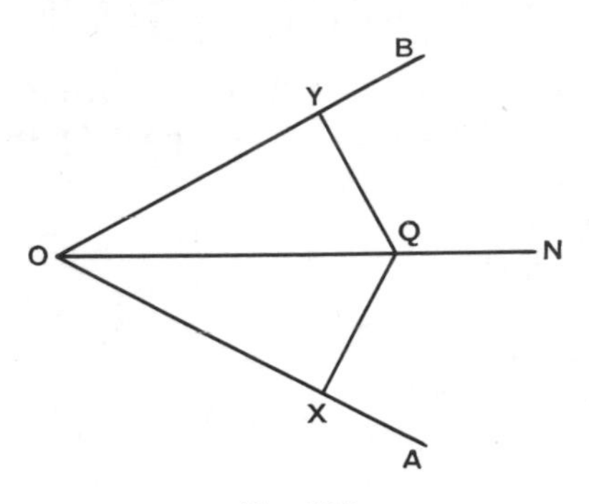

FIG. 206

Given an angle AOB, a point Q on the bisector ON of $\angle$AOB, and the perpendiculars QX, QY from Q to OA, OB.

To prove that QX = QY.

Proof In $\triangle$s QXO, QYO,

$\angle QOX = \angle QOY$ *given*,

$\angle QXO = \angle QYO$ *rt.* $\angle$*s, given*,

QO = QO

$\therefore \triangle$s QXO, QYO are congruent AAS.

$\therefore$ QX = QY.

The enunciations (1) and (2) together form the locus-theorem:

The locus of points which are equidistant from two given intersecting straight lines is the pair of lines which bisect the angles between the given lines.

Abbreviation for reference: ∠bisector locus.

EXERCISE 89 (Class Discussion)

1 The internal bisectors of ∠B, ∠C of △ABC meet at I; IP, IQ, IR are the perpendiculars from I to BC, CA, AB. Prove (i) IP = IQ = IR; (ii) **IA bisects ∠BAC.**

Prove (i) by using the locus-theorem (2), p. 180.

Prove (ii) by using the locus-theorem (1), p. 179.

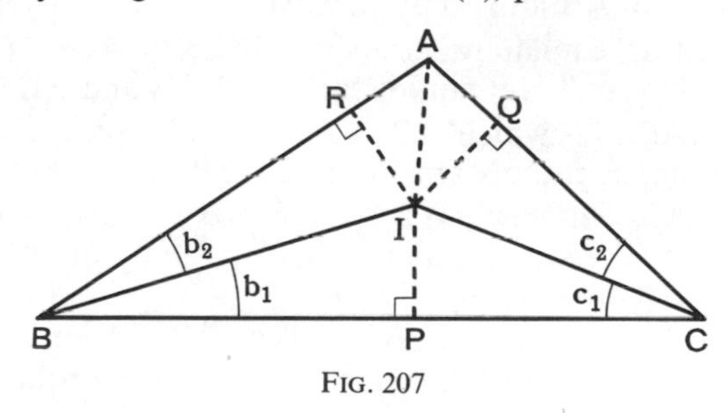

FIG. 207

2 The internal bisectors of the three angles of a triangle are concurrent.

External Bisector of an Angle If one arm AB of an angle ABC is produced to D, the bisector of the angle DBC is called the **external bisector** of the angle ABC.

3 The external bisectors of ∠B, ∠C of △ABC meet at I_1. I_1P_1, I_1Q_1, I_1R_1 are the perpendiculars from I_1, to BC, AC produced, AB produced. Prove

(i) $I_1P_1 = I_1Q_1 = I_1R_1$;

(ii) I_1A bisects ∠BAC.

Prove (i) by using the locus-theorem (2), p. 180.

Prove (ii) by using the locus-theorem (1), p. 179.

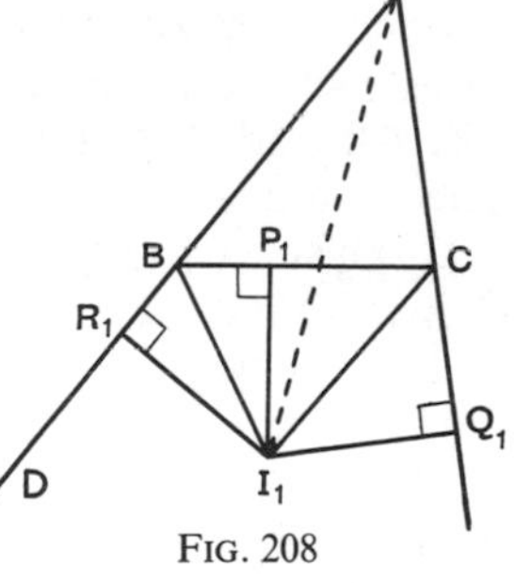

FIG. 208

4 The external bisectors of two angles of a triangle and the internal bisector of the third angle are concurrent.

Intersection of Loci

If the position of a point is given by two distinct conditions, it may be possible to construct the two corresponding loci and so fix the position, or the several possible positions, of the point by taking the intersection of these lines or curves.

EXERCISE 90

[*Use ruler and compasses only for the following constructions*]

1 Construct an equilateral triangle ABC, side 6 cm. What are the loci of points distant 4 cm from A, distant 2 cm from BC? Construct a point P 4 cm from A and 2 cm from BC.

[**2**] Construct an equilateral triangle ABC, side 4 cm. Construct points on AC and AC produced which are 2 cm from BC.

3 Construct an equilateral triangle ABC, side 4 cm. Construct a point P, 2 cm from AB and 3 cm from AC.

[**4**] Construct an equilateral triangle ABC, side 4 cm. Construct a point P, 2·5 cm from A and equidistant from BA and BC.

5 Draw △ABC in which AB = 4 cm, BC = 5 cm, CA = 4·5 cm. Construct a point P equidistant from CA, CB and 3 cm from B.

6 Draw △ABC in which AC = 3 cm, CB = 4 cm, ∠ACB = 90°. Construct the positions of all points which are 1 cm from AB, produced if necessary, and 4 cm from C.

7 Construct △ABC in which AB = 3 cm, BC = 5 cm, ∠ABC = 45° and construct a point P equidistant from AB, AC and equidistant from A, C.

[**8**] Draw △ABC in which BC = 7 cm, CA = 6 cm, AB = 5 cm. Construct a point P equidistant from BA, BC and 2 cm from AC.

[**9**] Draw △ABC in which AB = 6 cm, BC = 2 cm, CA = 7 cm. Construct points P, Q, on AB, AB produced, which are equidistant from CA, CB, produced if necessary. Measure PQ.

10 Draw △ABC in which BC = 6 cm, CA = 5 cm, AB = 7 cm. Construct the points P, Q which are equidistant both from the lines AB, AC and from the points B, C. Construct also the circle ABC; does it pass through P and Q?

11 Draw a triangle ABC in which BC = 5 cm, CA = 4 cm, AB = 6 cm. Construct the four points which are equidistant from the three sides of the triangle.

[**12**] Construct the triangle ABC in which BC = 8 cm, ∠ABC = 45°, ∠ACB = 60°. Construct a point P equidistant from AB, AC and such that BP, produced if necessary, is perpendicular to AC.

13 Construct the triangle ABC in which AB = 6 cm, ∠CAB = 30°, ∠CBA = 45°, and construct a point P which lies on the circle ABC and is equidistant from CA, CB.

14 Draw a line BC of length 5 cm. Construct the triangle ABC such that the median AA′ of △ABC is 4 cm and the perpendicular AD from A to BC is 3 cm. State shortly your method.

15 Draw two straight lines AOB, COD cutting at an angle of 60°. Sketch the form of the locus of a variable point P whose distance is always 2 cm from one of the lines AB, CD. Thicken the part of the locus for which P is not less than 2 cm from either of the lines AB, CD.

16 Draw a rectangle ABCD in which AB = 5 cm, BC = 4 cm. Sketch the form of the locus of a variable point P whose distance is always 1 cm from some side of the rectangle, produced if necessary. Thicken the part of the locus for which P is not less than 1 cm from any side of the rectangle.

[**17**] Draw a triangle ABC in which AB = 4 cm, BC = 5 cm, CA = 6 cm. Sketch the form of the locus of a variable point P whose distance is always 1 cm from some side of the triangle, produced if necessary, and never less than 1 cm from any side.

18 Draw a line AB such that AB = 5 cm. Draw the locus of a variable point P which is always 4 cm from one of the points A, B and never less than 4 cm from either point.

[**19**] Draw an equilateral triangle ABC, side 2 cm. Draw the locus of a variable point P which is always 2 cm from one of the points A, B, C and never less than 2 cm from any of these points.

20 Draw two perpendicular lines AB, AC. P is a variable point which is 2 cm nearer to AB than to AC. Draw the locus of P. [Draw a line DE parallel to AB such that P is equidistant from AC and DE.]

21 Draw on squared paper, with 2 cm as unit on each axis, the locus of a point whose coordinates (x, y) are subject to the laws:

(i) $y = 2x$; (ii) $y = \frac{1}{2}x+1$; (iii) $y = \frac{1}{5}x^2$; (iv) $y = \frac{1}{5}(x-1)^2$.

CHAPTER 19
INDICES AND LOGARITHMS

Positive Integral Indices

If n is any positive integer, a^n is an abbreviation for

$$a \times a \times a \times \ldots, n \text{ factors.}$$

Example 1 Simplify (i) $a^3 \times a^4$; (ii) $a^6 \div a^2$; (iii) $(a^2)^3$.

(i) $a^3 \times a^4 = (a \times a \times a) \times (a \times a \times a \times a) = \boldsymbol{a^7}$.

(ii) $a^6 \div a^2 = \dfrac{a \times a \times a \times a \times a \times a}{a \times a} = a \times a \times a \times a = \boldsymbol{a^4}$.

(iii) $(a^2)^3 = a^2 \times a^2 \times a^2 = (a \times a) \times (a \times a) \times (a \times a) = \boldsymbol{a^6}$.

EXERCISE 91 (Oral)

Obtain from first principles the simplest forms of:

1 $a^2 \times a^3$ **2** $b^5 \div b^2$ **3** $(c^3)^2$ **4** $d^9 \div d^3$

Write down the simplest forms of:

5 $a^4 \times a^2$ **6** $b^6 \div b^3$ **7** $c \times c^4$ **8** $d^5 \div d$
9 $(e^4)^2$ **10** $f^2 \times f^6$ **11** $g^5 \times g^3$ **12** $(h^5)^3$
13 $k^8 \div k^4$ **14** $(l^2)^4$ **15** $m^8 \div m^2$ **16** $n^5 \times n^5$

What general formulas include the following special cases?

17 $a^3 \times a^4 = a^7$; $a^2 \times a^6 = a^8$; $a^4 \times a^7 = a^{11}$.
18 $a^6 \div a^2 = a^4$; $a^8 \div a^3 = a^5$; $a^{10} \div a^4 = a^6$.
19 $(a^2)^3 = a^6$; $(a^5)^2 = a^{10}$; $(a^3)^4 = a^{12}$.
20 What is (i) the square of x^6; (ii) the cube of y^4?

The examples in Exercise 91 illustrate the important facts:

If m and n are positive integers,

$$\boldsymbol{a^m \times a^n = a^{m+n}};$$
$$\boldsymbol{a^m \div a^n = a^{m-n}}, \qquad \text{if } m > n;$$
$$\boldsymbol{(a^m)^n = a^{mn}}.$$

The symbol $\boldsymbol{a^m}$ has not yet been defined if m is not a positive integer, *e.g.* if $m = \frac{2}{3}$ or -4 or 0. We now proceed to show what meaning must be given to it if the law

$$\boldsymbol{a^m \times a^n = a^{m+n}}$$

remains true for *all* values of m and n.

Example 2 Find the meaning of $16^{\frac{1}{2}}$.

$$16^{\frac{1}{2}} \times 16^{\frac{1}{2}} = 16^{\frac{1}{2}+\frac{1}{2}} = 16^1 = 16;$$

$\therefore\ 16^{\frac{1}{2}}$ is a *square root* of 16;

$$\therefore\ 16^{\frac{1}{2}} = \mathbf{4} \text{ (or } -4).$$

Note Throughout this chapter we shall consider only the *positive* value of a square root.

Example 3 Find the meaning of $a^{\frac{1}{3}}$.

$$a^{\frac{1}{3}} \times a^{\frac{1}{3}} \times a^{\frac{1}{3}} = a^{\frac{1}{3}+\frac{1}{3}+\frac{1}{3}} = a^1 = a;$$

$\therefore\ a^{\frac{1}{3}}$ is the *cube root* of a;

$$\therefore\ \boldsymbol{a^{\frac{1}{3}} = \sqrt[3]{a}}.$$

Example 4 Find the meaning of $a^{\frac{3}{4}}$.

$$a^{\frac{3}{4}} \times a^{\frac{3}{4}} \times a^{\frac{3}{4}} \times a^{\frac{3}{4}} = a^{\frac{3}{4}+\frac{3}{4}+\frac{3}{4}+\frac{3}{4}} = a^3;$$

$\therefore\ \boldsymbol{a^{\frac{3}{4}}}$ is a *fourth root* of a^3;

$$\therefore\ a^{\frac{3}{4}} = \sqrt[4]{(\boldsymbol{a^3})}.$$

Or, by the method of Example 3, $a^{\frac{1}{4}} = \boldsymbol{\sqrt[4]{a}}$;

but $a^{\frac{3}{4}} = a^{\frac{1}{4}} \times a^{\frac{1}{4}} \times a^{\frac{1}{4}} = (a^{\frac{1}{4}})^3$; $\quad \therefore\ a^{\frac{3}{4}} = (\sqrt[4]{a})^3$.

Hence $\boldsymbol{a^{\frac{3}{4}} = \sqrt[4]{(a^3)} = (\sqrt[4]{a})^3}$.

Thus $16^{\frac{3}{4}} = \sqrt[4]{(16^3)} = \sqrt[4]{(2^4 \times 2^4 \times 2^4)} = \sqrt[4]{(2^{12})} = \mathbf{2^3}$.

But it is quicker to say,

$$16^{\frac{1}{4}} = \sqrt[4]{16} = 2,$$

$$\therefore\ 16^{\frac{3}{4}} = \mathbf{2^3}.$$

Example 5 Find the meaning of 10^{-3}.

$$10^5 \times 10^{-3} = 10^{5-3} = 10^2;$$

divide each side by 10^5,

$$\therefore 10^{-3} = \frac{10^2}{10^5} = \frac{\mathbf{1}}{\mathbf{10^3}}.$$

Example 6 Find the meaning of a^{-5}, given $a \neq 0$.

$$a^7 \times a^{-5} = a^{7-5} = a^2;$$

divide each side by a^7,

$$\therefore\ \boldsymbol{a^{-5}} = \frac{a^2}{a^7};$$

$$\therefore\ \boldsymbol{a^{-5} = \frac{1}{a^5}}.$$

Example 7 Find the meaning of a^0, given $a \neq 0$.

$$a^3 \times a^0 = a^{3+0} = a^3;$$

divide each side by a^3,

$$\therefore a^0 = \frac{a^3}{a^3} = 1.$$

$$\therefore \boldsymbol{a^0 = 1}.$$

By using the methods of Examples 2–6, it may be proved that,

(i) If p, q are any positive integers,

$$\boldsymbol{a^{1/q} = \sqrt[q]{a}; \qquad a^{p/q} = \sqrt[q]{a^p} = (\sqrt[q]{a})^p}.$$

(ii) If n is any number, and $a \neq 0$,

$$\boldsymbol{a^{-n} = \frac{1}{a^n}}.$$

Example 8 Find the value of (i) $(\frac{3}{4})^{-2}$; (ii) $9^{-1\cdot5}$.

(i) $(\frac{3}{4})^{-2} = 1 \div (\frac{3}{4})^2 = 1 \div \frac{9}{16} = \boldsymbol{\frac{16}{9}}$.

(ii) $9^{1\cdot5} = 9^1 \times 9^{0\cdot5} = 9 \times 9^{\frac{1}{2}} = 9 \times \sqrt{9} = 9 \times 3 = 27;$

$$\therefore 9^{-1\cdot5} = 1 \div 9^{1\cdot5} = \boldsymbol{\tfrac{1}{27}}.$$

EXERCISE 92 (Oral)

1 Express as powers of x the square roots of x^8, x^{10}, x^9.

2 Express as powers of a the cube roots of a^{12}, a^4, a.

3 Simplify $b^{\frac{2}{3}} \times b^{\frac{2}{3}} \times b^{\frac{2}{3}}$. What is the meaning of $x^{\frac{2}{3}}$?

4 Simplify $(b^{\frac{4}{3}})^3$. What is the meaning of $x^{\frac{4}{3}}$?

Show how to find the meanings of:

5 $a^{\frac{1}{4}}$ **6** $b^{\frac{2}{5}}$ **7** $c^{\frac{3}{2}}$ **8** $d^{1\frac{2}{3}}$ **9** e^0

Write down the numerical values of:

10 $16^{\frac{1}{2}}$ **11** $27^{\frac{1}{3}}$ **12** $8^{\frac{2}{3}}$ **13** $9^{1\frac{1}{2}}$ **14** $81^{\frac{1}{2}}$

15 $16^{1\frac{1}{4}}$ **16** $32^{\frac{3}{5}}$ **17** $81^{\frac{3}{4}}$ **18** $1000^{\frac{2}{3}}$ **19** $4^{2\frac{1}{2}}$

20 Write down as powers of x, that is in the form x^n, and also simplify in the ordinary way:

(i) $\frac{x^3}{x^5}$; (ii) $\frac{x^2}{x^6}$; (iii) $\frac{x}{x^6}$; (iv) $\frac{x^3}{x^3}$; (v) $\frac{x^5}{x^5}$.

21 Simplify $a^5 \times a^{-3}$; what is the meaning of a^{-3}?

22 Simplify $b^4 \times b^{-1}$; what is the meaning of b^{-1}?

23 Simplify $c^2 \times c^0$; what is the meaning of c^0?

Write down the numerical values of:

24 3^{-2} **25** 4^{-1} **26** 5^0 **27** 2^{-3} **28** $9^{-\frac{1}{2}}$ **29** $16^{-\frac{1}{2}}$
30 $8^{-\frac{1}{3}}$ **31** $27^{-\frac{1}{3}}$ **32** $8^{-\frac{2}{3}}$ **33** $(\frac{1}{3})^{-2}$ **34** $(\frac{1}{2})^{-1}$ **35** $(\frac{1}{4})^0$

Express with root signs:

36 $10^{\frac{1}{2}}$ **37** $10^{\frac{1}{4}}$ **38** $10^{0 \cdot 5}$ **39** $10^{0 \cdot 75}$ **40** $10^{1 \cdot 5}$

41 Find the value of n if $0{\cdot}058 = 5{\cdot}8 \times 10^n$.

42 Express without indices (i) $72{\cdot}1 \times 10^{-3}$, (ii) $0{\cdot}0081 \times 1000^{\frac{2}{3}}$.

Powers of 10 Since $10^{\frac{1}{2}} = \sqrt{10}$, the value of $10^{\frac{1}{2}}$ can be found to as many decimal places as desired by the square-root rule.

Thus $10^{\frac{1}{2}} = \sqrt{10} = 3{\cdot}162\ldots;$

similarly $10^{\frac{1}{4}} = \sqrt{10^{\frac{1}{2}}} = \sqrt{3{\cdot}162\ldots} = 1{\cdot}779\ldots$

and $10^{\frac{1}{8}} = \sqrt{10^{\frac{1}{4}}} = \sqrt{1{\cdot}779\ldots} = 1{\cdot}333\ldots.$

Also $10^{\frac{3}{4}} - \sqrt[4]{10^3} = \sqrt[4]{1000} = \sqrt{31{\cdot}62\ldots} = 5{\cdot}623\ldots,$

and in a similar way we can find approximate values of $10^{\frac{3}{8}}$, $10^{\frac{5}{8}}$, $10^{\frac{7}{8}}$; also $10^0 = 1$.

The results obtained may be tabulated and can be used to draw the graph of 10^x from $x = 0$ to $x = 1$.

x	0	0·125	0·25	0·375	0·5	0·625	0·75	0·875	1
10^x	1	1·33	1·78	2·37	3·16	4·22	5·62	7·50	10

The graph is shown in Fig. 209 on the next page.

Logarithms

If a number N is expressed in the form 10^x, the **index** x is called the common **logarithm** of the number N.

For example, from the graph on p. 188, $2 \simeq 10^{0 \cdot 30}$,

$\therefore$ the logarithm of 2 is approximately 0·30,

or more shortly $\log 2 \simeq 0{\cdot}30$.

The *object* of logarithms is to make common calculations less laborious, and the *method* consists in replacing multiplication by addition, and division by subtraction. This is done by expressing each number in the form 10^x and using the facts,

$$10^x \times 10^y = 10^{x+y} \quad \text{and} \quad 10^x \div 10^y = 10^{x-y}.$$

The graph could be used for this purpose, but it would only give results correct to 2 figures. For practical purposes, tables are used, which are arranged on the same general plan as those for squares, square-roots, etc.

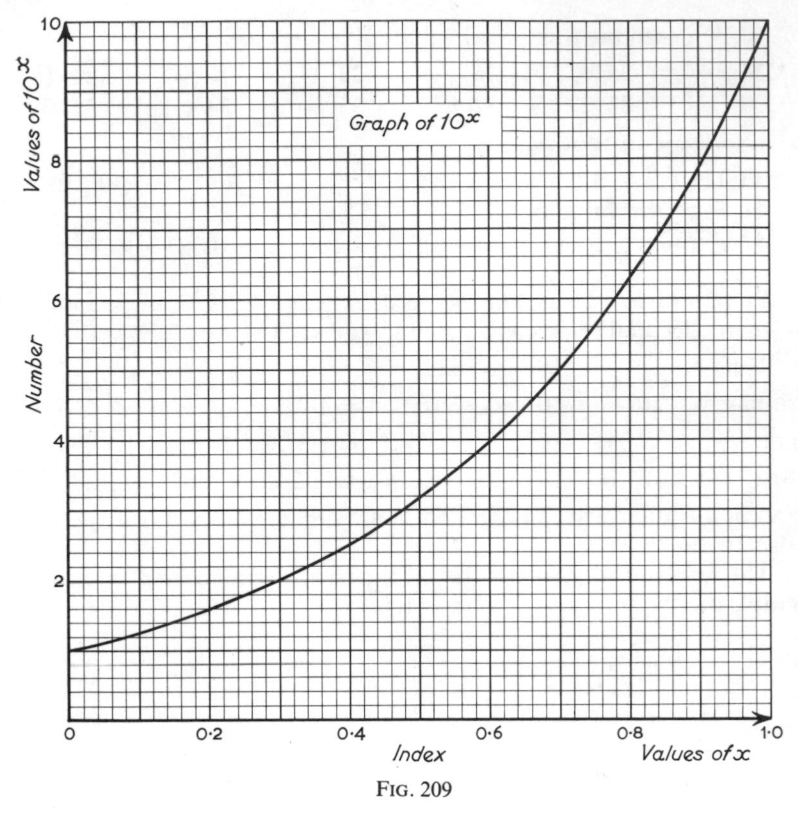

FIG. 209

Oral Examples on the graph of 10^x.

Write down approximate values of:

1 $10^{0\cdot2}$ **2** $10^{0\cdot3}$ **3** $10^{0\cdot7}$ **4** $10^{0\cdot42}$ **5** $10^{0\cdot84}$

Express in the form 10^x:

6 2 **7** 3 **8** 5 **9** 6 **10** 7·2 **11** 3·8

12 Find the values of a, b, c, N in the following:

$$2{\cdot}8 \times 3{\cdot}2 = 10^a \times 10^b = 10^c = N.$$

Numbers between 1 and 10

$$1 = 10^0 \quad \text{and} \quad 10 = 10^1,$$

$$\therefore \log 1 = 0 \quad \text{and} \quad \log 10 = 1;$$

therefore the logarithm of any number between 1 and 10 is between 0 and 1.

Logarithm tables give the logarithms of numbers between 1 and 10 **only**, and to save space *decimal points are omitted.*

The following extract from a four-figure table of logarithms will serve to show how the table is used.

	0	1	2		9	1 2 3	...	7 8 9
29	·4624	4639	4654	...	4757	1 3 4	...	10 12 13
30	·4771	4786	4800	...	4900	1 3 4	...	10 11 13

In this extract in the '29' row under the heading 0, we find ·4624, this means that $2{\cdot}90 = 10^{0 \cdot 4624}$; and in the '29' row under the heading 1, we find 4639, this means that $2{\cdot}91 = 10^{0 \cdot 4639}$; and so on.

Similarly $3{\cdot}00 = 10^{0 \cdot 4771}$, $3{\cdot}01 = 10^{0 \cdot 4786}$, $3{\cdot}02 = 10^{0 \cdot 4800}$, etc.

To find the logarithm of 2·918, we add to the value of log 2·91 the difference in the narrow column headed 8.

2·91	4639
Add diff. for 8	12
2·918	4651

Thus $2{\cdot}918 = 10^{0 \cdot 4651}$.

EXERCISE 93 (Oral)

Use tables to express in the form 10^x:

1 3·2 **2** 6·4 **3** 9·6 **4** 3 **5** 9
6 2·5 **7** 5 **8** 4·1 **9** 8·2 **10** 7·7
11 4·26 **12** 6·95 **13** 3·04 **14** 5·60 **15** 1·11
16 1·07 **17** 1·20 **18** 9·06 **19** 4·44 **20** 6·03
21 2·563 **22** 2·568 **23** 7·451 **24** 7·468 **25** 4·645
26 4·657 **27** 4·678 **28** 4·606 **29** 6·347 **30** 6·358
31 8·876 **32** 9·708 **34** 7·067 **34** 7·004 **35** 5·081
36 5·005 **37** 3·050 **38** 3·005 **39** 1·086 **40** 1·074
41 1·003 **42** 1·048 **43** 1·006 **44** 1·995 **45** 9·987

The same tables may be used by reversing the process to find the value of any number of the form 10^x.

Example 9 Find the value of

(i) $10^{0 \cdot 7076}$; (ii) $10^{0 \cdot 7143}$; (iii) $10^{0 \cdot 7148}$.

(i) Look for the figures 7076 in the four-figure columns; these occur in the '51' row in the column headed 0; therefore the significant figures in the value of $10^{0 \cdot 7076}$ are 510; but the value of $10^{0 \cdot 7076}$ lies between 10^0 and 10^1, that is between 1 and 10;

$\therefore 10^{0 \cdot 7076} =$ **5·100,** to 4 figures.

(ii) The figures 7143 occur in the '51' row in the column headed 8;

$\therefore 10^{0 \cdot 7143} =$ **5·180,** to 4 figures.

(iii) The actual figures 7148 do not occur in the four-figure columns;

$$10^{0\cdot7143} = 5\cdot180 \quad \text{and} \quad 10^{0\cdot7152} = 5\cdot190,$$

$\therefore$ the value of $10^{0\cdot7148}$ lies between 5·180 and 5·190.

The difference between 7143 and 7148 is 5, and in the column of mean differences for this row 5 occurs under the column headed 6;

$$\therefore 10^{0\cdot7148} = \mathbf{5\cdot186}, \text{ to 4 figures.}$$

Note Whenever mean differences are used in 4-figure tables, the fourth figure is not reliable; *this may be indicated by enclosing it in brackets.*

Example 10 Find the value of

$$\text{(i) } 10^{0\cdot0135}; \qquad \text{(ii) } 10^{0\cdot9543}.$$

(i) The figures less than 0135 which are nearest to it and occur in the four-figure columns are 0128; these occur in the '10' row in the column headed 3.

$$10^{0\cdot0128} = 1\cdot030, \text{ to 4 figures.}$$

The difference between 0128 and 0135 is 7; but 7 does not occur in the column of mean differences for this row, we therefore take the nearest difference, namely 8, which is in the column headed 2.

$$\therefore 10^{0\cdot0135} = \mathbf{1\cdot03(2)}, \text{ to 4 figures.}$$

(ii) From the tables, $10^{0\cdot9542} = 9\cdot000$, to 4 figures.

For this row, the difference 1 occurs in two columns, namely, those headed 2 and 3; we cannot therefore tell whether it is better to take the value as 9·002 or as 9·003; the result may be written:

$$10^{0\cdot9543} = \mathbf{9\cdot00(2)}.$$

EXERCISE 94 (Oral)

Use logarithm tables to find the values of:

1 $10^{0\cdot6232}$ **2** $10^{0\cdot7709}$ **3** $10^{0\cdot7782}$ **4** $10^{0\cdot5428}$
5 $10^{0\cdot5514}$ **6** $10^{0\cdot7059}$ **7** $10^{0\cdot9106}$ **8** $10^{0\cdot9713}$
9 $10^{0\cdot7016}$ **10** $10^{0\cdot574}$ **11** $10^{0\cdot781}$ **12** $10^{0\cdot9042}$
13 $10^{0\cdot4019}$ **14** $10^{0\cdot5258}$ **15** $10^{0\cdot7581}$ **16** $10^{0\cdot8642}$

Use logarithm tables to find the numbers whose logarithms are:

17 0·4835 **18** 0·7040 **19** 0·7801 **20** 0·9052 **21** 0·5255
22 0·6603 **23** 0·2753 **24** 0·1992 **25** 0·2011 **26** 0·6105
27 0·7138 **28** 0·7621 **29** 0·8455 **30** 0·9034 **31** 0·9701
32 0·0298 **33** 0·0314 **34** 0·0025 **35** 0·003 **36** 0·9546

Multiplication and Division

Example 11 Find the value of $3{\cdot}18 \times 2{\cdot}17$.

Rough Estimate: $3 \times 2 = 6$.

$$3{\cdot}18 \times 2{\cdot}17 = 10^{0{\cdot}5024} \times 10^{0{\cdot}3365} = 10^{0{\cdot}5024+0{\cdot}3365}$$
$$= 10^{0{\cdot}8389} = \mathbf{6{\cdot}90(1)} \quad \text{or} \quad \mathbf{6{\cdot}90(2)}.$$

0·5024
0·3365
0·8389

Example 12 Find the value of $8{\cdot}37 \div 5{\cdot}09$.

Rough Estimate: $8 \div 5 = 1{\cdot}6$.

$$8{\cdot}37 \div 5{\cdot}09 = 10^{0{\cdot}9227} \div 10^{0{\cdot}7067} = 10^{0{\cdot}9227-0{\cdot}7067}$$
$$= 10^{0{\cdot}2160} = \mathbf{1{\cdot}64(4)} \quad \text{or} \quad \mathbf{1{\cdot}64(5)}.$$

0·9227
0·7067
0·2160

Note To obtain the 4th figure accurately, it is necessary to use 5-figure tables; the error in the 4th figure obtained from 4-figure tables will usually be small. But in the exercises that follow, the reader is asked to give the answer to *four* figures, 'as given by 4-figure tables', in order to make it easier to check the accuracy of his use of the tables.

The fourth figure should be enclosed in brackets to show its approximate character.

EXERCISE 95

Find the value to 4 figures, as given by 4-figure tables, of:

1 2·36 × 3·24 **[2]** 2·73 × 3·18 **3** 4·19 × 1·84
4 3·624 × 2·315 **5** 5·278 × 1·406 **[6]** 2·073 × 4·108
7 2·086 × 4·104 **[8]** 7·263 × 1·173 **9** 3·708 × 2·046
10 8·67 ÷ 3·29 **[11]** 7·49 ÷ 4·08 **12** 9·07 ÷ 2·61
13 7·408 ÷ 2·165 **[14]** 9·032 ÷ 4·705 **15** 8·007 ÷ 6·503
16 9·804 ÷ 5·073 **17** 7·006 ÷ 1·088 **18** 6·704 ÷ 5·908
19 1·78 × 2·04 × 2·56 **20** 7·03 × 6·74 ÷ 8·26

Numbers greater than 10

Number		expressed as power of 10
7·243		$10^{0{\cdot}8599}$
72·43	$= 7{\cdot}243 \times 10^1$	$10^{0{\cdot}8599} \times 10^1 = 10^{1{\cdot}8599}$
724·3	$= 7{\cdot}243 \times 10^2$	$10^{0{\cdot}8599} \times 10^2 = 10^{2{\cdot}8599}$
7243	$= 7{\cdot}243 \times 10^3$	$10^{0{\cdot}8599} \times 10^3 = 10^{3{\cdot}8599}$
72430	$= 7{\cdot}243 \times 10^4$	$10^{0{\cdot}8599} \times 10^4 = 10^{4{\cdot}8599}$

Thus, so long as the order of the digits of a number is unaltered, the decimal portion of the logarithm, called its **mantissa**, *remains the same.*

The integral portion of the logarithm, called its CHARACTERISTIC, is obtained by counting the number of digits between the decimal point and where it would be in the standard form.

For example, for 7|24·3, two digits, the characteristic is 2;
for 7|2430, four digits, the characteristic is 4.

Example 13 Express 72 428 as a power of 10.

$$72\,428 \simeq 72\,430, \text{ to 4 figures,}$$
$$\simeq 10^{4\cdot8599}.$$

Example 14 Find the value of $10^{5\cdot5748}$.

From the tables, $10^{0\cdot5748} = 3\cdot757$.
Move the figures 5 places to the left. ∴ $10^{5\cdot5748}$ = **375 *7*00.**
The 4th figure in 375 700, here printed in italics, may be underlined, when written, to show that it is only approximate.

EXERCISE 96 (Oral)

Read off the characteristics of the logarithms of the numbers:

1 72 **2** 694 **3** 8850 **4** 11·17 **5** 7·916
6 15720 **7** 823000 **8** 61944·72 **9** 12 000 000

State the number of digits to the left of the decimal point in the numbers whose logarithms are:

10 1·8627 **11** 3·0145 **12** 2·6158 **13** 4·2717 **14** 3·162
15 5·7018 **16** 0·9824 **17** 0·0104 **18** 3·85 **19** 4·2

EXERCISE 97 (Oral)

Use tables to write down to 4 figures the logarithms of:

1 26 **2** 380 **3** 62·4 **4** 7480 **5** 41·72
6 3212 **7** 814·7 **8** 10·04 **9** 200·3 **10** 53 700
11 428 300 **12** 2 007 000 **13** 1000 **14** 100 000
15 10^7 **16** 1 **17** 63 748 **18** 371·62
19 800 735 **20** $5\cdot163 \times 10^2$ **21** $7\cdot04 \times 10^3$ **22** $9\cdot65 \times 10^8$

Use tables to write down to 4 figures the numbers whose logarithms are:

23 1·5843 **24** 2·7275 **25** 3·8865 **26** 2·4771 **27** 4·9304
28 5·9562 **29** 7·847 **30** 6·932 **31** 2·5256 **32** 1·6687
33 3·4777 **34** 1·3018 **35** 4·6029 **36** 2·0025 **37** 5·702
38 3·787 **39** 2·0000 **40** 5·0000 **41** 1·1 **42** 4·4

Example 15 Find the value of $816{\cdot}3 \times 37{\cdot}42$.

Rough Estimate: $800 \times 40 = 32000$.

$$816{\cdot}3 \times 37{\cdot}42 = 10^{2{\cdot}9119} \times 10^{1{\cdot}5731} = 10^{4{\cdot}4850} = \mathbf{30550}.$$

$$\begin{array}{r} 2{\cdot}9119 \\ 1{\cdot}5731 \\ \hline 4{\cdot}4850 \end{array}$$

Example 16 Find the value of $816{\cdot}3 \div 37{\cdot}42$.

Rough Estimate: $800 \div 40 = 20$.

$$816{\cdot}3 \div 37{\cdot}42 = 10^{2{\cdot}9119} \div 10^{1{\cdot}5731} = 10^{1{\cdot}3388} = \mathbf{21{\cdot}8(2)}.$$

$$\begin{array}{r} 2{\cdot}9119 \\ 1{\cdot}5731 \\ \hline 1{\cdot}3388 \end{array}$$

Example 17 Find the value of $(37{\cdot}06)^4$.

Rough Estimate: $40^4 = 2\,560\,000$.

$$(37{\cdot}06)^4 = (10^{1{\cdot}5689})^4 = 10^{1{\cdot}5689 \times 4} = 10^{6{\cdot}2756} = \mathbf{1\,886\,000}.$$

$$\begin{array}{r} 1{\cdot}5689 \\ 4 \\ \hline 6{\cdot}2756 \end{array}$$

Example 18 Find the value of $\sqrt[3]{561{\cdot}4}$.

$$\sqrt[3]{561{\cdot}4} = (10^{2{\cdot}7493})^{\frac{1}{3}} = 10^{2{\cdot}7493 \times \frac{1}{3}} = 10^{0{\cdot}9164} = \mathbf{8{\cdot}24(9)}.$$

$$3\overline{)2{\cdot}7493} \quad 0{\cdot}9164$$

Rough Check: $8^3 = 512$; $9^3 = 739$.

We shall now repeat Example 15, to show how the work may be arranged, after the principles have been grasped; for other examples of this method of arrangement, see p. 194.

Example 19 Find the value of $816{\cdot}3 \times 37{\cdot}42$.

Rough Estimate: $800 \times 40 = 32\,000$.

$$\therefore\ 816{\cdot}3 \times 37{\cdot}42 = 3{\cdot}055 \times 10^4 = \mathbf{30550}.$$

Number	Logarithm
816·3	2·9119
37·42	1·5731
Product	4·4850

Logarithmic Notation

If $x = 10^p$, p is the logarithm of x, and we write

$$\log x = p.$$

Examples 15–18 illustrate the following general properties:

$$\mathbf{\log (x \times y) = \log x + \log y;}$$
$$\mathbf{\log (x \div y) = \log x - \log y;}$$
$$\mathbf{\log (x^n) = n \log x.}$$

EXERCISE 98

Find the value to 4 figures, as given by 4-figure tables, of:

1 $53{\cdot}2 \times 67{\cdot}4$ [**2**] $724 \times 15{\cdot}3$ **3** $6070 \times 1{\cdot}08$
[**4**] $84{\cdot}3 \div 9{\cdot}15$ **5** $387 \div 71{\cdot}6$ **6** $4130 \div 8{\cdot}62$
[**7**] $268{\cdot}3 \times 37{\cdot}4$ [**8**] $409{\cdot}3 \times 64{\cdot}9$ **9** $20{\cdot}72 \times 18{\cdot}37$
[**10**] $627{\cdot}4 \div 8{\cdot}36$ **11** $5070 \div 21{\cdot}68$ [**12**] $61320 \div 483{\cdot}4$
13 $36828 \times 4{\cdot}073$ **14** $50 \div 3{\cdot}0643$ [**15**] 4527×3406
16 $1000 \div 28{\cdot}447$ **17** $10{\cdot}73 \times 2004$ [**18**] $73600 \div 909$
19 $(21{\cdot}76)^2$ [**20**] $(7{\cdot}294)^3$ **21** $(10{\cdot}71)^4$
22 (i) $\sqrt{7{\cdot}34}$; (ii) $\sqrt{73{\cdot}4}$ **23** (i) $\sqrt[3]{618}$; (ii) $\sqrt[3]{61{\cdot}8}$; (iii) $\sqrt[3]{6{\cdot}18}$
[**24**] $\sqrt{453{\cdot}6}$ [**25**] $\sqrt[3]{507{\cdot}3}$ [**26**] $\sqrt[4]{8672}$ [**27**] $\sqrt[3]{10{\cdot}43}$
28 $\sqrt[4]{200{\cdot}7}$ [**29**] $\sqrt[5]{40{\cdot}78}$ **30** $\sqrt[3]{100}$ [**31**] $\sqrt[4]{1000}$

If the expression is complicated, the working may be arranged in the more concise form illustrated by the following examples:

Example 20 Find the value of $\dfrac{65130 \times 37{\cdot}14}{793{\cdot}2 \times 4{\cdot}186}$.

Rough Estimate: $\frac{60\,000 \times 40}{800 \times 4} = \frac{600\,000}{800} \simeq 700.$

Number	Logarithm
65130	4·8138
37·14	1·5699
Numerator	6·3837
Denominator	3·5212
Expression	2·8625

Number	Logarithm
793·2	2·8994
4·186	0·6218
Denominator	3·5212

$\therefore$ expression $= 7{\cdot}286 \times 10^2 =$ **728·(6).**

Example 21 Find the value of $\dfrac{(35{\cdot}41)^3 \times 4{\cdot}783}{34{\cdot}65 \times \sqrt{41{\cdot}8}}$.

Rough Estimate: $\frac{30^3 \times 5}{30 \times 6} \simeq 30^2 = 900.$

Number	Logarithm	
$(35{\cdot}41)^3$	$1{\cdot}5491 \times 3$	4·6473
4·783		0·6797
Numerator		5·3270
34·65		1.5397
$\sqrt{41{\cdot}8}$	$1{\cdot}6212 \times \frac{1}{2}$	0·8106
Denominator		2·3503
Expression		2·9767

5·3270
2·3503
2·9767

$\therefore$ expression $= 9{\cdot}478 \times 10^2$
$=$ **947·(8).**

EXERCISE 99

Find the value to 4 figures, as given by 4-figure tables, of:

1 $4{\cdot}62 \times 20{\cdot}7 \times 35{\cdot}8$ [**2**] $83{\cdot}6 \times 1{\cdot}07 \times 155$

[3] $1732 \times 8{\cdot}04 \times 23{\cdot}69$ **4** $80{\cdot}72 \times 11{\cdot}01 \times 3{\cdot}141$

[5] $\dfrac{31{\cdot}73 \times 6{\cdot}482}{7{\cdot}918}$ [6] $\dfrac{520{\cdot}4 \times 8{\cdot}065}{97{\cdot}53}$ **7** $\dfrac{7324}{16{\cdot}42 \times 39{\cdot}81}$

[8] $\dfrac{800{\cdot}6}{7{\cdot}152 \times 9{\cdot}038}$ **9** $\dfrac{4{\cdot}045 \times 1760}{49{\cdot}13 \times 50{\cdot}5}$ [10] $\dfrac{76{\cdot}03 \times 908}{101{\cdot}2 \times 6{\cdot}317}$

11 $\dfrac{(19{\cdot}42)^2}{8{\cdot}73 \times 6{\cdot}04}$ [12] $\dfrac{7073 \times 8500}{(362{\cdot}5)^2}$ **13** $\dfrac{(161{\cdot}3)^3 \times 4{\cdot}03}{(285{\cdot}1)^2 \times 30}$

14 $\dfrac{(27{\cdot}3)^2 \times (4{\cdot}17)^3}{(11{\cdot}01)^3}$ [15] $\left(\dfrac{17{\cdot}46 \times 10{\cdot}8}{13{\cdot}47}\right)^2$ [16] $\left(\dfrac{7134}{85{\cdot}07 \times 26{\cdot}92}\right)^3$

17 $\sqrt{\dfrac{793{\cdot}4}{86{\cdot}13}}$ [18] $\sqrt[3]{\dfrac{8{\cdot}427}{5{\cdot}913}}$ **19** $\dfrac{\sqrt{874{\cdot}2}}{\sqrt[3]{75{\cdot}13}}$ [20] $\dfrac{\sqrt{635{\cdot}4}}{11{\cdot}6 \times 1{\cdot}07}$

[21] $1\frac{1}{3} \times 3{\cdot}142 \times (4{\cdot}102)^3$ **22** $2 \times \sqrt[3]{100} \div \sqrt[4]{1000}$

23 $\dfrac{\sqrt[3]{100}}{2{\cdot}085}$ [24] $\dfrac{39{\cdot}6 \times 4005}{\sqrt[3]{83470}}$ **25** $\sqrt[3]{\dfrac{71{\cdot}36 \times 80}{46{\cdot}07}}$

[26] $\dfrac{2\frac{1}{3} \times (3{\cdot}054)^3}{(4{\cdot}103)^2}$ **27** $\sqrt[3]{\{(2{\cdot}608)^3 - 10\}}$ **28** $\dfrac{(7{\cdot}165)^2 + 40}{(7{\cdot}165)^2 - 40}$

EXERCISE 100

[*Take* $\log \pi = 0{\cdot}4971$. *Give answers to 3 significant figures*]

1 Find the volume of a rectangular block 17·3 cm long, 12·8 cm wide, 10·4 cm high.

2 The weight of an iron cylinder, height h cm and radius r cm, is w grammes, where $w = 7{\cdot}6\pi r^2 h$. Evaluate w when $r = 3{\cdot}94$, $h = 8{\cdot}65$.

[3] If 1 cm^3 of lead weighs 11·4 g, find the weight of a cuboid of lead, 7·45 cm by 5·08 cm by 3·16 cm.

4 If a cuboid of iron 6·13 cm by 4·95 cm by 2·05 cm weighs 462 g, find the weight of 1 cm^3 of iron.

5 The tax on matches yielded £7 949 280 in one year and £9 080 457 in the next year. Find the increase per cent.

6 National Health prescriptions fell from 211 232 428 in three years to 203 071 240. Find the decrease per cent.

7 The area of a circle, radius r cm, is A cm^2, where $A = \pi r^2$, and $r = \sqrt{\left(\dfrac{A}{\pi}\right)}$. (i) Find A if $r = 4{\cdot}163$; (ii) find r if $A = 10$.

[8] The time of a complete oscillation of a pendulum l metres long is t seconds, where $t = 2\pi\sqrt{\left(\dfrac{l}{g}\right)}$ and $l = \dfrac{gt^2}{4\pi^2}$ and $g = 9{\cdot}81$. (i) Find l if $t = 2{\cdot}35$; (ii) find t if $l = 12$.

9 The weight of a brass cylinder, height h cm, base-diameter d cm, is w grammes, where $w = 2{\cdot}09d^2h$. (i) Find w if $d = 2{\cdot}516$, $h = 127{\cdot}4$; (ii) find h if $w = 856{\cdot}4$, $d = 3{\cdot}85$.

10 Find k from the formula, $w = \pi kl(D+d)(D-d)$ if $w = 4528$, $l = 10{\cdot}7$, $D = 6{\cdot}31$, $d = 4{\cdot}92$.

11 Find in metres the diameter of a circular enclosure of area $10\,890\,\text{m}^2$. [Radius of circle of area $A\,\text{m}^2 = \sqrt{\left(\dfrac{A}{\pi}\right)}$ metres.]

***12** If $\frac{4}{3}\pi r^3 = 100$, find the value of (i) r, (ii) $4\pi r^2$.

***13** The volume of a cube is $7\,\text{cm}^3$; find the area of its surface.

14 Evaluate $\sqrt{\{s(s-a)(s-b)(s-c)\}}$ if $a = 3{\cdot}15$, $b = 4{\cdot}27$, $c = 5{\cdot}14$, and $s = \frac{1}{2}(a+b+c)$.

Positive Numbers less than 1

Example 22 Find the logarithm of 0·0648.

$$0{\cdot}0648 = 6{\cdot}48 \div 10^2 = 10^{0\cdot8116} \div 10^2$$
$$= 10^{0\cdot8116-2} = 10^{-2+0\cdot8116}.$$

Similarly, $\quad 0{\cdot}648 = 6{\cdot}48 \div 10 = 10^{0\cdot8116} \div 10^1 = 10^{-1+0\cdot8116}$

and $\quad 0{\cdot}000648 = 6{\cdot}48 \div 10^4 = 10^{0\cdot8116} \div 10^4 = 10^{-4+0\cdot8116}$.

Logarithms of numbers between 0 and 1 are **negative,** *but are always written so that the decimal portion is positive*; for example, the logarithm of 0·0648 is taken as $-2+{\cdot}8116$, instead of $-1{\cdot}1884$, and for brevity is written $\bar{2}{\cdot}8116$, the 'minus' being placed *above* the 2 to show that it refers only to the 2 and not to ·8116. But at first it is best to write out such logarithms in full when making any calculations and to speak of 'minus 2 plus point 8116', although later the shorter phrase 'bar 2 point 8116' will be used.

Thus for any numbers between 0 *and* 1, *the* **characteristic** *is* **negative** *and is obtained by* **counting the number of digits between the decimal point and where it would be in the standard form.**

For example, for 0·6|48, one digit, the characteristic is -1;
for 0·0006|48, four digits, the characteristic is -4.

Notice that this is the same rule as that given on p. 192 for numbers greater than 10.

There is *no* logarithm for a *negative* number.

Example 23 Find the number whose logarithm is

(i) $\bar{3}{\cdot}6749$; (ii) $\bar{1}{\cdot}3027$.

(i) $10^{0\cdot6749} = 4{\cdot}730$.

Move the figures 3 places to the right, $\therefore\ 10^{\bar{3}\cdot6749} =$ **0·004730.**

(ii) $10^{0\cdot3027} = 2{\cdot}008$.

Move the figures 1 place to the right, $\therefore\ 10^{\bar{1}\cdot3027} =$ **0·2008.**

EXERCISE 101 (Oral)

[*It is suggested that the characteristics of the logarithms for Nos.* 1–16 *should be read off, before Exercise* 101 *is taken.*]

Find the logarithms of:

1 0·342 **2** 0·0483 **3** 0·0076 **4** 0·902
5 2·04 **6** 0·0075 **7** 10·01 **8** 0·0003
9 0·0025 **10** 0·101 **11** 0·001 **12** 0·0704
13 0·04503 **14** 0·4013 **15** 0·007138 **16** 0·01101
17 $\frac{4}{5}$ **18** $\frac{3}{400}$ **19** $\frac{1}{300}$ **20** $\frac{7}{250}$
21 $4{\cdot}03\times10^{-5}$ **22** $6{\cdot}104\times10^{-6}$ **23** 428×10^{-10}

Find the numbers whose logarithms are:

24 $-1+0{\cdot}5922$ **25** $-2+0{\cdot}8645$ **26** $-4+0{\cdot}9085$
27 $-1+0{\cdot}4771$ **28** $-2+0{\cdot}6085$ **29** $-3+0{\cdot}7067$
30 $-1+0{\cdot}8476$ **31** $-5+0{\cdot}9703$ **32** $-1+0{\cdot}5229$
33 $-4+0{\cdot}2939$ **34** $-2+0{\cdot}1650$ **35** $-3+0{\cdot}1315$
36 $\bar{2}{\cdot}3016$ **37** $\bar{1}{\cdot}0425$ **38** $\bar{3}{\cdot}0055$ **39** $\bar{2}{\cdot}0028$
40 $\bar{1}{\cdot}0000$ **41** $\bar{2}{\cdot}0000$ **42** $\bar{4}{\cdot}44$ **43** $\bar{3}{\cdot}044$

The working in the following illustrative examples may be abbreviated as soon as the processes are understood.

Example 24 Express with the decimal portion positive:

(i) $2{\cdot}89+\bar{5}{\cdot}47$; (ii) $\bar{3}{\cdot}76+\bar{1}{\cdot}58$.

(i)
$$\begin{array}{r} 2+0{\cdot}89 \\ -5+0{\cdot}47 \\ \hline -3+1{\cdot}36 \\ \hline \mathbf{\bar{2}{\cdot}36} \end{array}$$

(ii)
$$\begin{array}{r} -3+0{\cdot}76 \\ -1+0{\cdot}58 \\ \hline -4+1{\cdot}34 \\ \hline \mathbf{\bar{3}{\cdot}34} \end{array}$$

Example 25 Express with the decimal portion positive:

(i) $\bar{4}{\cdot}21-1{\cdot}73$; (ii) $\bar{2}{\cdot}63-\bar{5}{\cdot}87$.

(i)
$$\begin{array}{r} -4+0{\cdot}21 \\ 1+0{\cdot}73 \\ \hline \end{array}$$
becomes

Equal Additions
$$\begin{array}{r} -4+1{\cdot}21 \\ 2+0{\cdot}73 \\ \hline -6+0{\cdot}48 \\ \hline \mathbf{\bar{6}{\cdot}48} \end{array}$$
or

Decomposition
$$\begin{array}{r} -5+1{\cdot}21 \\ 1+0{\cdot}73 \\ \hline -6+0{\cdot}48 \\ \hline \mathbf{\bar{6}{\cdot}48} \end{array}$$

In the subtraction, the integer -6 is obtained by changing the sign of the lower line and adding: $-2-4=-6$; $-1-5=-6$.

(ii) $\begin{array}{r} -2+0{\cdot}63 \\ -5+0{\cdot}87 \\ \hline \end{array}$ becomes

	Equal Additions		Decomposition
	$-2+1{\cdot}63$	or	$-3+1{\cdot}63$
	$(-5+1)+0{\cdot}87$		$-5+0{\cdot}87$
	$2+0{\cdot}76$		$2+0{\cdot}76$
	2·76		**2·76**

In the 'equal additions' method, $(-5+1) = -4$, change the sign of the lower line and add: $+4-2 = 2$.

Example 26 Express with the decimal portion positive:

(i) $\bar{2}{\cdot}76\times4$; (ii) $\bar{6}{\cdot}42\div3$; (iii) $\bar{4}{\cdot}52\div3$.

(i) $(-2+0{\cdot}76)\times4 = -8+3{\cdot}04 = -5+0{\cdot}04 = \mathbf{\bar{5}{\cdot}04}$.

(ii) $(-6+0{\cdot}42)\div3 = -2+0{\cdot}14 = \mathbf{\bar{2}{\cdot}14}$.

(iii) $(-4+0{\cdot}52)\div3$; since -4 is not exactly divisible by 3, it must be expressed as $-6+2$:

$$(-4+0{\cdot}52)\div3 = (-6+2{\cdot}52)\div3 = -2+0{\cdot}84 = \mathbf{\bar{2}{\cdot}84}.$$

EXERCISE 102

Express with the decimal portion positive:

1 $\bar{2}{\cdot}3+\bar{1}{\cdot}4$	**2** $\bar{5}{\cdot}2+3{\cdot}6$	**3** $\bar{3}{\cdot}6+\bar{2}{\cdot}8$	**4** $\bar{4}{\cdot}7+2{\cdot}8$
5 $\bar{2}{\cdot}7+4{\cdot}8$	**6** $\bar{3}{\cdot}9+2{\cdot}7$	**7** $\bar{3}{\cdot}8-\bar{1}{\cdot}5$	**8** $\bar{4}{\cdot}7-1{\cdot}2$
9 $4{\cdot}6-\bar{3}{\cdot}2$	**10** $2{\cdot}7-5{\cdot}4$	**11** $\bar{4}{\cdot}5-\bar{2}{\cdot}8$	**12** $\bar{3}{\cdot}4-1{\cdot}6$
13 $3{\cdot}2-\bar{2}{\cdot}5$	**14** $4{\cdot}2-6{\cdot}7$	**15** $1-\bar{1}{\cdot}32$	**16** $0-2{\cdot}8$
17 $0-\bar{3}{\cdot}4$	**18** $\bar{3}{\cdot}1-\bar{7}{\cdot}8$	**19** $\bar{2}{\cdot}3\times2.$	**20** $\bar{2}{\cdot}4\times3$
21 $\bar{3}{\cdot}9\times4$	**22** $\bar{6}{\cdot}8\div2$	**23** $\bar{9}{\cdot}6\div3$	**24** $\bar{3}{\cdot}4\div2$
25 $\bar{5}{\cdot}5\div3$	**26** $\bar{1}{\cdot}4\div3$	**27** $\bar{7}{\cdot}7\div3$	**28** $\bar{1}{\cdot}6\div4$
29 $\bar{6}{\cdot}8\div4$	**30** $\bar{2}{\cdot}4\div5$	**31** $\bar{3}\div5$	**32** $\bar{4}{\cdot}5\div7$
33 $\bar{2}{\cdot}2\div9$	**34** $3{\cdot}7\times(-2)$	**35** $\bar{2}{\cdot}6\times(-3)$	**36** $3{\cdot}6\div(-2)$
37 $\bar{3}{\cdot}6\div(-2)$	**38** $\bar{5}{\cdot}8\div(-3)$	**39** $3\div(-5)$	**40** $\bar{2}{\cdot}2\div(-4)$

Example 27 Evaluate $0{\cdot}000645\times82{\cdot}3$.

Rough Estimate: $0{\cdot}0006\times80 = 0{\cdot}048$.

Number	Logarithm
0·000645	$\bar{4}{\cdot}8096$
82·3	$1{\cdot}9154$
Expression	$\bar{2}{\cdot}7250$

$$\begin{array}{r} -4+0{\cdot}8096 \\ 1+0{\cdot}9154 \\ \hline -3+1{\cdot}7250 \\ \hline -2+0{\cdot}7250 \\ \hline \end{array}$$

$\therefore$ expression = **0·0530(9).**

Example 28 Evaluate $429{\cdot}3 \div 0{\cdot}00736$.

Rough Estimate: $\dfrac{400}{0{\cdot}007} = \dfrac{400\,000}{7} \simeq 50\,000.$

Number	Logarithm
429·3	2·6328
0·00736	$\bar{3}{\cdot}8669$
Expression	4·7659

$$\begin{array}{r} 2+0{\cdot}6328 \\ -3+0{\cdot}8669 \\ \hline 4+0{\cdot}7659 \end{array}$$

$\therefore$ expression = **58 330.**

Example 29 Evaluate $(0{\cdot}08644)^3$.

Rough Estimate: $(0{\cdot}09)^3 = 0{\cdot}000729.$

Number	Logarithm	
$(0{\cdot}08644)^3$	$\bar{2}{\cdot}9367 \times 3$	$\bar{4}{\cdot}8101$

$$\begin{array}{r} -2+0{\cdot}9367 \\ 3 \\ \hline -6+2{\cdot}8101 \\ \hline \bar{4}{\cdot}8101 \end{array}$$

$\therefore$ expression = **0·000645(8).**

Example 30 Evaluate $\sqrt[4]{0{\cdot}5173}$.

Number	Logarithm	
$\sqrt[4]{0{\cdot}5173}$	$\bar{1}{\cdot}7138 \div 4$	$\bar{1}{\cdot}9284$

$$\begin{array}{r} -1+0{\cdot}7138 \\ 4\,\underline{|\,-4+3{\cdot}7138} \\ -1+0{\cdot}9284 \\ \hline \bar{1}{\cdot}9284 \end{array}$$

$\therefore$ expression = **0·848(0).**

Rough Check: $(0{\cdot}8)^4 = (0{\cdot}64)^2 \simeq (0{\cdot}6)^2 = 0{\cdot}36$
$(0{\cdot}9)^4 = (0{\cdot}81)^2 \simeq (0{\cdot}8)^2 = 0{\cdot}64.$

EXERCISE 103

Find the value to 4 figures, as given by 4-figure tables, of:

1 $0{\cdot}243 \times 3{\cdot}12$ **2** $0{\cdot}743 \times 0{\cdot}814$ **3** $0{\cdot}729 \times 56{\cdot}2$
[**4**] $0{\cdot}0863 \times 0{\cdot}924$ **5** $0{\cdot}0072 \times 0{\cdot}091$ [**6**] $0{\cdot}000389 \times 47{\cdot}4$
7 $614{\cdot}3 \times 0{\cdot}03617$ [**8**] $0{\cdot}8172 \times 0{\cdot}5049$ **9** $0{\cdot}06194 \times 1{\cdot}032$
10 $0{\cdot}542 \div 3{\cdot}67$ [**11**] $0{\cdot}618 \div 8{\cdot}04$ **12** $0{\cdot}847 \div 0{\cdot}623$
13 $6{\cdot}71 \div 0{\cdot}426$ [**14**] $83{\cdot}7 \div 0{\cdot}00217$ **15** $729 \div 0{\cdot}0564$
16 $42{\cdot}71 \div 6832$ [**17**] $0{\cdot}3007 \div 0{\cdot}0942$ **18** $1 \div 0{\cdot}03716$
19 $(0{\cdot}284)^2$ [**20**] $(0{\cdot}0846)^2$ **21** $(0{\cdot}372)^3$ **22** $(0{\cdot}006)^4$
[**23**] $\sqrt{(0{\cdot}0863)}$ **24** $\sqrt{(0{\cdot}5167)}$ [**25**] $\sqrt[3]{(0{\cdot}615)}$ **26** $\sqrt[3]{(0{\cdot}0485)}$
[**27**] $\sqrt[4]{(0{\cdot}732)}$ **28** $\sqrt[5]{(0{\cdot}00704)}$ [**29**] $(0{\cdot}473)^{\frac{2}{3}}$ **30** $(0{\cdot}0066)^{\frac{3}{5}}$
[**31**] $0{\cdot}387 \times 0{\cdot}473 \times 7{\cdot}08$ **32** $0{\cdot}01031 \times 0{\cdot}2074 \times 1{\cdot}702$

33 $\dfrac{6.314 \times 0.7285}{37.62}$ [**34**] $\dfrac{0.5178}{4.917 \times 14.97}$ **35** $\dfrac{17.35 \times 3.628}{0.4271 \times 0.00726}$

[**36**] $\dfrac{4007 \times 0.0275}{0.9193 \times 1.063}$ **37** $\dfrac{(0.7356)^2}{3.142 \times 0.0867}$ [**38**] $\dfrac{\sqrt{(0.07324)}}{16.81 \times 53.46}$

39 $\sqrt{\dfrac{6.372 \times 15.08}{8307}}$ [**40**] $\sqrt[3]{\dfrac{713.6 \times 84}{0.00525}}$ **41** $\dfrac{\sqrt[3]{100}}{(0.06182)^2}$

EXERCISE 104

[*Take log* $\pi = 0.4971$. *Give answers to 3 figures*]

1 A rectangular tank, 1·72 m long, 0·94 m wide, contains 3000 litres of water. Find the depth of water.

[**2**] The rainfall one day was 1·17 cm. Find the weight of rain which fell on a courtyard 7·05 m long, 5·2 m wide.

3 Wireless licences increased in 15 years from 9 034 187 to 13 455 061. Find the increase per cent.

[**4**] From the formula, $V = \pi r^2 h$, for the volume of a cylinder, find V if $r = 2.35$, $h = 0.73$.

5 From the pendulum formula, $t = 2\pi\sqrt{(l \div g)}$, find t if $l = 2.86$, $g = 9.81$.

6 From the formula $l = \dfrac{gt^2}{4\pi^2}$, find l if $g = 9.81$, $t = 1.87$.

7 A copper pipe, internal diameter 2 cm, thickness 3 mm, weighs $\pi(1.3+1)(1.3-1) \times 0.89$ kg per metre length. Evaluate this expression.

[**8**] Find the weight of a piece of glass tubing 12·5 cm long with external and internal diameters 4·82 mm, 3·54 mm, if it is equal to $\frac{1}{4}\pi(0.482+0.354)(0.482-0.354) \times 12.5 \times 2.36$ grammes.

9 The time for a carrier to pass through a tube, diameter d metres, length l metres, under pressure P kg per cm² is $0.000235\,(\sqrt{l^3}) \div (d\sqrt{P})$ seconds. Find the time for a tube, diameter 0·45 m, 1 km long, under pressure 0·846 kg per cm².

10 Express in the form $N \times 10^p$ where N is a number between 1 and 10 and p is integral the value of $x^2 y \div z$ when $x = 3.18 \times 10^{-4}$, $y = 4.27 \times 10^{-3}$, $z = 9.62 \times 10^{-1}$.

11 From the formula, $s = \frac{1}{2}gt^2$, find t if $s = 7.24$, $g = 9.81$.

[**12**] Find r if $0.72 \times \pi r^2 \times 11.3 = 6.55$.

13 Find d if $\frac{1}{6}\pi d^3 = 0.85$.

14 The diameter d mm of a tin wire which is just fused by a current c amperes is given by $d^3 = \left(\dfrac{c}{12.8}\right)^2$. Find d if $c = 35$.

Logarithms of Trigonometric Functions Numerical work necessary in applications of Trigonometry can generally be shortened by using logarithms, and, to save time, tables of logarithms of each trigonometrical ratio have been constructed. For example, from the tables we find sin 25° = 0·4226 and log 0·4226 = $\bar{1}$·6259; therefore log sin 25° = $\bar{1}$·6259; but if we use the table of log-sines we obtain this result by a *single* reading.

Tables of log-sines, log-cosines, log-tangents are arranged and used in the same way as those for natural sines, cosines and tangents.

It is not always easy to fix the characteristic by common sense; it is therefore always printed, but only at the beginning of each line; for example,

$$\log \tan 36° = \bar{1}{\cdot}8613; \quad \log \tan 36°\,18' = \bar{1}{\cdot}8660.$$

A change in the characteristic is indicated by a difference of type, *e.g.*

$$\log \tan 84° = 0{\cdot}9784; \quad \log \tan 84°\,18' = 1{\cdot}\mathit{0008}.$$

In using the cosine-table, figures in the difference-columns must be **subtracted;** the same is true for the table of log-cosines.

Example 31 Find log cos 66° 34′.

log cos 66° 30′	$\bar{1}$·6007
Subtract diff. for 4′	12
log cos 66° 34′	$\bar{1}$·5995

∴ log cos 66° 34′ = **$\bar{1}$·5995.**

Usually the difference-columns give *average* differences calculated over intervals of 1°; but if these differences are changing rapidly, more accurate results are obtained by taking the average difference for 1′ calculated over 12′ intervals:

Example 32 Find (i) log tan 85° 8′, (ii) log tan 85° 50′.

(i)

log tan 85° 6′	1·0669
Add diff. for 2′	30
log tan 85° 8′	**1·0699**

Diff. for 1′, interval 1′ to 11′ is 15,
∴ diff. for 2′ is 15 × 2 = 30.

(ii)

log tan 85° 48′	1·1341
Add diff. for 2′	36
log tan 85° 50′	**1·1377**

Diff. for 1′, interval 49′ to 59′ is 18,
∴ diff. for 2′ is 18 × 2 = 36.

Example 33 Find $x°$ if (i) log sin $x°$ = $\bar{1}$·7461,
(ii) log cos $x°$ = $\bar{1}$·5874.

(i)

$\bar{1}$·7453	log sin 33° 48′
8	diff. for 4′ *add*
$\bar{1}$·7461	log sin 33° 52′

∴ $x°$ = **33° 52′.**

(ii)

$\bar{1}$·5865	log cos 67° 18′
9	diff. for 3′ *subtract*
$\bar{1}$·5874	log cos 67° 15′

$\therefore x^\circ =$ **67° 15′.**

EXERCISE 105 (Oral)

Write down, by using tables, the logarithms of:

1 sin 23° **2** sin 23° 18′ **3** sin 23° 20′ **4** sin 23° 23′
5 sin 64° 42′ **6** sin 64° 45′ **7** sin 77° 18′ **8** sin 77° 21′
9 cos 40° 24′ **10** cos 40° 30′ **11** cos 40° 26′ **12** cos 52° 45′
13 cos 68° 8′ **14** cos 28° 13′ **15** cos 72° 52′ **16** cos 79° 5′
17 cos 84° 12′ **18** cos 84° 18′ **19** cos 85° 8′ **20** cos 85° 50′
21 cos 86° 15′ **22** cos 86° 51′ **23** cos 84° 23′ **24** cos 84° 53′
25 tan 31° 27′ **26** tan 56° 38′ **27** tan 72° 35′ **28** tan 84° 36′
29 tan 5° 48′ **30** tan 5° 50′ **31** tan 6° 10′ **32** tan 6° 52′
33 tan 87° 2′ **34** tan 87° 50′ **35** tan 87° 11′ **36** tan 87° 53′

Find the angles whose log-sines are:

37 $\bar{1}$·7854 **38** $\bar{1}$·7859 **39** $\bar{1}$·9335 **40** $\bar{1}$·9343
41 $\bar{1}$·9767 **42** $\bar{1}$·9768 **43** $\bar{1}$·0403 **44** $\bar{1}$·0436

Find the angles whose log-cosines are:

45 $\bar{1}$·4350 **46** $\bar{1}$·4359 **47** $\bar{1}$·7922 **48** $\bar{1}$·7929
49 $\bar{1}$·7214 **50** $\bar{1}$·8801 **51** $\bar{1}$·9310 **52** $\bar{1}$·9509
53 $\bar{1}$·6424 **54** $\bar{1}$·8252 **55** $\bar{1}$·0046 **56** $\bar{1}$·9970
57 $\bar{1}$·0756 **58** $\bar{1}$·0447 **59** $\bar{2}$·7690 **60** $\bar{2}$·6500

Find the angles whose log-tangents are:

61 $\bar{1}$·7907 **62** 0·0425 **63** 0·9932 **64** 1·0085
65 $\bar{1}$·7831 **66** 0·1572 **67** 0·5801 **68** 0·7105
69 0·9881 **70** 1·0898 **71** 0·0150 **72** 0·0732

Example 34 In △ABC, ∠A = 33°, ∠C = 90°, AC = 3·5 cm; find AB.

Let AB = x cm,
then $x \cos 33^\circ = 3{\cdot}5$,

$$\therefore x = \frac{3{\cdot}5}{\cos 33^\circ} = 4{\cdot}17(4),$$

Number	Log
3·5	0·5441
cos 33°	$\bar{1}$·9236
x	0·6205

∴ AB = **4·17(4) cm.**

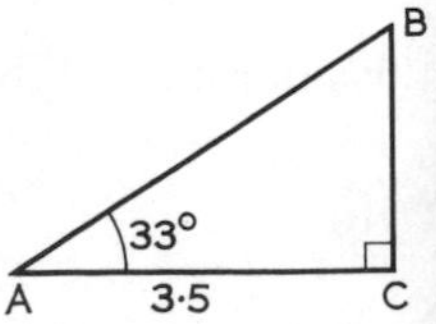

FIG. 210

Example 35 A level road runs 384 m in the direction N. 56° E. from P to Q and then 518 m from Q to a point R north of P and continues to T, where QT = 650 m. Find

(i) the direction of QR,

(ii) how far T is north of P.

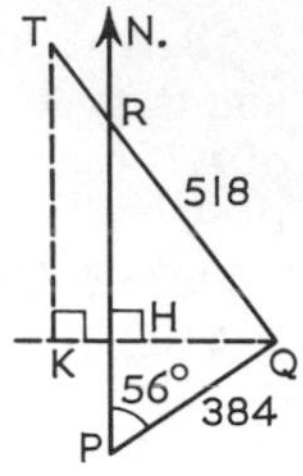

FIG. 211

Draw QH perpendicular to PR, draw TK perpendicular to QH.

(i) QH = 384 sin 56° m;

$$\therefore \sin HRQ = \frac{384 \sin 56°}{518}.$$

384	2·5843
sin 56°	$\bar{1}$·9186
Numerator	2·5029
518	2·7143
sin HRQ	$\bar{1}$·7886

$\therefore \angle HRQ = 37° 55'$,

$\therefore$ the direction of QR is **N. 37° 55′ W.**

(ii) PH = 384 cos 56° m

= 214·7 m;

KT = 650 cos 37° 55′ m

= 512·8 m.

$\therefore$ the distance T is north of P

= (214·7 + 512·8) m

= 727·(5) m.

384	2·5843
cos 56°	$\bar{1}$·7476
PH	2·3319
650	2·8129
cos 37° 55′	$\bar{1}$·8970
KT	2·7099

EXERCISE 106

1 The shadow on level ground of a vertical pole 4·14 m high is 5·68 m long. Find the angle of elevation of the sun.

2 A ladder 3·54 m long leans against a vertical wall with its foot on level ground 1·36 m from the wall. Find the angle which the ladder makes with the ground.

3 ABC is a triangular plot of land; AB = AC = 148 m, ∠BAC = 83° 40′. Find the length of BC.

[**4**] PQ is a side of a regular polygon with 7 sides inscribed in a circle. If PQ = 6·5 cm, find the radius of the circle.

5 Two points P and Q are 2·32 cm apart on a map, scale 1 cm to the km. P is on the 80 m contour line and Q is a spot height 201 m. Find the angle of elevation of Q from P.

6 Use the data and results of Example 35 to find (i) how far T is west of P, (ii) the bearing of T from P.

[**7**] Two men start from O; one walks 5 km to A in the direction N. 67° E.; the other walks 4 km to B in the direction S. 13° E. Find (i) how far A is north of B and east of B, (ii) the bearing of A from B.

8 An aeroplane starts from A and flies 30 km in direction N. 22° W. and then 20 km in direction S. 75° W. How far is it north of A and west of A? Find in what direction and how far it must fly to return direct to A.

9 AB, DC are the parallel sides of a trapezium ABCD, see Fig. 212; $\angle ABC = \angle CAD = 90°$, $\angle BAC = 35°$, AB = 3 cm. Find (i) BC, (ii) AC, (iii) CD, (iv) $\angle BDC$.

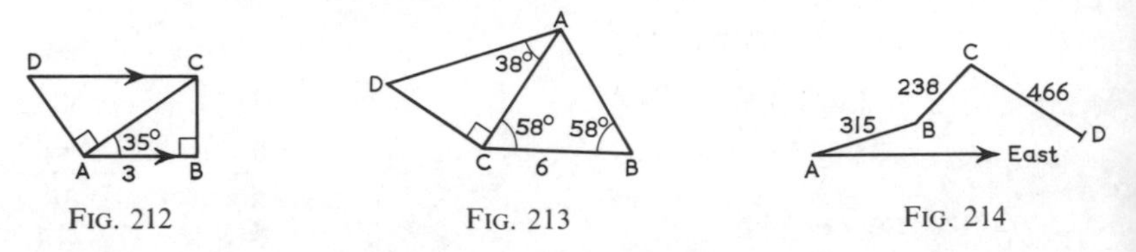

FIG. 212 FIG. 213 FIG. 214

10 In Fig. 213, BC = 6 cm, $\angle ABC = \angle ACB = 58°$, $\angle ACD = 90°$, $\angle CAD = 38°$. Find the lengths of (i) AC, (ii) CD, (iii) AD, (iv) $\angle BDC$. [Draw BN perpendicular to DC produced.]

11 Fig. 214 represents a hill-track ABCD running east; AB, BC slope upwards and CD downwards at angles 16° 30′, 22° 45′, 19° 10′ with the horizontal; AB = 315 m, BC = 238 m, CD = 466 m. Find (i) how far D is east of A, (ii) the height of D above A, (iii) the angle of slope of a tunnel direct from A to D.

12 The end A of a vertical flag-staff AB rests on level ground; BC is a wire fastening B to a point C on the ground; AC = 3·06 m, $\angle ABC = 14°$. Another wire is fixed joining C to a point of the flag-staff three-quarters of the way up AB. Find the length of this wire and the angle it makes with the vertical.

CHAPTER 20
CIRCLES AND CYLINDERS

Circumference of a Circle The relation between the lengths of the circumference and diameter of a circle should be illustrated experimentally.

EXERCISE 107 (Class Discussion)

1 A fine thread is wrapped 5 times round a circular cylinder, like a reel of cotton, of diameter 4 cm. When unwrapped, the thread measures 62·9 cm. Find the value of the ratio of the length of the circumference to the diameter.

Complete the following:

5 times the circumference = 62·9 cm, $\therefore$ circumference = ... cm.

$$\therefore \frac{\text{circumference}}{\text{diameter}} = \frac{\ldots\text{ cm}}{4\text{ cm}} = \ldots$$

Note A better approximation would be obtained by wrapping the thread 10 times or 20 times round the cylinder.

2 Take a cylinder and a long piece of thread. Use the process of No. 1 to find the ratio, circumference: diameter, for this cylinder.

3 Measurements in cm of various cylindrical objects are as follows:

	Reel	Tumbler	Bottle	Jar	Tin	Disc
Diameter	3	6	7	8	9	10
Circumference	9·4	18·8(5)	22·0	25·1	28·3	31·4

(i) Represent the results by a graph.
Scales: 1 cm represents 1 cm diameter; 1 cm represents 2 cm circumference

(ii) Is the graph a *straight line through the origin*? If so, what does this mean?

(iii) Plot the results obtained in No. 2 in your figure.

(iv) Add to the table a row which gives the value of the ratio, circumference: diameter, to 2 places of decimals for the cylinders.

Any two circles are of the same shape, the larger is an enlargement of the smaller; this suggests that the ratio of the circumference to the diameter of any circle is a constant. The discussion in Exercise 107 shows that a rough approximation of this constant is 3·14. The value can in fact be calculated to as many places of decimals as desired, for example to 9 places of decimals it is 3·141592653 . . . ; it is always denoted by the Greek letter π. We therefore write

$$\frac{\text{circumference}}{\text{diameter}} = \pi.$$

$\therefore$ **the circumference of a circle, diameter d units, is πd units.**

If the radius of a circle is r units, its diameter is $2r$ units,

$\therefore$ **the circumference of a circle, radius r units, is $2\pi r$ units.**

A convenient approximation for π is $\mathbf{\frac{22}{7}}$ or $3\frac{1}{7}$, since $3\frac{1}{7} = 3{\cdot}1428\ldots$; and a closer approximation is **3·1416.**

It is useful to remember that $\log \pi = \mathbf{0{\cdot}4971}$, to 4 figures.

Approximate results should be expressed as **decimals.**

Example 1 Find the circumference of a circle, radius 4 cm.

$$\text{Circumference} = \text{radius} \times 2\pi = 4 \times 2\pi \text{ cm}$$
$$\simeq 8 \times 3{\cdot}142 \text{ cm} = \mathbf{25{\cdot}1\ cm}, \text{ to 3 figures.}$$

Example 2 Find the radius of a circle, circumference 10 cm.

If the radius is r cm, the circumference is $2\pi r$ cm;

$$\therefore 2\pi r = 10; \quad \therefore r = \frac{10}{2\pi} = \frac{5}{\pi} = 1{\cdot}592;$$

$$\therefore \text{radius} = \mathbf{1{\cdot}59\ cm}, \text{ to 3 figures.}$$

0·6990
0·4971
0·2019

Example 3 The length of the minute-hand of a clock is 10·25 cm. Find the distance which the tip of the minute-hand moves in 22 minutes.

In 60 minutes, the tip moves $2\pi \times 10{\cdot}25$ cm.

$\therefore$ in 22 minutes, it moves $\frac{22}{60} \times 2\pi \times 10{\cdot}25$ cm.

$$= \frac{44 \times \pi \times 10{\cdot}25}{60} \text{ cm}$$

$$= \mathbf{23{\cdot}6\ cm}, \text{ to 3 figures.}$$

1·6435
0·4971
1·0107
3·1513
1·7782
1·3731

Do not substitute for π its approximate numerical value before it is necessary to do so.

Example 4 The diameter of the wheel of a car is 76 cm. Find the number of revolutions made by the wheel per minute when the car is travelling at 64 km an hour.

In 60 min the car travels 64 km = 64 000 m.

$\therefore$ in 1 min the car travels $\frac{64\,000}{60}$ m.

The circumference of the wheel, diameter 0·76 m, is $\pi \times 0{\cdot}76$ m.

$\therefore$ the car travels $0{\cdot}76\pi$ metres for each revolution of the wheel;

$\therefore$ the number of revolutions made in 1 min

$$= \frac{64\,000}{60 \times 0{\cdot}76\pi}$$

$$= 446{\cdot}8;$$

$\therefore$ the wheel makes **447 revs. per min.**

4·8062	1·7782
2·1561	$\bar{1}$·8808
	0·4971
2·6501	2·1561

EXERCISE 108

[*Give answers to* 3 *figures. Take* $\pi = 3{\cdot}142$, *log* $\pi = 0{\cdot}4971$, *unless otherwise stated.*]

Find the lengths of the circumferences of the circles, Nos. 1–6:

1 Radius 5 cm **2** Diameter 2·4 cm **3** Diameter 165 m

[**4**] Radius 3 cm [**5**] Diameter 100 m [**6**] Diameter 345 m

Find the radii of the circles, Nos. 7–10:

7 Circumference 15·4 m. [Take $\pi = 3\frac{1}{7}$.]

8 Circumference 250 m. [**9**] Circumference 7·4 cm.

[**10**] Circumference 1760 m.

11 A 500 m track is a circle. Find the radius in m. [Take $\pi = 3\frac{1}{7}$.]

[**12**] The minute-hand of a clock is 8 cm long. What distance does the tip of the hand move in 1 h? in 20 min?

13 The diameter of a semicircular protractor is 7 cm; find its perimeter. [Take $\pi = 3\frac{1}{7}$.]

[**14**] AB is a diameter of a circular pond of radius 100 m. How much farther does a man go who walks from A to B round the edge than a man who rows straight across the pond?

15 A bicycle wheel is 70 cm in diameter. How many revolutions does it make per km? [Take $\pi = 3\frac{1}{7}$.]

16 A boy finds the value of π experimentally by wrapping a piece of thread 5 times round a cylinder of diameter 5 cm. He finds that the length of the thread when unwrapped is 78·65 cm. What value for π should he obtain?

[**17**] A bicycle wheel, diameter 70 cm, is making 25 revolutions in 10 seconds. At what speed in km per hour is the bicycle travelling?

[**18**] The radii of the inner and outer edges of a circular running-track are 70 m and 75 m. What is the difference between the lengths of the two edges? Would it be the same if the two radii were 100 m and 105 m?

19 An arc AB of a circle, of radius 10 cm, subtends an angle 144° at the centre O of the circle. Find the length of the arc AB. [Arc = $\frac{144}{360}$ of circumference.]

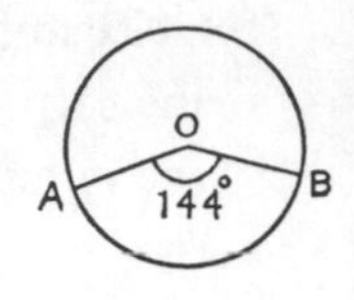

FIG. 215

20 An arc AB of a circle subtends an angle 144° at the centre O of the circle. If the length of the arc AB is 4·4 cm, find the radius of the circle.

21 A dog-cart is being driven at 16 km an hour. If each wheel is 84 cm in diameter, find the number of revolutions made by each wheel per minute. [Take $\pi = 3\frac{1}{7}$.]

[**22**] A wheel of diameter 7 cm is rotating at 3000 revolutions per minute. Find the speed of a point on the rim in cm per second.

***23** A piece of wire is in the form of an arc of a circle of radius 10 cm, subtending an angle 150° at the centre of the circle. It is bent into the form of a complete circle; find the radius of this circle.

***24** A piece of wire 1 m long is bent into the form of a semicircular arc and its diameter. Find the radius.

***25** A bucket is raised from the bottom of a well 16·5 m deep by a rope wound on an axle of diameter 45 cm. How many turns of the axle are required to bring the bucket up to the top? [Neglect the thickness of the rope.]

***26** The statement that the 'gear' of a bicycle is 80 means that each revolution of the pedals makes the bicycle move a distance equal to the circumference of a circle of diameter 80 cm. How many revolutions of the pedals are made per km for this bicycle?

Area of a Circle An approximate value of the area of a circle can be found by drawing the circle on squared paper and counting the number of small squares it contains.

EXERCISE 109 (Class Discussion)

1 Find approximately the area of a circle, radius 3 cm, and the ratio of the area of the circle to the area of the square on a radius.

Draw on paper ruled in centimetres and millimetres a circle, radius 3 cm, and a square whose side is a radius. Count the number of small squares inside the circle. The quickest way is to mark out slabs of rectangles inside the circle and to count individually only the small squares in the margins. Disregard a small square if less than half of it is inside the circle and count as whole squares those for which more than half is inside the circle. Your answer should be about 2828 small squares. Complete the following:

There are ... small squares in 1 cm^2,

$\therefore$ the area of 2828 small squares is ... cm^2.

$\therefore$ the area of a circle, radius 3 cm, $\simeq$ ... cm^2.

The square whose side is a radius contains 900 small squares;

$$\therefore \frac{\text{area of circle}}{\text{area of square on radius}} \simeq \frac{2828}{900} \simeq \ldots, \text{ to 3 figures.}$$

Suppose now you look at your figure through a magnifying glass so that each length appears doubled. You will still see 2828 small squares inside the circle and 900 small squares inside the square whose side is a radius of the circle, although the area of each small square appears 4 times as large as before. This shows that

$$\frac{\text{area of circle, radius 6 cm}}{\text{area of square, side 6 cm}} \simeq \frac{2828}{900}, \textit{ as before.}$$

The same argument may be used for any magnification. This result should be verified by using the method of No. 1 for other circles.

2 Repeat No. 1 for a circle of radius 2 cm.

3 Draw a *quadrant* of a circle, radius 10 cm. Count the number of small squares inside the quadrant and then find the area of the complete circle and the ratio of the area of the circle to the area of the square on a radius.

The calculations in Nos. 1–3 show that an approximation for the ratio $\frac{\text{area of circle}}{\text{area of square on radius}}$ is 3·14, and this suggests that the actual value is the number denoted by π. Thus $\frac{\text{area of circle}}{\text{area of square on radius}} = \pi$ or area of circle = (square on radius) $\times \pi$.

$\therefore$ **the area of a circle, radius r units, is πr^2 units of area.**

The re-appearance of the number π in the expression for the area of a circle may be illustrated experimentally as follows.

4 Draw a large circle on *stiff* paper and then draw radii all round the circle so that each radius makes an angle 15° with the next radius. The quickest method is as follows:

(i) Draw a diameter AB and place a semicircular protractor with its centre at the centre of the circle and with graduations 0°, 180° on AB; mark by dots the graduations 15°, 30°, 45°, . . . , 165°, see Fig. 216.

(ii) Draw the diameters of the circle through these dots.

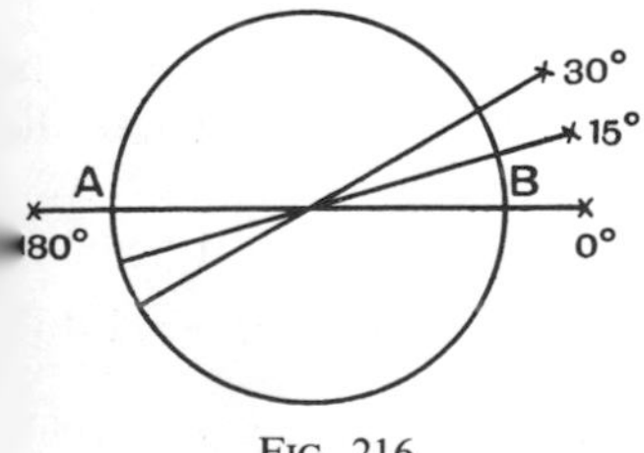

FIG. 216

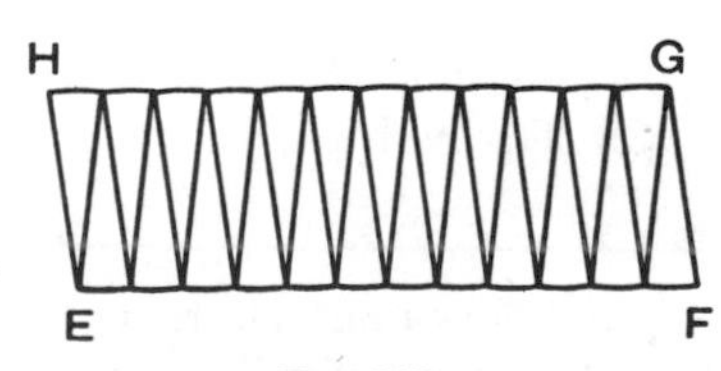

FIG. 217

Cut out the circle and cut along AB; then cut up each semicircle into 12 sectors by cutting along the radii. Arrange the sectors so that alternate sectors point in reverse directions as shown in Fig. 217; then they form a figure which is roughly a rectangle EFGH; the length of the base EF is approximately half the circumference of the circle and the length of the height EH is approximately the radius of the circle. The more sectors there are, the closer is this approximation.

Since area rect. EFGH = EF × EH, this experiment suggests that area of circle = length of half-circumference × length of radius;

$\therefore$ **area of circle, radius r cm,** $= \frac{1}{2}$ of $2\pi r \times r$ cm^2 $= \boldsymbol{\pi r^2}$ **cm^2**.

5 Show that the area of a circle, diameter d cm, is $\frac{1}{4}\pi d^2$ cm^2.

The formulas, $\boldsymbol{C = 2\pi r}$ and $\boldsymbol{A = \pi r^2}$,
for the circumference, C units of length, and the area, A units of area, of a circle, radius r units, illustrate the idea of variation.

If r is multiplied by $2, 3, 4, \ldots, k, \ldots,$
then C is multiplied by $2, 3, 4, \ldots, k, \ldots,$
and A is multiplied by $2^2, 3^2, 4^2, \ldots, k^2, \ldots.$

We say that C *varies directly as* r and that A *varies directly as the square of* r, that is, r^2. The graph which shows the relation between A and r is shaped as in Fig. 181, p. 151. If however values of A are plotted against values of r^2, the graph is a *straight line through the origin.*

Example 5 Find the area of a circle whose diameter is 9 cm. [Take $\log \pi = 0{\cdot}4971$.]

Since the diameter = 9 cm, the radius = 4·5 cm,

$\therefore$ area of circle $= \pi(4{\cdot}5)^2$ cm^2

$=$ **63·6 cm^2.**

0·4971
0·6532
0·6532
1·8035

Example 6 Find the radius of a circle whose area is 132 cm^2, (i) taking $\pi = 3\frac{1}{7}$, (i) taking $\log \pi = 0{\cdot}4971$.

Let the radius be r cm; then

$$\pi r^2 = 132, \quad \therefore r^2 = \frac{132}{\pi}, \quad \therefore r = \sqrt{\frac{132}{\pi}}.$$

(i) Taking $\pi = \frac{22}{7}$, $r = \sqrt{(132 \times \frac{7}{22})} = \sqrt{42}$,

$\therefore r = 6{\cdot}481$; $\therefore$ radius = **6·48(1) cm.**

(ii) Taking $\log \pi = 0{\cdot}4971$, $r = 6{\cdot}482$;

$\therefore$ radius = **6·48(2) cm.**

2·1206
·4971
2)1·6235
0·8117(5)

The figure bounded by two concentric circles, see Fig. 218, is often called an **annulus.**

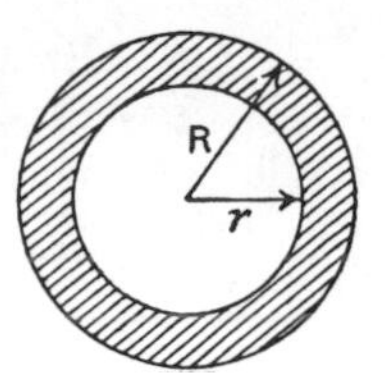

FIG. 218

If the radii of the outer and inner circles are R cm, r cm,

$$\text{area between circles} = (\pi R^2 - \pi r^2)\ \text{cm}^2$$
$$= \pi(R^2 - r^2)\ \text{cm}^2.$$

In numerical examples, use factors:

$$\textbf{area of annulus} = \boldsymbol{\pi(R+r)(R-r)}\ \textbf{cm}^2.$$

For example, if the radii are 8·6 cm, 7·9 cm,

$$\text{area of annulus} = \pi(8{\cdot}6+7{\cdot}9)(8{\cdot}6-7{\cdot}9)\ \text{cm}^2$$
$$= \pi \times 16{\cdot}5 \times 0{\cdot}7\ \text{cm}^2$$
$$= \mathbf{36{\cdot}3\ cm^2}, \text{ to 3 figures.}$$

0·4971
1·2175
$\bar{1}$·8451
1·5597

EXERCISE 110

[*Give answers to* 3 *figures. Take* $\pi = 3{\cdot}1416$, $\log \pi = 0{\cdot}4971$, *unless otherwise stated.*]

Find the areas of the circles, Nos. 1–6:

1 Radius 6 cm　　**2** Diameter 3·5 cm　　**3** Diameter 85 m

[**4**] Radius 4 cm　　[**5**] Diameter 10·5 cm　　[**6**] Diameter 250 m

Find the radii of the circle whose areas are given in Nos. 7–12:

7 22 cm^2 [$\pi = 3\frac{1}{7}$]　　**8** 1000 m^2　　**9** 7744 m^2

[**10**] 38·5 cm^2 [$\pi = 3\frac{1}{7}$]　　[**11**] 13·5 m^2　　[**12**] 2420 m^2

[**13**] Find the area of the top of a coin, diameter 3·3 cm.

14 Find the area of a semicircle of diameter 42 cm. [$\pi = 3\frac{1}{7}$.]

[**15**] A circle of diameter 3 cm is cut out of a sheet of paper 3 cm square; what is the area of the remainder?

16 A circle, radius 4 cm, is drawn on squared paper ruled in millimetres, and the number of small squares inside the circle is counted. What should be the result?

[**17**] 440 m of fencing are available for enclosing part of a field. Find the area enclosed if (i) square, (ii) circular. [Take $\pi = 3\frac{1}{7}$.]

18 Find the area of the ring between two concentric circles of radii 3 cm and 4 cm. [Take $\pi = 3\frac{1}{7}$.]

19 From a metal sheet, 4 cm square, four quadrants of a circle, each radius 2 cm, are cut away at the corners. Sketch the shape of what remains, and find its area.

[**20**] A circular pond of diameter 15 m is surrounded by a path 2 m wide. Find the area of the path.

21 A circular metal plate of radius 20 cm weighs 5 kg; find the weight of a plate of the same material and thickness, radius 24 cm. [There is no need to substitute for π.]

22 OA, OB are two radii of a circle, centre O, radius 4 cm; if the angle between OA and OB is 144°, find the area of the part of the circle between OA and OB. [This area is called a **sector** of the circle of angle 144°. What fraction is the area of the sector of the area of the circle?]

Area of Surface of a Circular Cylinder

A solid, whose cross-section is uniform and circular, such as a curtain rod or a pencil of circular section, is called a **circular cylinder,** and the radius of the cross-section is called the *radius of the cylinder*, and the straight line which passes through the centre of every cross-section is called the *axis of the cylinder*. The length of the axis is often called the *length* of the cylinder, as for a pencil, but this depends to some extent on the kind of cylindrical object. Thus for a jug of circular cross-section, the length of the axis is called the *height* of the cylinder, while for a circular disc, like a penny, the length of the axis is called the *thickness* of the cylinder, and for a garden roller the length of the axis is called the *width* of the cylinder. The name used is merely a matter of common sense.

A rectangular sheet of paper can be rolled up into the shape of a circular cylinder with open ends. Therefore

the area of the curved surface of a cylinder

= perimeter of base × height of cylinder.

If the radius of a cylinder is r units and if the height of the cylinder is h units, the perimeter of the base is $2\pi r$ units,

$\therefore$ **area of curved surface of cylinder = $2\pi rh$ units of area.**

In order to find the area of the *total* surface of a solid circular cylinder, it is necessary also to calculate the area of each end.

The area of each end is the area of a circle of radius r units, and is therefore πr^2 units of area.

$$\therefore \textit{total} \text{ area of surface} = (2\pi rh + 2\pi r^2) \text{ units of area}$$
$$= 2\pi r(h+r) \text{ units of area.}$$

Example 7 Find the total area of the surface of a solid cylinder, base-radius 3 cm, height 4 cm.

$$\text{Perimeter of base} = (2\pi \times 3)\text{ cm} = 6\pi \text{ cm};$$
$$\therefore \text{area of curved surface} = (6\pi \times 4)\text{ cm}^2 = 24\pi \text{ cm}^2.$$
$$\text{and area of base} = (\pi \times 3^2)\text{ cm}^2 = 9\pi \text{ cm}^2.$$
$$\therefore \textit{total} \text{ area of surface} = (24\pi + 9\pi \times 2)\text{ cm}^2$$
$$= 42\pi \text{ cm}^2$$
$$= \mathbf{132\ cm^2}, \text{ to 3 figures.}$$

1·6232
0·4971
2·1203

Notice that the substitution for π is done as late as possible.

EXERCISE 111

[*Nos.* 1 *and* 2 *are intended for oral work*]

1 A rectangular sheet of paper 11 cm wide, 5 cm high, is rolled into the form of an open hollow cylinder, 5 cm high. What is the greatest possible diameter of the cylinder (take $\pi = \frac{22}{7}$)? What is the area of its curved surface? What is the area of the smallest piece of paper which will cover one end of it?

2 The radius of an open hollow cylinder is 3·5 cm and the height of the cylinder is 6 cm. A cut is made straight down the surface so that the cylinder can be unwrapped to form a rectangle. Find the breadth, height, and area of the rectangle.

Find the areas of the curved surfaces of the cylinders, Nos. 3–6:

3 Radius 4 cm, height 5 cm
4 Diameter 5·8 cm, width 16 cm
[**5**] Radius 3 cm, length 1 m
[**6**] Diamcter 36 cm, 0·5 cm thick

Find the *total* areas of the surfaces of the solid cylinders, Nos. 7–10:

7 Radius 2 cm, height 5 cm
[**8**] Radius 10 cm, thickness 5 mm
[**9**] Diameter 1 m, width 3 m
10 Diameter 5 cm, length 12 cm

Find the total area of the external surface of a cylinder closed at one end and open at the other end, with the given external measurements, Nos. 11, 12:

11 Radius 3 cm, height 2 cm
[**12**] Diameter 6 cm, height 10 cm

13 Find the area of thin tin-sheeting required for making a tin cylinder, radius 4 cm, height 6 cm, with a slip-on lid which overlaps 5 mm.

14 A garden roller is 75 cm in diameter and is 1·05 m wide. What area does it cover in 40 rcvolutions?

15 Fig. 219 shows the section of a closed tin by a plane through the common axis of the two cylinders which compose it; dimensions in centimetres. Find the area of thin tin-sheeting required for making it.

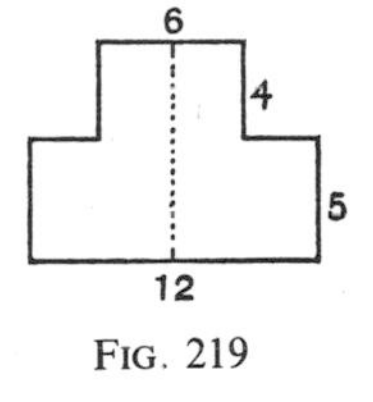

FIG. 219

[**16**] A cylindrical tank, 2·1 m in diameter, contains water to a depth of 1·2 m. Find the total area of the wetted surface.

Find the radius of each cylinder, Nos. 17, 18:

17 Area of curved surface 210 cm^2, height 5 cm.
[**18**] Area of curved surface 79·5 cm^2, height 3 cm.

Find the height of each cylinder, Nos. 19, 20:

19 Area of curved surface 100 cm^2, radius 6 cm.
[**20**] Total area of surface of solid cylinder 675 cm^2, radius 5 cm.

Volume of Circular Cylinder The volume of any solid of uniform cross-section is given by the rule,

area of cross-section × length.

If a circular cylinder is of radius r units and of height h units, the area of its cross-section is πr^2 units of area,

$$\therefore \textbf{volume of cylinder} = \boldsymbol{\pi r^2 h} \textbf{ units of volume.}$$

Also if a hollow tube is l cm long, and if its outer and inner radii are R cm, r cm,

volume of material composing tube

$$= (\pi R^2 l - \pi r^2 l)\text{ cm}^3 = \pi l(R^2 - r^2)\text{ cm}^3$$
$$= \boldsymbol{\pi l(R+r)(R-r)}\ \textbf{cm}^3.$$

Example 8 Find the number of litres of water in a cylindrical tank, 1·8 m in diameter, if the water is 60 cm deep.

The radius = 90 cm and the depth of water = 60 cm.

$$\therefore \text{volume of water} = \pi \times 90^2 \times 60\text{ cm}^3$$
$$= 1\,526\,000\text{ cm}^3$$
$$= \textbf{1530 litres}, \text{ to 3 figures.}$$

0·4971
3·9084
1·7782
6·1837

Example 9 A cylindrical tank, 90 cm in diameter, contains 600 litres of water. Find the depth of the water.

$$\text{Radius} = 45\text{ cm}, \therefore \text{area of base} = \pi \times 45^2\text{ cm}^2.$$
$$600\text{ litres} = 600\,000\text{ cm}^3.$$
$$\therefore \text{depth of water} = \frac{\text{volume of water}}{\text{area of base}}$$
$$= \frac{600\,000}{\pi \times 45^2}$$
$$= 94{\cdot}34\text{ cm}$$
$$= \textbf{94·3 cm}, \text{ to 3 figures.}$$

5·7782	0·4971
3·8035	3·3064
1·9747	3·8035

Example 10 Find the volume of metal in a hollow pipe, 60 cm long, of internal diameter 12 cm, made of metal 0·5 cm thick.

Internal radius = 6 cm, ∴ external radius = (6 + 0·5) cm = 6·5 cm.

The cross-section is the area between two concentric circles, radii 6·5 cm, 6 cm;

$\therefore$ area of cross-section $= \pi(6{\cdot}5^2 - 6^2)\text{ cm}^2$
$= \pi(6{\cdot}5+6)(6{\cdot}5-6)\text{ cm}^2$
$= \pi \times 12{\cdot}5 \times 0{\cdot}5\text{ cm}^2.$

But length of pipe $= 60$ cm.

$\therefore$ volume of metal $= \pi \times 12{\cdot}5 \times 0{\cdot}5 \times 60\text{ cm}^3$
$=$ **1180 cm³**, to 3 figures.

0·4971
1·0969
$\bar{1}$·6990
1·7782
3·0712

EXERCISE 112

[*Give answers to* 3 *figures*]

Find the volumes of the circular cylinders, Nos. 1–4:

1 Radius 4 cm, height 6 cm.
2 Diameter 1·3 m, 1 cm thick.
[**3**] Radius 9 cm, length 1 m.
[**4**] Diameter 3 cm, length 24 cm.

Find the heights of the circular cylinders, Nos. 5, 6:

5 Volume 4 litres, radius 8·2 cm.
[**6**] Volume 50 cm^3; radius 1·6 cm.

Find the diameters of the circular cylinders, Nos. 7, 8:

7 Volume 100 cm^3, height 6·8 cm.
[**8**] Volume 380 cm^3, height 10 cm.

9 A telegraph pole is 7 m high and 20 cm in diameter. Find its weight if the wood weighs 0·8 g per cm^3.

[**10**] Find the weight of 10 m of silver wire, diameter 4 mm, if 1 cm^3 of silver weighs 10·5 g.

11 A cylindrical tank is 1·2 m in diameter and 1·5 m high, internal measurements. How many litres will it hold?

12 A cylindrical tank, 1·8 m in diameter, contains 2000 litres of water. Find the depth of the water.

Find the volume of metal in the hollow pipes, Nos. 13, 14:

13 Internal diameter 7 cm, metal 5 mm thick, length 6 cm.
[**14**] External diameter 6 cm, metal 0·3 cm thick, length 60 cm.

15 How many cylindrical glasses, diameter 8 cm, height 12·5 cm, can be filled from a cylindrical vessel, diameter 30 cm, height 75 cm, full of milk? [There is no need to substitute for π.]

[**16**] A cylindrical jar, diameter 6 cm, depth 8 cm, is full of water. If this water is poured into an empty cylindrical jar of diameter 4 cm, find the depth of the water. [There is no need to substitute for π.]

17 Find the weight per metre of wire of sectional area 0·13 cm^2, if 1 cm^3 of the material weighs 7·5 g.

18 Find the volume of the closed tin whose dimensions are given in Ex. 111, No. 15.

19 Water is pouring into a cylindrical tank of diameter 7 m at the rate of 20 000 litres per minute. Find the rise of water-level per minute.

[**20**] Find the number of litres discharged in an hour from a pipe 15 cm in diameter through which water is flowing at 105 cm per second, if the pipe remains full.

Miscellaneous Examples

EXERCISE 113

[*Give answers to* 3 *figures, unless otherwise stated.*]

1 The diameter of the clock-face on a town-hall is 2·7 m. Find the area of the clock-face.

2 The hour-hand of a clock is 10 cm long; find the distance the tip of this hand travels in 45 min.

[**3**] The circumference of a circle is 1000 m long. Find the area of the circle.

4 The wheel of a vehicle is 70 cm in diameter. Find the number of revolutions made by the wheel for each kilometre travelled.

5 ABCD is a rectangular plate, AB = 6 cm, BC = 3 cm; two quadrants, centres A, B, each of radius 3 cm, are cut away from the plate. Sketch the shape of what remains, and find its area.

[**6**] The perimeter of a semicircular window is 2·7 m; find (i) the diameter, (ii) the area of the window.

7 A uniform metal circular plate of radius 15 cm weighs 6 kg. Three circular holes, each of radius 5 cm, are pierced in it; find the weight of the remainder.

8 The girth of a solid circular cylinder is 6 cm, and its length is 11 cm; find its volume.

9 The wheel of a bicycle, 70 cm in diameter, is making two revolutions per second. At what speed in kilometres per hour is the bicycle travelling?

[**10**] The base of a metal plate, 0·5 cm thick, is a rectangle 6 cm by 5 cm, from which quadrants each of radius 2 cm have been removed at each of the four corners. Find the weight of the plate if the metal weighs 8·9 g per cm^3.

11 A cylindrical well, internal diameter 3·15 m, contains 70 000 litres of water. Find the depth of the water.

12 An archway is formed by two vertical walls, 1·5 m high, 2·4 m apart, surmounted by a roof of semicircular section. The length of the archway is 6 m; find the area of its internal surface.

13 A cylindrical tin is 15 cm high and of radius 3·75 cm. There is a strip of paper covering the whole of the curved surface and a ring 1 cm wide round the edge of the top. Find the total area of the paper.

[**14**] The cross-section of a tube is the space between two concentric circles, radii 5 cm, 2 cm, and the length of the tube is 8 cm. Find the *total* area of its surface.

15 The cross-section of a thin sheet of corrugated iron is composed of 30 semicircles each of diameter 5 cm, and the sheet is 2·4 m long. Find the area of the upper surface of the sheet.

[**16**] The cross-section of a barn is a rectangle with a semicircle above it. The width of the barn is 8·4 m, its greatest height is 9 m, and its length is 18 m. Find its volume.

[**17**] An open cylindrical jar is 17·5 cm in diameter and 27·5 cm high, external measurements. The material is 1 cm thick. Find in cubic centimetres the amount of water it will hold. Find also the volume of the material used in making the jar.

18 A cylindrical measure which holds 4 litres is 21 cm in diameter. Find its height.

19 The area of a sector of a circle, radius 7 cm, is 55 cm^2; find the angle of the sector.

20 A bar 60 cm long of square cross-section, side 1 cm, is recast so as to have a circular cross-section of diameter 0·5 cm. Find its new length.

CHAPTER 21

SIMPLE INTEREST

If a man occupies a house he does not own, he pays the owner money, called the *rent* of the house, for being allowed to use it. If you put money into the National Savings Bank, you are lending money to the Government, the Government will pay you for being allowed to make use of your money, and this payment is called *interest*. As the house for which a man pays rent remains the property of the landlord, so the money you have put in the bank remains your property; a year's rent of a house and a year's interest on money deposited in the bank are merely fair payments for being allowed to use what has been borrowed for the year.

Any money lent or borrowed is called the **Principal,** and the charge made for its use is called the **interest,** and depends on how much has been borrowed and the length of time for which it is borrowed.

If the charge for borrowing £100 for 1 year is £4, we say that *interest is reckoned at the rate of* 4% *per annum*, or merely at 4%, as the words *per annum* are implied if not actually stated.

Interest is usually paid at fixed intervals, yearly, half-yearly, or quarterly, just as rent is so paid; and in this case the principal is said to be lent at **simple interest.** If the simple interest for any given time is added to the principal, the sum is called the **amount** at simple interest for that time. Thus if a man borrows £400 for 3 years at 5 per cent per annum, the *simple interest* for 1 year is $\frac{5}{100}$ of £400, that is £20, and therefore for 3 years is £60; and the *amount* at simple interest for 3 years is £460. The abbreviation for *per annum* is p.a.; thus 4% p.a. means 4 per cent per annum.

In transactions where the interest as it falls due is not paid to the lender but is added on to, *i.e.* compounded with, the principal, the money is said to be lent at **compound interest.** In the present chapter we shall be dealing only with simple interest.

EXERCISE 114 (Oral)

Find the simple interest on, and the amount at simple interest of:

1 £100 for 1 year at 3% p.a.

2 £100 for 3 years at 4% p.a.

3 £100 for 2 years at $3\frac{1}{2}$% p.a.

4 £100 for $\frac{1}{2}$ year at 8% p.a.

5 £200 for 1 year at 4% p.a.

6 £300 for 1 year at 7% p.a.

7 £400 for 2 years at 3% p.a.

8 £800 for 3 years at 4% p.a.

9 £200 for 4 years at $2\frac{1}{2}$% p.a.

10 £400 for 5 years at $3\frac{1}{2}$% p.a.

11 £5000 is borrowed at 4% p.a., payable yearly. What is the total debt at the end of 1 year? If no repayment is made, what is the total debt at the end of 2 years?

Calculation of Simple Interest

Method I. Direct Proportion

Study the following simple interest table:

Principal	Rate	Time in years	Interest	Amount
£300	4% p.a.	1	£12	£312
£300	4% p.a.	2	£24	£324
£300	4% p.a.	3	£36	£336

This illustrates the fact that

the simple interest is proportional to the time;

but it is important to note that *the* **Amount** *is* **not** *proportional to the time.*

Example 1 Find the simple interest on £285 for $2\frac{1}{2}$ years at 6% p.a. Find also the amount after $2\frac{1}{2}$ years.

The interest on £285 for 1 year is $\frac{6}{100}$ of £285,

$\therefore$ the interest on £285 for $2\frac{1}{2}$ years is ($\frac{6}{100}$ of £285) $\times 2\frac{1}{2}$.

$\therefore$ interest $= £(285 \times \frac{6}{100} \times \frac{5}{2}) = £\frac{171}{4} =$ **£42·75.**

$\therefore$ the amount $=$ (£285 + £42·75) $=$ **£327·75.**

Method II Simple Interest Formula

The method of Example 1, which should first be worked, may be used to obtain a general formula:

To find the simple interest on £*P* for *T* years at *R*% per annum.

The interest on £P for 1 year is $\frac{R}{100}$ of £P, $= £\frac{P \times R}{100}$;

$\therefore$ the interest on £P for T years is $£\left(\frac{P \times R}{100}\right) \times T, = £\frac{P \times R \times T}{100}$.

Hence we have the following formula:

If the simple interest on £P for T years at R% per annum is £I,

$$\boldsymbol{I = \frac{P \times R \times T}{100}}.$$

Example 2 Find the simple interest on £213·60 for 15 months at 3% per annum.

Principal = £213·60; time = 15 months = $\frac{5}{4}$ years;

$\therefore$ in the formula, $P = 213{\cdot}60, \quad T = \frac{5}{4}, \quad R = 3$;

$\therefore$ the interest £I is given by $I = \dfrac{213{\cdot}6 \times 3 \times \frac{5}{4}}{100} = \dfrac{213{\cdot}6 \times 3 \times 5}{400} = 8{\cdot}01$.

$\therefore$ the simple interest is **£8·01.**

Note As, in the formula, T is years, we must express 15 months as $\frac{5}{4}$ years.

Example 3 Find to the nearest penny the simple interest on £239·22 for $2\frac{3}{4}$ years at $4\frac{1}{2}\%$ p.a.

With the notation of the formula,

$$P = 239{\cdot}22, \quad R = \tfrac{9}{2}, \quad T = \tfrac{11}{4};$$

$$\therefore I = 239{\cdot}22 \times \tfrac{9}{2} \times \tfrac{11}{4} \times \tfrac{1}{100}$$

$$= \frac{2{\cdot}3922 \times 99}{8} = 29{\cdot}6034.$$

$$\begin{array}{r} 2{\cdot}3922 \\ 99 \\ \hline 21{\cdot}5298 \\ 215{\cdot}298 \\ \hline 8\overline{)236{\cdot}8278} \\ \hline 29{\cdot}6034 \end{array}$$

$\therefore$ the interest is **£29·60,** to the nearest penny.

N.B. 4-figure log tables cannot be used if the result is required to the nearest penny.

In calculating the interest for a period between two given dates, assume that the year contains 365 days. *Do not count the day when the money is deposited*, but count the day when it is removed.

Example 4 Find to the nearest penny the simple interest on £450 at 6% p.a. from July 4 to August 16.

Number of days

July	27	(31–4)
Aug.	16	
	43	

$\therefore$ with the notation of the formula,

$$P = 450, \quad R = 6, \quad T = \frac{43}{365};$$

$$\therefore I = 450 \times 6 \times \frac{43}{365} \times \frac{1}{100} = \frac{54 \times 43}{730} = \frac{2322}{730} \simeq 3{\cdot}1808.$$

$\therefore$ the interest is **£3·18,** to the nearest penny.

EXERCISE 115

Find the simple interest on, and the amount at simple interest of:

1 £350 for 3 years at 6% p.a.

[**2**] £420 for 5 years at $2\frac{1}{2}\%$ p.a.

[**3**] £184 for 2 years at 5% p.a.

4 £845 for $3\frac{1}{2}$ years at 6% p.a.

[**5**] £765 for $2\frac{1}{2}$ years at $5\frac{1}{3}\%$ p.a.

6 £62·50 for $2\frac{1}{4}$ years at $3\frac{1}{5}\%$ p.a.

7 £375 for 2 years 8 months at $4\frac{1}{2}\%$ p.a.

[**8**] £192 for 20 months at $2\frac{3}{4}\%$ p.a.

9 £168·50 for 3 years at 4% p.a.

10 £48 for 2 years at 4% p.a.

[**11**] £272 for $2\frac{1}{2}$ years at 3% p.a.

[**12**] £528 for $1\frac{1}{2}$ years at $3\frac{1}{2}\%$ p.a.
13 £416 for 16 months at $4\frac{1}{2}\%$ p.a.
[**14**] £632 for 14 months at $5\frac{1}{4}\%$ p.a.

Find to the nearest penny the simple interest on:

[**15**] £342·84 for $2\frac{1}{2}$ years at $3\frac{1}{2}\%$ p.a.
16 £47·57 for 4 years at $2\frac{1}{2}\%$ p.a.
17 £604·69 for $1\frac{1}{4}$ years at 6% p.a.
[**18**] £860·28 for 3 years at 4% p.a.
19 £724 from April 5 to June 10 at $4\frac{1}{2}\%$ p.a.
20 £848 from August 17 to November 4 at $5\frac{1}{2}\%$ p.a.

Inverse Problems on Simple Interest

Example 5 If the simple interest on £560 for 4 years is £78·40, find the rate per cent per annum.

With the notation of the formula, $P = 560$, $T = 4$, $I = 78{\cdot}4$ and it is required to find the value of R.

$$\frac{P \times R \times T}{100} = I; \qquad \therefore P \times R \times T = 100 \times I; \qquad \therefore R = \frac{100 \times I}{P \times T};$$

$$\therefore R = \frac{100 \times 78{\cdot}4}{560 \times 4} = \frac{7840}{560 \times 4} = \frac{7}{2};$$

$\therefore$ the rate of interest is **$3\frac{1}{2}\%$ p.a.**

Example 6 In what time will £640 amount to £684 at $2\frac{1}{2}\%$ p.a. simple interest?

$P = 640$, $I = 684 - 640 = 44$, $R = 2\frac{1}{2}$; find T.

$$P \times R \times T = 100 \times I; \qquad \therefore T = \frac{100 \times I}{P \times R};$$

$$\therefore T = \frac{100 \times 44}{640 \times 2\frac{1}{2}} = \frac{100 \times 44 \times 2}{640 \times 5} = 2\tfrac{3}{4};$$

$\therefore$ the time is **$2\frac{3}{4}$ years.**

Example 7 What sum of money will yield £81 interest in $2\frac{1}{2}$ years at $4\frac{1}{2}\%$ p.a. simple interest?

$I = 81$, $T = 2\frac{1}{2}$, $R = 4\frac{1}{2}$; find P.

$$P \times R \times T = 100 \times I; \qquad \therefore P = \frac{100 \times I}{R \times T};$$

$$\therefore P = \frac{100 \times 81}{4\frac{1}{2} \times 2\frac{1}{2}} = \frac{100 \times 81 \times 2 \times 2}{9 \times 5} = 720;$$

$\therefore$ the principal is **£720.**

EXERCISE 116

Find the unknown quantities in Nos. 1–16:

	Principal	Interest	Amount	Time	Rate % p.a.
1	£120	£18	...	3 years	...
2	£640	...	£696	$2\frac{1}{2}$ years	...
[**3**]	£240	...	£267	...	$4\frac{1}{2}$
4	£960	£198	...	...	$5\frac{1}{2}$
5	...	£48	...	$1\frac{1}{4}$ years	$5\frac{1}{3}$
[**6**]	...	£42	...	$1\frac{2}{3}$ years	$4\frac{1}{2}$
7	£360	...	...	$2\frac{1}{2}$ years	$4\frac{1}{2}$
[**8**]	£10	...	...	5 months	$4\frac{1}{2}$
9	£1560	£245·70	...	$3\frac{1}{2}$ years	...
[**10**]	£205	...	£207·05	146 days	...
11	£2000	£422·50	...	...	$3\frac{1}{4}$
[**12**]	...	£341·25	...	1 year	5
13	£487·50	...	...	5 years	3
[**14**]	£840	...	£913·50	...	$3\frac{1}{2}$
15	£834	£12·51	...	3 months	...
[**16**]	£280	...	£324·80	4 years	...

17 A man borrows £160 on condition that he pays back £169 after 9 months. At what rate % p.a. is interest charged?

18 After what time will £5 amount to £5·25 at $2\frac{1}{2}$% p.a.?

19 A savings bank pays interest at the rate of 5% p.a. How much have I in the Bank if my interest for 9 months is £6·30?

[**20**] A moneylender charges 1p per month interest on a loan of 25p. Find the rate % p.a.

21 If the rate of interest is reduced from 5% p.a. to $3\frac{1}{2}$% p.a., find the decrease in a half-year's interest on £540.

[**22**] A man lends £500 at 4% p.a., £800 at $3\frac{1}{4}$% p.a., and £700 at 6% p.a. Find how much interest he receives each year, and express this as a percentage of the total sum lent.

23 Find correct to the nearest penny the interest on £500 at 5% p.a. from June 1 to June 29.

24 A man borrowed £100 at $7\frac{1}{2}$% p.a.; when repaid the interest correct to the nearest penny was £2·46. How long did the loan run?

25 If the rate of interest on a loan is raised from 3% to $3\frac{1}{2}$%, the annual interest is increased by £2·75. What is the loan?

QUICK REVISION PAPERS 1–18

In the diagrams, lines shown as straight are given to be straight; parallel lines and right angles are indicated in the usual way.

Q.R. 1–6 (Ch. 1–6)

Q.R. 1

1 $2\frac{1}{4}-\frac{7}{8}$ **2** $1-(0{\cdot}63+0{\cdot}36)$ **3** $0{\cdot}01 \div 0{\cdot}4$

4 $2\frac{1}{4}\times 24$ **5** Ratio, 7·5 cm : 10 cm

6 Decrease 30p in ratio 3 : 5

7 Value of e^3 if $e = -2$

8 $\frac{3}{c}\times\frac{c}{3}$ **9** $\frac{1}{x} = 2\frac{1}{2}$; find x

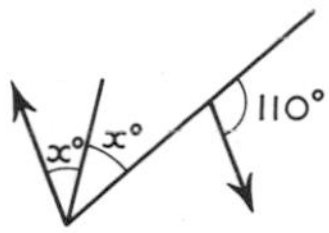

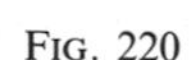

FIG. 220

10 Express $\frac{2}{5}n$ cm in mm

11 Find x in Fig. 220   **12** Cost of 21 kg if 14 kg cost 114p

Q.R. 2

1 Add: 2·73
0·69
1·08

2 $11\times 9{\cdot}19$ **3** $0{\cdot}23 \div 23$ **4** $b^4 \div \frac{1}{b^4}$

5 $\frac{3}{25}$ of 1 m in cm

6 Ratio, 150 m : 1 km

7 Decrease 20 g by 30 per cent

8 Cost of 2 m if $1\frac{1}{2}$ m costs 30p   **9** $5-k-2(1-k)$

10 Value of $2x^3+17x^2+5x+9$ if $x = 10$.

11 $bc^2 = 20$ and $c = -2$, find b and bc.

12 $x°$, $5x°$, $6x°$ are angles of a triangle; find $6x°$.

Q.R. 3

1 $(\frac{1}{3}\times\frac{1}{2})+\frac{1}{2}$ **2** $\frac{0{\cdot}2}{0{\cdot}4}+\frac{0{\cdot}3}{0{\cdot}6}$ **3** $(0{\cdot}03)^2$ **4** $\frac{11}{200}$ of £2

5 What is 10% of $2\frac{1}{2}$ kg?   **6** Divide 24 cm in ratio 5 : 7

7 $\frac{4}{9}$ of a number is 36; what is the number?

8 Value of $6ab^2+2c^3$ if $a = -1$, $b = 2$, $c = 0$.

9 $0{\cdot}2y = 0{\cdot}01$, find y **10** $3x^3y^3 \times 2x^2y^2$

11 How long to travel $\frac{1}{2}s$ km at $\frac{1}{2}v$ km per hour?

12 70°, 85°, 150°, $x°$ are angles of a quadrilateral; find x.

Q.R. 4

1 $\frac{3}{5} \div 1\frac{1}{2}$ **2** $0{\cdot}0709 \times 1{\cdot}2$ **3** $0{\cdot}1 \div 0{\cdot}001$ **4** $\frac{1}{7}$ of £3·78

5 Ratio, 750 m : 1 km.

6 R.F. of a map, 1 cm to 2 km.

7 What % is 18p of 30p?

8 $\frac{3}{n} = \frac{4}{5}$; find n **9** $\frac{a^4b^6}{a^2b^3} \div \frac{1}{2}$

10 In Fig. 221, if $x = 80$, $y = 34$, find z.

11 Find $2a - b - c$ if $a = 5$, $b = -3$, $c = 1$.

12 Cost in £ of $200x$ articles at $1\frac{1}{2}$p each.

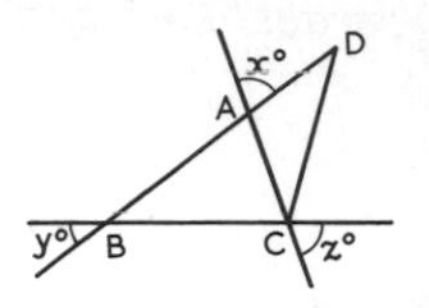

Fig. 221

Q.R. 5

1 Add:

1·617
5·42
3·504
2·059

2 $1 - \frac{1}{2\frac{1}{2}}$ **3** $(0{\cdot}12)^3$

4 $0{\cdot}028 \div 14$ **5** £1·25 × 48

6 $x - \frac{3}{4}(x - y)$

7 Ratio of areas of two squares, sides 8 cm and 12 cm.

8 18 litres cost £6; find the cost of 27 litres.

9 40% of N is 12; find N **10** Find $x - z + z^2$ if $x = 2$, $z = -3$.

11 In Fig. 221, if AB = BC and $z = 70$, find y.

12 $a = 5b = \frac{1}{3}c = 10$; find $a + b - c$.

Q.R. 6

1 $\frac{1}{3\frac{3}{4}} \times 3$ **2** $(0{\cdot}2)^4$ **3** $156 \div 1{\cdot}2$ **4** 99p × 36 in £

5 Share £1·20 in the ratios 2 : 3 : 7

6 Find area of border 1 m wide all round the edge of the floor of a room, 7 m long, 5 m wide.

7 Length in km represented by length 12·6 cm on a map, scale 1 : 100 000.

8 Value of $\frac{t^2}{x^3}$ if $t = -12$, $x = -2$

9 $\frac{a}{ab} + \frac{c}{bc}$

10 Increase $20n$ by 20%

11 $p^2 - 1 = 80$; find p.

12 In Fig. 221, if AB = BC = CD and $y = 40$, find ∠ACD.

Q.R. 7–12 (Ch. 1–14)

Q.R. 7

1 Subtract: 3·001
2·207

2 $0{\cdot}75 \times 1\frac{1}{3}$

3 $1{\cdot}01 \times 0{\cdot}11$

4 £2·20 × 35

5 80% of 30p

6 What % is 14 of 35?

7 Decrease £P by 15%. Ans. in p.

8 Weight in kg of 5 litres of spirit, if 1 cm^3 weighs 0·78 g.

9 $p+q = 97$, $p-q = 53$; find p and q.

10 Expand $(3a-5b)^2$

11 Factorise t^2-t-72

12 In Fig. 222, ABCD is a rectangle. If $y = 110$, find x.

Q.R. 8

1 $1 \div (\frac{1}{2}+\frac{1}{3})$

2 $0{\cdot}129 \times 0{\cdot}12$

3 $\frac{7}{300}$ of 3 m, in cm

4 $3 \div 0{\cdot}01$

5 17% of 50

6 Ratio, 1p per g : £1 per kg.

7 What % is 350 mm of 1 m?

8 Complete: $x^2+7x+\ldots = (x+\ldots)^2$.

9 Factorise $4b^2-36c^2$.

10 $y = 3z$, $y+z = 12$; find y, z.

11 C.P. £4; S.P. £5; find gain %.

12 In Fig. 222, ABCD is a rectangle. If $z = 35$, find y.

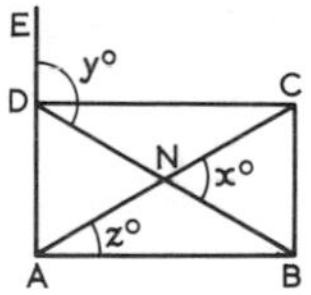

FIG. 222

Q.R. 9

1 $1\frac{3}{4}-(1\frac{1}{2}\div 2)$

2 $0{\cdot}3 \times 0{\cdot}4 \times 0{\cdot}5$

3 $17 \div 0{\cdot}17$

4 Express $7\frac{1}{2}$% as a fraction

5 $10{\cdot}2^2-9{\cdot}8^2$

6 What % is 36 of 80?

7 Increase £350 by 10%

8 S.P. £3; gain 20%; find C.P.

9 How many metres travelled in $15t$ min at $\frac{1}{2}n$ km per hour?

10 Factorise $2x^2-7x+6$

11 Expand $(7p-4q)^2$

12 In Fig. 222, ABCD is a rectangle. If AN = 2·5 cm, find BD.

Q.R. 10

1 Complete addition: £3·58
£4·89
– –
£10·00

2 $(\frac{1}{4}\times\frac{1}{2})+(\frac{1}{4}\div\frac{1}{2})$

3 $5{\cdot}05 \times 0{\cdot}2$

4 80% of 75p.

5 24 is 60% of N; find N.

6 Divide £48 in ratios 7:3:2.

7 $(a-b)^2-(a^2-b^2)$

8 Find error % if a distance of 2·5 cm is measured as 2·3 cm.

9 $c^2-\dfrac{1}{d}$ if $c=-\frac{1}{2}$, $d=-3$

10 $\dfrac{a-b}{a+b}$ if $\dfrac{a}{b}=\dfrac{3}{2}$

11 Factorise y^3-y^2-6y.

12 In Fig. 223, find BC, PQ.

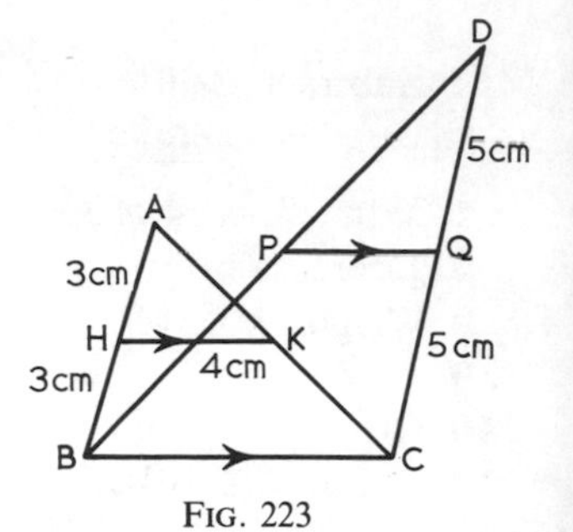

FIG. 223

Q.R. 11

1 $(0{\cdot}7)^2+(0{\cdot}1)^2$ **2** $(1+\frac{1}{2})\div(1-\frac{1}{3})$ **3** 9·99 m × 15

4 Decrease 16 kg by $12\frac{1}{2}$% **5** Express $\frac{1}{30}$ as a percentage

6 $\cos x° = 0{\cdot}4483$, find $x°$ **7** $s=\frac{1}{2}gt^2$, find t in terms of s, g

8 Find 3 integers proportional to $\frac{1}{36}, \frac{1}{63}, \frac{1}{28}$.

9 S.P. £100, loss £10; find loss % **10** $(x+3)^2-(x-3)^2$

11 Find b if $x+4$ is a factor of $x^2+bx-24$.

12 In Fig. 223, prove that HP is parallel to KQ.

Q.R. 12

1 $8{\cdot}4\div0{\cdot}012$ **2** $(1-\frac{1}{4})(1-\frac{5}{9})$ **3** $8\frac{1}{3}$% as a fraction

4 16 kg cost £45·36. Find the cost of 6 kg.

5 Length in m represented by 0·8 cm on map, scale 2 cm to 1 km.

6 Increase N by 50% **7** $(x+4y)^2-(x+3y)(x+5y)$

8 Value of y^2-z^2 if $y=5{\cdot}15$, $z=4{\cdot}85$.

9 $7\tan x° = 5$; find $x°$ and $\cos x°$.

10 Factorise $1+2t-15t^2$. **11** Factorise $3(2-p)+q(p-2)$

12 In Fig. 223, prove HQ, PK bisect each other.

Q.R. 13–18 (Ch. 1–21)

Q.R. 13

1 $0{\cdot}002\div0{\cdot}4$ **2** $2\frac{1}{2}$% of £12 **3** $(2\frac{1}{3}\times1\frac{1}{2})-(3\frac{1}{2}\div1\frac{1}{3})$

4 What % is 8p of 30p?

5 $\sqrt{(3{\cdot}7)}$ to 4 figures, from tables.

6 $6 \sin x^\circ = 1$; find x°.

7 Factorise $5t^2 - 45$.

8 $(3C)^2 \times (2C^3)^3 \div (6C^2)^2$.

9 $n(R-r) = 2R$; find R.

10 $5780 = 5{\cdot}78 \times 10^n$; find n.

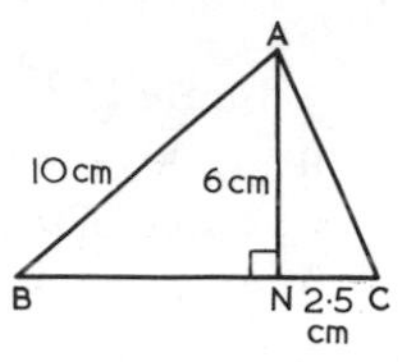

FIG. 224

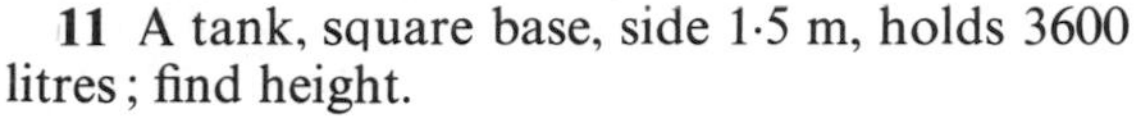
11 A tank, square base, side 1·5 m, holds 3600 litres; find height.

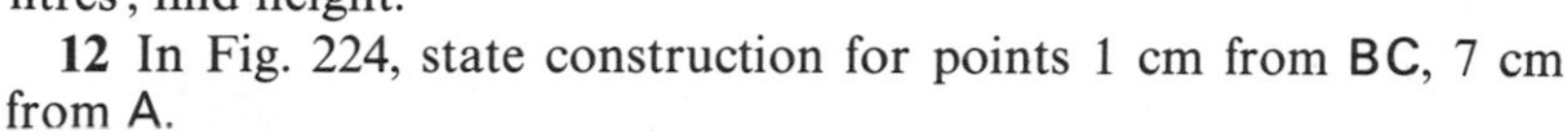
12 In Fig. 224, state construction for points 1 cm from BC, 7 cm from A.

Q.R. 14

1 $\frac{1}{2} + \frac{11}{12} - (1\frac{2}{3} \div 1\frac{1}{4})$

2 Express $0{\cdot}3\ \text{m}^2$ in cm^2

3 $0{\cdot}2849 \div 92{\cdot}5$

4 $0{\cdot}1 \times 0{\cdot}2 \times 0{\cdot}3$

5 $\sqrt{(0{\cdot}29)}$ to 4 figures, by tables

6 $0{\cdot}036 = 3{\cdot}6 \times 10^n$; find n

7 $d + \frac{1}{2}\pi d = 45$; find d, taking $\pi = 3\frac{1}{7}$. Interpret result.

8 In Fig. 224, calculate ∠ABC and ∠ACB.

9 Factorise $6x^2 + xy - 5y^2$

10 $(z+1)^3 - 9(z+1)$ when $z = -4$

11 $1 \div 38{\cdot}46$, to 4 figures

12 In Fig. 224, find AC, BC

Q.R. 15

1 $(\frac{3}{4} \times 2\frac{1}{2}) - (1\frac{1}{6} \div 1\frac{1}{3})$

2 Cost of 145 tonnes at £2·36 per tonne

3 $1{\cdot}3677 \div 194$

4 $85{\cdot}86^2$ (tables)

5 $\log x = 1{\cdot}825$, find x

6 35% of x is 42; find x

7 Factorise $6ab + 4a - 3b - 2$

8 Simple interest on £840 for 3 years at 4% p.a.

9 *Sketch* the graphs of $2-x$ and $x-1$ with the same axes.

10 Area of circle is $16{\cdot}5\ \text{cm}^2$; find radius to 3 figures. [$\pi = 3\frac{1}{7}$.]

11 If $x+5$ is a factor of $x^2 + bx - 30$, find b.

12 In Fig. 224, state how to construct centre of circle ABC.

Q.R. 16

1 $36\text{p} \times 47$, in £

2 $1{\cdot}1951 \div 8{\cdot}5$

3 $\frac{2}{3}$ of $(4 - \frac{3}{10} - 1\frac{8}{15})$.

4 Factorise $4x^4 - 16x^2$.

5 In Fig. 225, find ∠ABC.

6 C.P. £4, S.P. £3; find loss %.

7 Find the error % if 25 m is taken as 20 m.

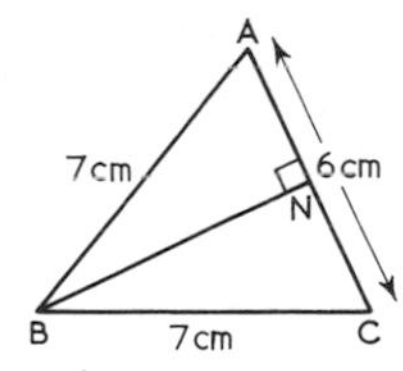

FIG. 225

8 Area of a circle, circumference 440 m. [$\pi = 3\frac{1}{7}$.]

9 Distance between points $(-1, -2)$, $(2, 2)$

10 Value of $16^{\frac{3}{4}}$

11 Express with decimal portion positive, (i) $\bar{1}{\cdot}64 \times 4$, (ii) $\bar{1}{\cdot}64 \div 3$.

12 In Fig. 225, state construction for point equidistant from CA, CB and equidistant from A, B.

Q.R. 17

1 Share £1 in ratio $1:1\frac{1}{2}$ 2 $2\frac{1}{2}\%$ of £14·80 3 $\sqrt{(0{\cdot}0256)}$
4 $\sqrt[3]{100}$, to 3 figures 5 $\cos x° = 2\cos 72°$; find $x°$
6 Value of $a^3 - 8b^3$ if $a = -\frac{1}{2}$, $b = -\frac{3}{4}$.
7 $p+2q = 4$ and $4p-4q = 7$; find $p:q$.
8 If $y = 1-2x$, express $2x^2-5xy-3y^2$ in terms of x.
9 If $x-2$ is a factor of x^2+3x+c, find c.
10 Find the simple interest on £700 for 2 years at 7% p.a.
11 If $V = \pi r^2 h$ and $S = 2\pi rh$, find r in terms of V, S. Interpret.
12 In Fig. 225, find position of centre of circle ANB.

Q.R. 18

1 $245{\cdot}83 \div 0{\cdot}305$ 2 $\sin x° = 0{\cdot}88$, find $\cos x°$ 3 $(6\frac{1}{4})^{-1\frac{1}{2}}$
4 Decrease £1·60 by 10%.
5 0·0896 km in metres and centimetres.
6 Solve the inequality $x+4 > 3x-2$.
7 In Fig. 225, find ∠BAN and BN 8 $\sqrt[3]{0{\cdot}1}$, to 3 figures
9 *Sketch* the graphs of $3-\frac{1}{2}x$ and $2x$ from $x = -1$ to $x = 4$.
10 In Fig. 225, state construction for finding all points 3 cm from BC and 2 cm from AC, produced if necessary.
11 Factorise $6t^2+14t-12$.
12 Simple interest on £64 for 3 years is £4·80. Find rate % p.a.

TESTS IN COMPUTATION 1–40

TESTS 1–10 (Volume I)

Test 1

1 Find in prime factors the square root of 9801.
2 $5\frac{7}{10} - 1\frac{23}{36} - 1\frac{11}{45}$ 3 $0{\cdot}0998 \times 0{\cdot}085$
4 $\frac{p}{q} \times \frac{q}{r} \times \frac{r}{p}$ 5 Solve $\frac{t+1}{5} = \frac{t+3}{6}$
6 Two angles of a triangle are $2(n+6)$ degrees, $3(n-12)$ degrees. Find the third angle in terms of n.

Test 2

1 $\frac{2}{5} \times \frac{3}{4}$ of $\frac{5}{12}$ 2 $0{\cdot}2967 \div 69$ 3 Square $3c^3$
4 $(0{\cdot}12)^3$ 5 Solve $0{\cdot}4(2y-1) = 0{\cdot}6$
6 A boy is now $2t$ years old and his father is $2\frac{1}{2}$ times as old as his son. Find the sum of their ages in 5 years' time.

Test 3

1 What must be added to $4\frac{2}{5}+3\frac{1}{10}$ to make 10?

2 $0{\cdot}6315 \times 124$

3 $3x(3x-2y)-2y(2x-3y)$

4 How many rectangular pieces of cloth 5 cm by 4 cm can be cut from a piece 60 cm long, 24 cm wide?

5 $\dfrac{3a}{2b}+\dfrac{2a}{3b}+\dfrac{5a}{6b}$

6 $F = \frac{9}{5}(C+40)-40$; find F if $C = 25$

Test 4

1 Find in prime factors the L.C.M. of 35, 75, 98.

2 $(\frac{1}{2}+\frac{1}{3})(\frac{1}{4}+\frac{1}{6})-\frac{1}{3}$

3 $0{\cdot}124 \div 0{\cdot}155$

4 $\dfrac{9a^4b^6c^2}{3a^2b^3c^6}$

5 $\dfrac{x+y}{4}-\dfrac{x-y}{6}$

6 If $y^2-2yz = z+1$, find z when $y = 4$.

Test 5

1 $1\frac{13}{15}+(\frac{5}{7}\div 2\frac{1}{7})-(\frac{8}{9}\times 1\frac{4}{5})$

2 $1{\cdot}2 \times 0{\cdot}05 \times 0{\cdot}4$

3 A car travels 550 m in 25 seconds; find its average speed in km per hour.

4 Add $8ab$ to $12bc$ and divide the sum by $4b$.

5 Solve $\dfrac{t-1}{4}-\dfrac{t-2}{5} = 1$.

6 One of the base angles of an isosceles triangle is $(\frac{1}{2}n+15)$ degrees. Find the vertical angle.

Test 6

1 Find the perimeter of a tile, 4·3 cm long, 3·8 cm wide.

2 Express 46 000 cm^2 in square metres.

3 Find the cost of 52 articles at £3·36 each.

4 $2b^2 \times 3b^3 \times 4b^4$

5 $\frac{4}{5}(t+2r)-\frac{3}{5}(r-2t)$

6 A pencil costs $1\frac{1}{2}x$ pence; how many can be bought for 30p?

Test 7

1 $\sqrt{(189 \times 84)}$

2 $0{\cdot}061103 \div 0{\cdot}301$

3 The volume of a rectangular solid, 12 cm long, 9 cm wide, is 540 cm^3; find the total area of its surface.

4 If $p = \frac{1}{2}, q = \frac{3}{4}, r = \frac{1}{4}$, evaluate $(q-r)^2-p^2$.

5 Solve $0{\cdot}11r = 2{\cdot}2$

6 $12a-2\{a+3b-(b-2a)\}$

Test 8

1 $\{(\frac{1}{2}\times\frac{3}{4})+\frac{5}{6}\}\div(\frac{2}{3}+\frac{1}{4}+\frac{7}{6})$

2 $(0{\cdot}02)^3 \times 0{\cdot}5 \times (0{\cdot}1)^2$

3 Divide 10·948 by 0·23.

4 $\dfrac{a^4 \times a^5}{a^{10}}$ **5** Solve $\dfrac{x-3}{5} = 8 - \dfrac{x-7}{4}$.

6 By what must $\dfrac{2x^2}{y}$ be multiplied to make $\frac{1}{2}x^6y$?

Test 9

1 $187{\cdot}5 \times 0{\cdot}0256$ **2** Divide $(6a^2b^3)^3$ by $(3ab^3)^2$

3 Express 13 m 9 cm as a decimal of 24 m, to 3 figures.

4 Find the total area of the walls of a room, 4·6 m long, 3·75 m wide, 2·5 m high.

5 If $a = \frac{2}{3}$, $b = \frac{1}{6}$, $c = \frac{1}{9}$, evaluate $(a+b)^2 - (a-b-3c)^2$.

6 For what value of x is $(5x - 0{\cdot}1)$ equal to twice $(3x - 0{\cdot}7)$?

Test 10

1 Divide 7·525 by 0·043.

2 The perimeter of a rectangle is 17 cm; its length is 5·65 cm; find its breadth.

3 Find the cost of 136 articles at £0·66 each.

4 If $a = 3b = 4c = 60$, evaluate $\dfrac{a-b-c}{a-b+c}$.

5 Solve $\dfrac{x+5}{2} - \dfrac{2}{3}(x-4) + 1 = 0$.

6 The vertical angle of an isosceles triangle is $(3x+40)$ degrees. Find a base angle.

TESTS 11–20 (Volume II, Ch. 1–5)

Test 11

1 $(2\frac{1}{4} - 3\frac{2}{3} + 5\frac{1}{6}) \div (2\frac{1}{4} - 3\frac{5}{6} + 2\frac{1}{3})$ **2** $6r - \{5r - 3(r-2) - 8\}$

3 Find the cost of 73 articles at £2·73 each.

4 Express $\frac{3}{8}$ of 4·544 litres in cm^3.

5 If $x = 3$, $y = -2$, $z = 0$, evaluate $x - 2y + 3z$ and $\sqrt{(x^2 + y^4)}$.

6 If $s = \frac{2}{3}nt$, find t in terms of n, s.

Test 12

1 $0{\cdot}309 \times 0{\cdot}01015$ **2** Increase £7 in the ratio 11:4

3 A wooden cuboid 30 cm by 14 cm by 5 cm weighs 1·281 kg; find in grammes the weight of 1 cm^3 of the wood.

4 $\frac{3}{4}(3x+2) - \frac{2}{5}(4x+5)$.

5 What must be added to $3x - 2y + 5z$ to make $5x - y - z$?

6 If cx^2y equals 240 when $x = -4$, $y = 5$, find its value when $x = 2$, $y = -3$.

Test 13

1 Express 3·36 m as a percentage of 15 m.

2 $1{\cdot}72 \div 0{\cdot}1075$ **3** $\frac{1}{a} \times \frac{1}{a^2} \div \frac{b^2}{a}$

4 The scale of a map is 2 cm to the km. How many square metres are represented by an area 2·35 cm^2 on the map?

5 If $a = 2$, $b = 1$, $c = -3$, evaluate $a - 2c$ and $(b+c)^3$.

6 £5n per kg is the same price as $(n-3)$ pence per g; find n.

Test 14

1 38p × 12 000 **2** Express $\frac{7}{130}$ as a decimal to 4 places

3 A hectare is 10 000 m^2. How many hectares are there in a square kilometre?

4 $p\left(\frac{1}{p}+\frac{1}{q}\right) - q\left(\frac{1}{q}-\frac{1}{p}\right)$ **5** Solve $\frac{4x+1}{3} - \frac{2-3x}{4} = \frac{26-x}{6}$

6 If $a(x^2-1)+x(a+1)$ is zero when $x = 2$, find the numerical value of a.

Test 15

1 $(4\frac{2}{5} \times \frac{1}{2}) - (6\frac{1}{5} \div 3)$ **2** Subtract $(r-s)$ from 75% of $(r+s)$

3 Express £1·16$\frac{1}{2}$ as a decimal of £2·40, to 3 figures.

4 Find the cost of 3·6 tonnes at £2·15 per tonne.

5 Solve $x - \frac{x-1}{3} = \frac{x+1}{4}$.

6 Make n the subject of the formula, $l = a+(n-1)d$.

Test 16

1 $0{\cdot}01 \times 1{\cdot}01 \times 0{\cdot}11$ **2** Decrease £8 by $3\frac{3}{4}$%

3 Evaluate $1 \div 0{\cdot}914$, correct to 3 figures.

4 Divide £100 into 3 shares in the ratios 8:5:3.

5 If $a = \frac{1}{2}$, $b = -2$, $c = 1$, $d = 0$, find the value of $a^2(b-c)+2d^2(a-c)$.

6 If $2y+z+10 = 0$ and $y = \frac{1}{2}z$, find y.

Test 17

1 $9{\cdot}9176 \div 0{\cdot}01771$ **2** $h - \frac{1}{2}(h+k)$

3 Find the cost of 670 articles at 89p each.

4 $\frac{3}{7}$ of a number is 63; what is $\frac{4}{7}$ of this number?

5 Subtract $1-3x(1-3x+x^2)$ from $x(2x^2-x-2)$.

6 Solve $1{\cdot}2t - 0{\cdot}034 = 0{\cdot}05t + 0{\cdot}012$.

Test 18

1 Express 30% of £0·88 to the nearest penny.

2 Evaluate $0{\cdot}76 \times 1093{\cdot}6$ to the nearest whole number.

3 Find the cost of cloth per m^2 if a metre length, 64 cm wide, costs £1·12.

4 If $p = 5, r = -3, t = -4$, evaluate $p-2t-r$ and $\sqrt{(p^2-t^2)}$.

5 $(12xy^3-9x^3y)\div(-6xy)$.

6 If $x:y = 7:3$, find $(x-y):(x+3y)$.

Test 19

1 $1\frac{1}{4}+[\frac{3}{20}(4\frac{1}{3}-2\frac{1}{2})\times 5]$

2 $(0{\cdot}04)^3 \times (0{\cdot}15)^2$

3 Evaluate $\frac{11}{16}$ of 277·3, to the nearest whole number.

4 $\dfrac{r-2s}{4r}-\dfrac{2r-3s}{6s}+1\frac{1}{4}$

5 $\dfrac{(x-y)(y-z)}{x-z}$ if $x = 1, y = -2, z = 7$.

6 If $p = n(l+b)$ and $A = lb$, express A in terms of n, p, b only.

Test 20

1 Express $\frac{4}{7}+\frac{7}{17}$ as a decimal, correct to 2 places.

2 Express 0·5673 km in m and cm.

3 Find the ratio of 7·5 cm to 20 cm in simplest form.

4 $(8x^4-6x^2y^2)\div(-2x^2)$.

5 If $s = ut+\frac{1}{2}ft^2$, find f when $s = 54, u = 30, t = 6$.

6 If x exceeds y by 20%, find by how much % $2y$ exceeds x.

TESTS 21–30 (Ch. 1–16)

Test 21

1 $4\frac{1}{2}$% of £15

2 Solve $4(t+1) = 3-3(2t-7)$

3 $\frac{2}{7}$ of £3·22$+\frac{11}{25}$ of £4$+\frac{3}{4}$ of £1·08.

4 Cost price £1·60, gain $22\frac{1}{2}$%; find the selling price.

5 If $a = -2, b = 3, c = -4$, evaluate $(b-c)(c-a)(a-b)$.

6 Factorise $3p^2+5p-2$

7 $3 \tan x° = 5$; find $x°$

Test 22

1 $0{\cdot}3728 \times \frac{17}{25}$

2 What percentage is $4\frac{1}{2}$p of 60p?

3 Cost of 12 000 articles at 99p each.

4 If $\frac{1}{2}x-\frac{1}{3}y = 1$, find y in terms of x.

5 Solve $3y-z = 15, \quad 7z+14y = 35$.

6 Factorise $ab-ac-b^2+bc$

7 $\cos x° = 0{\cdot}5660$; find $x°$

Test 23

1 $26{\cdot}44528 \div 0{\cdot}004195$ **2** Increase £248 by $1\frac{1}{4}\%$

3 A pond of surface area $30\,000\text{ m}^2$ is frozen over with ice of average thickness 6 cm; 1 m^3 of ice weighs 900 kg. Find the total weight of the ice in tonnes.

4 $2(p-q-r)-\{3p-5(q-r)\}$ **5** $1-\dfrac{5b-10c}{5b}$

6 Multiply $2c^2-3c-5$ by $3c+2$. **7** $\tan x^\circ = 2\tan 40^\circ$; find x°

Test 24

1 Express $\frac{29}{40}$ as a percentage **2** Factorise $1-t-12t^2$

3 Find the volume in litres of a tank 125 cm by 84 cm by 50 cm.

4 Selling price £2·94, gain 40%; find the cost price.

5 $(3x)^3(4y)^2 \div (6xy)^2$ **5** Solve $\dfrac{p-7}{4} = 8 - \dfrac{2p-1}{3}$

7 AB = AC = 6 cm, ∠BAC = 108°; find BC.

Test 25

1 $0{\cdot}908 \times 0{\cdot}071$.

2 Increase 304 by 125%.

3 Find the rate payable on property assessed at £1230, if the rate is 6p in the £.

4 Factorise $x^2-(y-z)^2$.

5 Subtract $\dfrac{a}{5}-\dfrac{3b}{10}+\dfrac{2c}{15}$ from $\dfrac{b}{5}-\dfrac{c}{3}+\dfrac{a}{10}$.

6 If $x:y = 2\frac{1}{2}:1\frac{2}{3}$ and $y:z = 1\frac{3}{4}:1\frac{2}{5}$, find $x:z$.

7 PQ = QR = 7 cm, PR = 12 cm; find ∠PQR.

Test 26

1 $17\frac{5}{6}-10\frac{1}{3}+(3\frac{1}{7}\text{ of }8{\cdot}4)$.

2 Factorise $6-54z^2$.

3 By how much per cent to 2 figures does 8435 exceed 7942?

4 A poultry-run, 26 m by 9·6 m, is enclosed with netting 1·3 m high. Find to the nearest penny the cost of the netting at 7p per m^2.

5 If $\dfrac{1}{b}+\dfrac{1}{c} = 1$ and if $b = 1\frac{1}{2}$, find c.

6 $\dfrac{6r^6-9r^3s^3}{-3r^2}$.

7 ABCD is a rhombus; AC = 7 cm, BD = 5 cm; find ∠BAD.

Test 27

1 $0{\cdot}1702 \div 0{\cdot}46$.

2 $1 - \dfrac{2c-1}{15} + \dfrac{2-c}{6}$.

3 $67\tfrac{1}{2}\%$ of N is 81; find N.

4 $\cos x^\circ = 0{\cdot}7640$; find $\cos(2x^\circ)$.

5 An article is sold for £37·80 at a loss of 16%; find the loss.

6 Add together $2x(x^2-3x+6)$, $3x(5-4x+x^2)$, $5x^2(1-x)$.

7 Solve $11y-3z+5 = 7y-z+3 = 4z-5y+6$.

Test 28

1 $84\text{p} \times 724$ in £.

2 Simplify $\dfrac{a^6}{a^2}$ and $\dfrac{5b}{6c} - \dfrac{3b}{4c} + \dfrac{b}{3c}$.

3 Express 58p as a decimal of £5.

4 Find the cost of matting for a floor, 9·4 m long, 6·7 m wide, at 24p per m², to the nearest penny.

5 If $t = 3r$, express $(t-r)(t^2-r^2)$ in terms of r.

6 Factorise $(3x+1)(x+2)^2-(x+2)(x+3)^2$.

7 B is 750 m N. 52° E. from A. How far is B north of A?

Test 29

1 What percentage is £2·52 of £3·36?

2 Divide £9·10 in the ratios $\tfrac{1}{2}:\tfrac{1}{3}:\tfrac{1}{4}$.

3 If $7\tfrac{1}{2}\%$ is deducted from a bill, £7·40 remains to be paid. How much was the bill?

4 Factorise $3x^2-36xy+108y^2$.

5 Subtract $5y(5y^2-2y+3)$ from $4y(4y^2-2y-1)+2$.

6 Solve $5x+6y = \tfrac{3}{4}x+\tfrac{1}{3}y = 4$.

7 Q is 480 m S. 24° W. from P. How far is Q west of P?

Test 30

1 $0{\cdot}65 \times 1{\cdot}08 \div 3{\cdot}9$.

2 $37\tfrac{1}{2}\%$ of 28·64 m.

3 $4\tfrac{1}{2}$ kg of tea cost £3·78; find the cost per kg.

4 $x = 2a^2+b^2$ and $y = 3ab-b^2$; find the value of $(x+2y)(x-y)$ if $a = 2$, $b = -1$.

5 Factorise $(3x-4y)^2-4(x+y)^2$.

6 Simplify $\dfrac{9n^9}{3n^3}$ and $2x\left(\dfrac{3}{x}+3\right)-3y\left(\dfrac{2}{y}+2\right)$.

7 $6\sin x^\circ = 7\sin 43^\circ$; find x°.

TESTS 31–40 (Ch. 1–21)

[Take $\log \pi = 0{\cdot}4971$]

Test 31

1 Arrange in *ascending* order of magnitude: $\frac{5}{16}, \frac{9}{32}, \frac{2}{7}$, 0·28.

2 $\dfrac{18{\cdot}92 \times 7{\cdot}064}{563{\cdot}7 \times 0{\cdot}808}$ to 3 figures.

3 $\dfrac{1+a}{3} - \dfrac{1+a}{2} + \dfrac{2-a}{4}$.

4 Volume of cylinder, diam. 7·35 cm, height 4·8 cm, to nearest cm^3.

5 Simplify $N - (40\,\%$ of $N)$.

6 Factorise $x^2 - 2xy - 15y^2$.

5 $\log \sin x° = \bar{1}{\cdot}7678$; find $x°$ and $\log \cos x°$.

Test 32

1 Simplify $(2\frac{1}{2})^2 \times 2\frac{4}{5}$.

2 56 % of a debt is £47·88; what is 18 % of the debt?

3 $\sqrt{(0{\cdot}0949)}$, to 4 figures.

4 Factorise $1 + a^2b^2 - a^2 - b^2$.

5 Simplify $2x\left(x^2 - 7x + 1 - \dfrac{4}{x}\right) - x^2(5-x) + 3x\left(x + \dfrac{1}{x} - 3\right)$.

6 Simplify $(x^6 - x^2) \div (-x^2)$ and $(n+2)(n-7) - (n-3)^2$.

7 $743 \cos x° = 386$; find $x°$.

Test 33

1 S.P. £15·84, loss 12 %. Find the C.P.

2 Find the radius of a circle whose area is $1000\ \text{cm}^2$. (Take $\log \pi = 0{\cdot}4971$.)

3 $\dfrac{(0{\cdot}7246)^3}{(0{\cdot}8103)^2}$, to 3 figures.

4 $\dfrac{3b-1}{10} - \dfrac{2b-4}{15}$.

5 Simplify $(c-d)(d-c) - (c+d)(d+c)$

6 Solve $4x - y - 3 = 3y - 2x + 4 = 10x - 8y - 4$.

7 $4{\cdot}87 \tan x° = 5{\cdot}93$; find $x°$.

Test 34

1 Express 87p as a percentage of 116p.

2 $\sqrt{(2{\cdot}87^2 + 1{\cdot}69^2)}$, to 3 figures.

3 Factorise $1 + c - 6c^2$.

4 A tank containing water is 1·4 m long, 1·2 m wide. How many cm does the water-level fall when 84 litres are drawn off?

5 $4\pi r^3 = 3$; find r.

6 Complete $a^2 - 5ab + \ldots = (a \ldots)^2$.

7 AB = AC = 7·84 cm, BC = 9·45 cm; find ∠BAC.

Test 35

1 $\dfrac{0{\cdot}64 \times 0{\cdot}035}{0{\cdot}14}$.

2 $\dfrac{0{\cdot}9172 \times 0{\cdot}0604}{10{\cdot}79 \times 0{\cdot}5147}$, to 3 figures.

3 The population of a town was 42 732 in 1945 and increased by $8\frac{1}{3}$ per cent in the next 10 years. What was it in 1955?

4 Simplify $(a-2b)(a+3b)-2(a-b)(a-2b)+3(2a-b)(2b-a)$.

5 Solve $2r+p+10 = 0$, $3r-2p+1 = 0$.

6 If $x-3$ is a factor of $3x^2+bx-3$, find b.

7 ∠PQR = 52°, ∠PRQ = 90°, QR = 7·85 cm; find PQ.

Test 36

1 Find $\sqrt[4]{0{\cdot}1656}$, to 3 figures.

2 How many pieces, each 15 cm long, can be cut from 4·25 m of tape, and how much remains over?

3 Find to 3 figures the radius of a circle, area 5 cm^2.

4 Solve $\dfrac{2t-7}{5} - \dfrac{t-2}{2} = 1$.

5 Factorise $4a^2 - 14a - 18$.

6 Find x if the square of $3x-2$ is equal to 25.

7 $5{\cdot}69 \sin x° = 4{\cdot}73 \sin 50°$; find $x°$.

Test 37

1 Find the rent for 52 weeks at 69p a week.

2 Coal production in Great Britain in a certain year was 224 190 000 tonnes, valued at the pit-head at £643 287 515; find the value per tonne to the nearest penny.

3 $\sqrt[3]{(0{\cdot}6218)}$, to 3 figures.

4 $x^2 \times (2x)^3 \div (8x^6)$.

5 Evaluate $\dfrac{1}{x} - \dfrac{1}{y}$ if $x = -4\frac{1}{2}$, $y = -1\frac{1}{2}$.

6 $\left(\dfrac{8}{125}\right)^{-\frac{2}{3}}$

7 Find x to 3 figures if $x = x \cos 68°\,40' + 5$.

Test 38

1 36% of a load is $4\frac{1}{2}$ tonnes; how much is 64% of the load?

2 An allotment 25 m long, 9 m wide, yields 423 kg of potatoes; find the average yield in tonnes per 10 000 m^2.

3 $\dfrac{9{\cdot}5\times 0{\cdot}7073}{(0{\cdot}3174)^2\times 0{\cdot}1}$, to 3 figures.

4 Factorise $5-9y-2y^2$.

5 $\pi r^2 h = 20$; find r if $h = \frac{3}{4}r$.

6 Solve $\dfrac{x-1}{3}+\dfrac{y+1}{2} = 1$, $\dfrac{2x+1}{5}-\dfrac{3y+1}{4} = 5$.

7 Find $17{\cdot}24 \sin 62° 20' \div \sin 37° 10'$, to 3 figures.

Test 39

1 $22\frac{1}{2}$% of £1·40.

2 $\dfrac{1}{4{\cdot}72}+\dfrac{1}{7{\cdot}36}$, to 3 figures.

3 A garden roller is 80 cm in diameter and 1 m wide. What area, to nearest m^2, does it roll over in 100 revolutions?

4 Factorise $4(2a-b)^2-(a+3b)^2$.

5 $10^x = \sqrt[3]{0{\cdot}683}$; find x.

6 Find b if $x^2+bx+2\frac{1}{4}$ is a perfect square.

7 If $c^2 = a^2+b^2-2ab \cos C$, where $a = 7{\cdot}8$, $b = 6{\cdot}4$, $c = 5{\cdot}5$, find C.

Test 40

1 Express $\dfrac{103}{125}$ as a percentage.

2 $\dfrac{8{\cdot}672\times 10{\cdot}18}{634{\cdot}7\times 7\frac{1}{3}}$, to 3 figures.

3 A cylindrical tank, diameter 1·35 m, is 1·05 m high, internal measurements; find its capacity in litres, to 3 figures.

4 If $\pi r^2 h = 465$, find r when $h = 8{\cdot}15$, to 3 figures.

5 $\dfrac{a^3b^3}{a^3c^3}\div(b^3c^3)$.

6 Factorise $m(p+q)-p-q$.

7 If $\frac{1}{2}bc \sin A = \sqrt{\{s(s-a)(s-b)(s-c)\}}$, where $a = 4{\cdot}6$, $b = 5{\cdot}9$, $c = 6{\cdot}3$, $s = \frac{1}{2}(a+b+c)$, find A, given A is acute.

REVISION PAPERS 1–81

PAPERS 1–20 (Volume I)

Paper 1

1 Express 7546 in prime factors. Find the least integral value of x if $7546x$ is (i) a perfect square, (ii) a perfect cube.

2 195 men can make a track in 17 days. How many more men must be employed to reduce the time by 2 days?

3 A photograph, 6 cm by 5 cm, is mounted on a rectangular card; there is a margin 1·5 cm wide all round. Find area of margin.

4 What must be added to $2(2l-m-2n)$ to make $5(l+2m-n)$?

5 A workman is paid 48p an hour ordinary time and 72p an hour overtime. He receives £25·20 for 49 hours' work; how many of these hours are overtime?

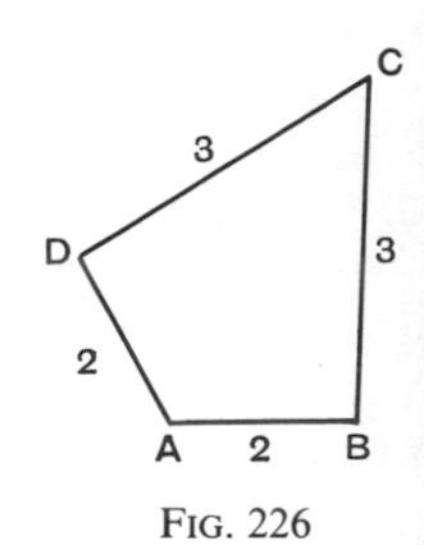

FIG. 226

6 In Fig. 226, prove $\angle ABC = \angle ADC$, and AC bisects $\angle BCD$.

Paper 2

1 Divide £75·68 by 55, to the nearest penny.

2 A car which runs 11 km to the litre of petrol uses 48 litres for a journey. How much petrol is used for this journey by a car which runs 8 km to the litre?

3 If $x = 2y$ and $3y = z$, evaluate $xy-2z$ when $z = 9$.

4 Square $6a^2b^3$ and divide the result by $4b^2$.

5 What angle is equal to (i) $\frac{1}{5}$ of its supplement, (ii) $\frac{1}{8}$ of its complement?

6 AB, DC are the parallel sides of the trapezium ABCD. If AD = DC and $\angle BAC = 33°$, find $\angle ADC$.

Paper 3

1 How many coins weigh 1 tonne, if 5 of them weigh 28·4 g?

2 A box without a lid is 1·8 m long, 1·2 m wide, 90 cm deep, internally. What area of tin sheeting is required to line inside the base and sides? If the box contains 27 litres of water, find the depth of water.

3 Find the L.C.M. of $4abc^2$, $6a^2b^2c$, $8a^3c^3$.

4 Solve $\frac{1}{3}h+12 = h$.

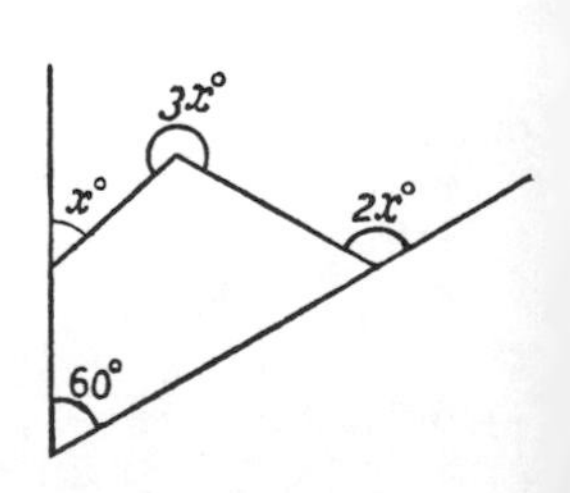

FIG. 227

5 How many minutes are there between p minutes past nine and $2p$ minutes to twelve on the same morning? Find p if the number of minutes is $12p$.

6 In Fig. 227, find x.

Paper 4

1 Evaluate $0{\cdot}7363+0{\cdot}805-0{\cdot}0909-0{\cdot}8124$.

2 A steel rail is 9·6 m long and weighs 140 kg. Find the weight per km in tonnes.

3 Simplify $(5x+y)\div(x-y)$ if $x = 7y$.

4 What must be added to $10-a$ to make 9?

5 A man buys $18n$ eggs for £4 and sells them at 4p each; his profit is £2·48. Find n.

6 Two circles have the same centre; a straight line XABY cuts the inner at A, B and the outer at X, Y; prove BX = AY. [XABY is *not* a diameter.]

Paper 5

1 Multiply 0·039 by 0·6065, giving the answer to 3 figures.

2 A window, 90 cm by 60 cm, contains 9 panes of glass, each 27 cm by 17·5 cm. Find in cm^2 the area not occupied by glass.

3 By what must a^2b^2 be divided to give b^6?

4 What can you say about h if $10-h$ is less than $2h+1$?

5 The proper dose for a child n years old is either $\dfrac{n}{n+12}$ or $\dfrac{n}{24}$ of the adult dose. What fractions of the adult dose should be given to a child aged 8, according to the rules?

6 In Fig. 228, AB is parallel to EF. Find x. Give reasons.

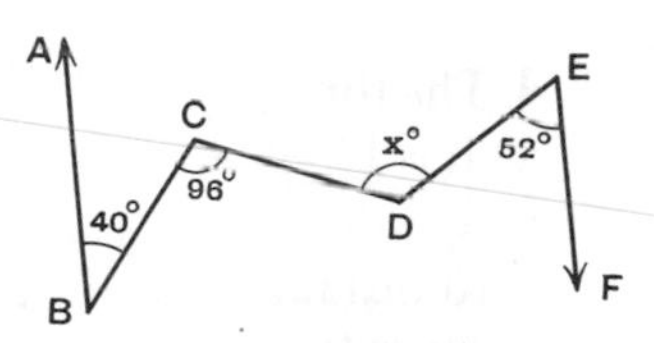

FIG. 228

Paper 6

1 Simplify $3\frac{3}{4}\times\frac{2}{2\frac{1}{2}}\div\left(\frac{1}{2\frac{1}{3}}+1\frac{1}{2}\right)$.

2 10 kg of tea cost £6·65. Find to the nearest penny the cost of $3\frac{1}{2}$ kg.

3 Simplify $2-4c-c(1+4c)-1+c^2$.

4 Solve $2x-5 = 1\frac{1}{2}x-\dfrac{2x+1}{3}$.

5 The marks in an examination run from 30 to 90; they are scaled so that an original mark n becomes N where $N = \frac{5}{3}(n-30)$. What is the new mark if the original mark is 48? What is the new top mark and the new bottom mark? What mark remains unchanged?

6 Draw two triangles ABC, PQR, which are *not* congruent, but are such that AB = PQ = 5 cm, AC = PR = 4 cm, $\angle$ABC = $\angle$PQR = 46°. Measure BC and QR.

Paper 7

1 Divide 31·11 by 0·0305.

2 Simplify $(3b^3)^3 \div 3b^2$.

3 The area of a rectangular field, 240 m wide, is 84 000 m^2. Find in metres the length of the fence which encloses it.

4 Subtract $4x+2y-\{5x-3(x-3y)\}$ from $14xy$.

5 A pan size N is of diameter $5(N+3)$ cm and costs $\frac{5}{2}(3N+10)$ pence. Find the size number and diameter if the cost is 70p.

6 P is a point inside $\triangle$ABC such that $\angle$PBC = $\angle$PCA. Prove that $\angle$BPC + $\angle$ACB equals 2 right angles.

Paper 8

1 A grocer buys 50 kg of tea at 62p per kg, 45 kg at 56p per kg and 24 kg at 60p per kg. He mixes it all together and sells it at 68p per kg. What is his profit?

2 A man bicycles 49 km in $2\frac{1}{3}$ hours. How long does he take to go $17\frac{1}{2}$ km at the same rate?

3 By what must $\dfrac{3}{p}$ be multiplied to give $\dfrac{p}{3}$?

4 The total value of c coins of 10p each, d coins of 5p, and e coins of 1p each is £1·50. Find c in terms of d, e.

5 A car travelling at v km per hour can be stopped in an emergency within a distance of s metres where $s = 0{\cdot}02v^2 + 0{\cdot}6v$. Find the distance if the car is travelling at 90 km an hour.

6 ABCD is a straight line such that AB = CD; K is a point outside the line such that KB = KC. Prove $\triangle$KCA $\equiv$ $\triangle$KBD.

Paper 9

1 What number exceeds 72·84 by the same amount that 72·84 exceeds 49·17?

2 The upright portion of a metal cross, 1 cm thick, is 75 cm long, 12 cm wide; the horizontal portion is 50 cm long, 10 cm wide. Find (i) the volume, (ii) the *total* area of the surface of the solid.

3 Simplify $\dfrac{10xyz^2}{6x^2yz} + \dfrac{xz}{3x^2}$.

4 Simplify $a(b-c)+b(c-a)-c(b-a)$.

5 The ordinary price of a mat is P pence. In a sale 20p in the £ is deducted, making the price 84p. Find P.

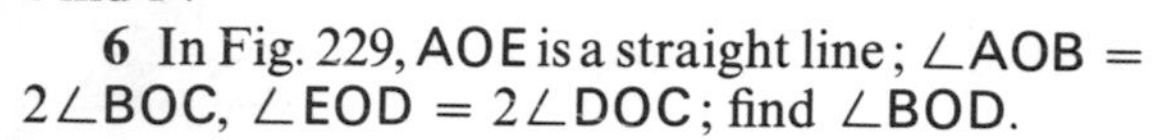

6 In Fig. 229, AOE is a straight line; $\angle AOB = 2\angle BOC$, $\angle EOD = 2\angle DOC$; find $\angle BOD$.

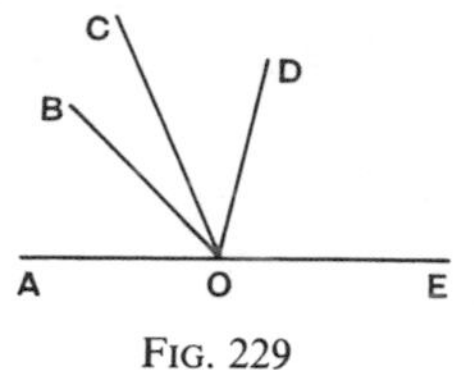

FIG. 229

Paper 10

1 Find in prime factors the L.C.M. of 24, 42, 63, 105, 135.

2 It requires 4 complete turns of the handle to wind up a bucket from the bottom of a well 12 m deep. Through what angle must the handle be turned to raise the bucket 5 m?

3 Simplify $2a+3-\dfrac{a}{2}-\dfrac{3b}{b}$.

4 Solve $6-\frac{3}{4}(r-6) = \frac{1}{2}(r+1)$.

5 A man walks at a certain rate for the first 2 hours of a journey and then increases his speed by 2 km per h for the next 30 minutes. He walks 16 km in all; find his first speed.

6 P is a point on the side BC of quad ABCD. If $\triangle ABP$ and $\triangle ACD$ are both equilateral, prove that $PD = BC$.

Paper 11

1 Multiply £87·37 by 73.

2 How many sheets of paper, each 12 cm by 16 cm, can be made from a strip of paper, 48 cm wide, 1 km long?

3 Simplify $\dfrac{3r^2}{s^2}\times\dfrac{s^2}{3r^2}$.

4 If $a = 3b = 4c = 60$, find the values of $a-b+c$ and abc.

5 A bottle weighs $(W-k)$ grammes when empty and weighs $(W+k)$ grammes when it contains v cm^3 of oil. Find the weight of 1 cm^3 of oil.

6 The bisector of $\angle A$ in $\triangle ABC$ meets BC at P. Given that AP = 5 cm, $\angle B = 28°$, $\angle C = 68°$, draw $\triangle ABC$; measure BC.

Paper 12

1 Simplify $4\frac{2}{21}-2\frac{13}{14}+(1\frac{5}{7}\div 5\frac{1}{7})$.

2 An engine travelling at 80 km per h picks up water from a tank between the rails, 200 m long, at the rate of 800 litres per second. Find the volume of water picked up.

3 Simplify $\frac{1}{2}(a+2b+3)-\frac{1}{4}(2a-b-c)$.

4 Solve $4-\frac{3}{t}=\frac{18}{t}-3$.

5 A rectangle is l cm long, b cm broad. If $\frac{l}{9}=\frac{b}{4}=\frac{c}{6}$, find (i) the perimeter, (ii) the area of the rectangle in terms of c.

6 In Fig. 230, find e in terms of b, c, d.

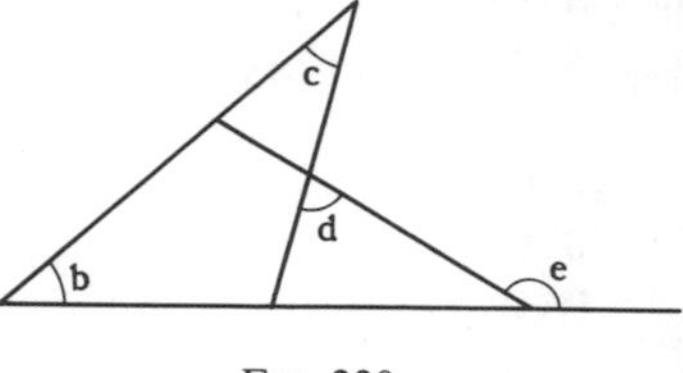

FIG. 230

Paper 13

1 Divide 0·12903 by 25·5.

2 A rectangular piece of wood, 14 cm by 11 cm, has two rectangular holes, each 5 cm by 4 cm, pierced in it; the wood is 2 cm thick. Find its weight, if 1 cm^3 of wood weighs 0·8 g.

3 How many metres at 36p per metre can be bought for £A?

4 Solve $\frac{7x}{5}-\frac{2x+1}{2}=\frac{x-1}{3}+\frac{1}{2}$.

5 A man starts from P at 9 a.m. and walks to Q, 18 km away. He walks at 6 km per h for 9 km, then rests for 30 min, and then walks on to Q at 6 km per h. A cyclist leaves Q at 11 a.m. and rides at 18 km per h to P by the same road. Find from a graph at what time and how far from P they meet.

6 *Using ruler and compass only*, construct $\triangle$ABC so that $\angle$ACB = 90°, $\angle$ABC = 60°, BC = 4 cm. Measure AB.

Paper 14

1 Simplify $6^3 \times 8^2 \div (12^2 \times 4^3)$.

2 A train travelling at 60 km per h takes 9 seconds to pass a telegraph pole. Find the length of the train in metres.

3 Simplify $13(y-x)-5\{2y-3(x+y)\}$.

4 Find the sum of 5 consecutive even numbers if the middle number is $8n$.

5 In a sale the price of a chair is reduced from £P to £$(P-\frac{1}{5}P)$; the reduced price is £1·40. Find the reduction in price.

6 In $\triangle$ABC, AB = AC and $\angle$A = 20°; B is a point on AC such that $\angle$DBC = 60°. Prove AD = DB.

Paper 15

1 Simplify $(\frac{1}{4}\times 2\frac{2}{3})-(1\frac{3}{5}\times\frac{1}{24})$.

2 The internal area of the base of a rectangular tank which can hold 1400 litres is 1·75 m^2. Find the internal depth.

3 Express $4a-2(3b-2c)$ in terms of n if $a = 4n+3$, $b = 3n-4$, $c = 2n-7$. Find the value of n if $4a-2(3b-2c) = 35$.

4 Divide $10t^8$ by $5t^4$.

5 A submarine can travel at 10 knots below the surface and at 14 knots on the surface. [1 knot means 1 sea-mile per hour.] If the submarine takes 8 hours to travel 98 sea-miles, find the distance travelled below the surface.

6 ACB is a straight line; ABX, ACY are equilateral triangles on opposite sides of AB. Prove CX = BY.

Paper 16

1 Multiply 0·5075 by 0·0408.

2 Fig. 231 represents the cross-section of a solid, 0·5 cm thick. The corners are right-angled and dimensions are shown in cm. Find (i) the area of the cross-section, (ii) the weight of the solid if the material weighs 10·5 g per cm³.

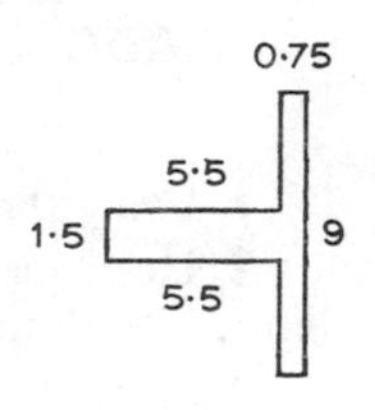

FIG. 231

3 A sheet of paper is $\dfrac{3}{2n}$ cm thick; how many sheets are there in a pile 6 cm high?

4 Which of the numbers 1, 2, 3, 4, 5, 6 are roots of the equation, $x^2(11-x) = 36(x-1)$?

5 I think of a number, halve it and subtract 1; the result is double the amount obtained by dividing the number by 3 and subtracting 4. Find the number.

6 ABCD is a quadrilateral in which AD is parallel to BC. R is a point inside ABCD such that AR = AB and ∠BAR = 40°. If ∠RBC = 30°, calculate ∠ABC and prove that AR bisects ∠BAD.

Paper 17

1 Evaluate $0{\cdot}16 \times 0{\cdot}015 \times 0{\cdot}0625$.

2 Simplify $3\frac{1}{7} \times 1\frac{2}{3} \div 1\frac{4}{21}$.

3 Simplify $(3xy^3)^3 \div (6xy^2)^2$.

4 If $8(R-2r) = 3R$, find R when $r = \frac{3}{4}$.

5 £30 are shared between A, B, C so that A gets £6 more than B, and C gets $\frac{3}{7}$ of the amount A and B together get. How much does each receive?

6 Draw △ABC in which AB = 4 cm, AC = 3 cm, ∠BAC = 35°. *Construct* equilateral triangles ABX, ACY, each *outside* △BAC. Prove that △XAC ≡ △BAY.

Paper 18

1 Find the duty on $4\frac{1}{2}$ kg of tobacco at £5·80 per kg.

2 It takes me 28 minutes to walk from my house to the station if I walk at $6\frac{1}{2}$ km per h. How long do I take at 7 km per h?

3 Simplify $1\frac{3}{4}(a+b)-1\frac{1}{4}(a-b)$.

4 Simplify $\dfrac{2{\cdot}7\times 0{\cdot}04}{0{\cdot}9\times 0{\cdot}8}$.

5 Solve $\dfrac{3x}{2}-\dfrac{x-1}{3}=\dfrac{1}{2}(3x-2)$.

6 *Using ruler and compasses only*, construct $\triangle$ABC so that AB = 3 cm, BC = 5 cm, $\angle$ABC = 90°. Construct a point D on the same side of AC as B such that AD = 5 cm, CD = 3 cm. Prove that $\angle$ADC = 90°.

Paper 19

1 Evaluate, to 3 places of decimals, $108{\cdot}36 \div 173{\cdot}04$.

2 How many litres of water must be pumped into a swimming bath, 25 m long, 18 m wide, to raise the water-level 20 cm?

3 Simplify $6xyz\times(2x^2y)^3\div(6xy^3)^2$.

4 Find the increase in value of $5x^2-7x-3$ when x increases from 2 to 4.

5 I walk for $2\frac{1}{2}$ hours in the morning at x km an hour and for $\frac{1}{2}x$ hours in the afternoon at 6 km an hour; this makes $38\frac{1}{2}$ km altogether. How far did I walk in the morning?

6 Fig. 232 shows in degrees the angles at which a line is cut by two parallel lines. Find x.

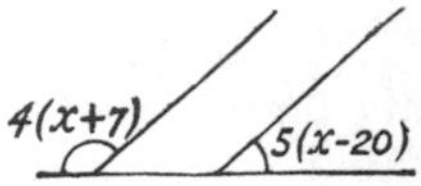

4(x+7)
5(x-20)

Fig. 232

Paper 20

1 Divide $0{\cdot}80376$ by $63{\cdot}04$.

2 A man buys 6 dozen articles at 4 for 3p and 6 dozen at 3 for 2p; he sells them all at 6 for $5\frac{1}{2}$p. Find his profit.

3 Simplify $\{(a-b+c)-(b-c+a)\}-\{(c-a+b)-(a-b-c)\}$.

4 The sum of the first n numbers of the set

$$a,\ a+d,\ a+2d,\ a+3d,\ \ldots$$

is s where $s=\frac{1}{2}n\{2a+d(n-1)\}$. (i) Find d if $a=9$, $n=16$, $s=48$. (ii) Find the sum of the first 20 numbers of the set 1, 3, 5, 7, 9, ...

5 The average speed of a motorist for the first $1\frac{1}{2}$ hours of a journey of 108 km is 16 km an hour less than his average speed for the remaining 20 minutes. Find his average speed for the first part.

6 In $\triangle$ABC, $\angle$C = 36°; a line parallel to BC cuts AB, AC at H, K; $\angle$BHK = 108°. Calculate $\angle$A and prove that CA = CB.

Paper 21

1 Evaluate (i) $(0{\cdot}25)^2 \times (0{\cdot}12)^3$; (ii) $0{\cdot}022344 \div 73{\cdot}5$.

2 After paying two-fifths of his whole income in taxes, a man has £1020 left. Find his income.

3 If $p = 6$, $q = -3$, $r = -12$, evaluate $2p^2 - 3p(r-q) - q^2$.

4 I travel 40 km in $2\frac{2}{15}$ hours, walking part of the way at 6 km per h and driving the rest at 40 km per h. How far do I walk?

5 If $t(R-2r) = 3R$, express R in terms of t, r.

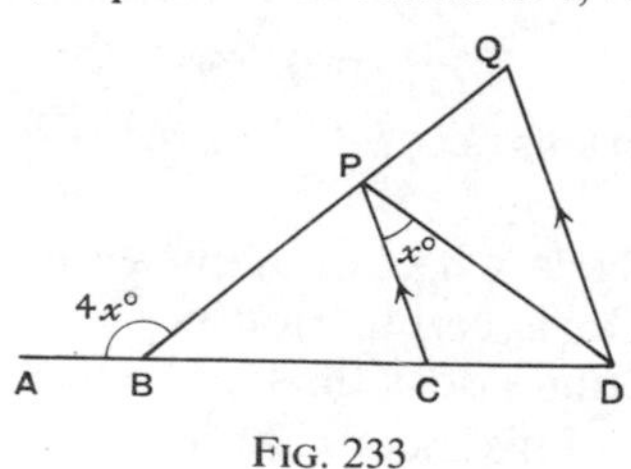

FIG. 233

6 In Fig. 233, ABCD and BPQ are straight lines; BP = BC and DQ is parallel to CP. Prove (i) CP = CD, (ii) DP bisects ∠CDQ.

Paper 22

1 Simplify $(1\frac{5}{8} - \frac{3}{4}) \div (1\frac{5}{12} - \frac{1}{4})$.

2 A tin sheet 75 cm by 1 m weighs 240 g. Find the weight of a tin sheet of the same thickness, 1·875 m by 1·05 m.

3 Simplify

$$3x\left(5x^2 - 7x + 4 - \frac{1}{x}\right) - 2x\left(4x^2 + 5x - 3 + \frac{1}{x}\right) - (x^3 - 7x^2 + 2x - 7).$$

4 45% of a debt is £1·53; what is 75% of the debt?

5 A man will be $3n$ years old in $(n+3)$ years' time. How old was he $(n-3)$ years ago?

6 In Fig. 234, ∠ACD = ∠ABC, and CP bisects ∠BCD. Prove that ∠APC = ∠ACP.

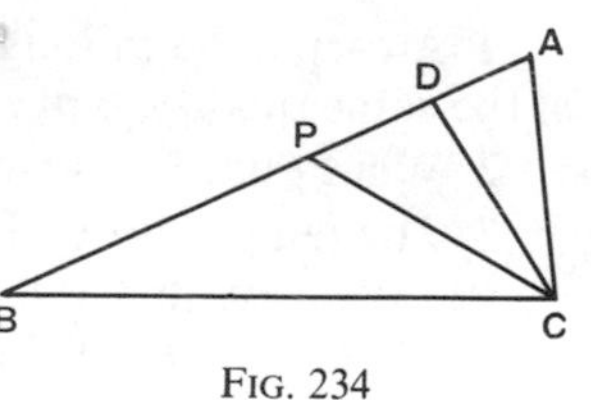

FIG. 234

Paper 23

1 How many allotments, each 50 m long, 30 m wide, can be made from 0·06 km^2?

2 The R.F. of a map is 1 : 20 000. Find the area of an estate which is represented on the map by 13·5 cm^2.

3 Subtract $1-5c+3c^2$ from $2c-1$.

4 Solve $\frac{1}{3}(9-2p) = 6-\frac{1}{2}(p+4)$.

5 The price of a bottle of wine in Portugal is 28 escudos; the same bottle in England costs 95p. Express the Portuguese price as a percentage of the English price, taking £1 to be worth 68 escudos. (Give the answer to 1 decimal place.)

6 ABCDE is a regular pentagon; CB, EA are produced to meet at X. Find ∠AXB.

Paper 24

1 A stack of 4500 bricks occupies $3{\cdot}82\ \text{m}^3$. Find the size of a brick in cubic centimetres.

2 A bicycle has a back wheel of diameter 70 cm and a front wheel of diameter 65 cm. On a certain journey the back wheel revolves 728 times; how many times does the front wheel revolve?

3 A man spends $\frac{4}{9}$ of his income at home, $16\frac{2}{3}\%$ of his income on tax and saves the rest; he saves £616. Find his income.

4 Simplify $(2x)^3 \times (3y)^3 \div (6x^2y)$.

5 Draw the travel graph of a man A who walks for 40 minutes at 6 km per h, then halts for half an hour and then bicycles on at 20 km per h for 24 km. Draw in the same figure the travel graph of a man B who starts from the same place as A but $1\frac{1}{2}$ hours later and drives along the same road at 48 km per h. Find the distance B has gone when he passes A.

6 ABPQ, ACXY are squares on the sides AB, AC of △ABC, drawn outside the triangle. Prove BY = CQ.

Paper 25

1 A carpet 7·5 m long, 4·2 m wide, costs £252. What will a carpet of the same quality, 5 m square, cost?

2 What number, increased by $\frac{4}{7}$ of itself, makes 99?

3 The temperature of water in a boiler rises at a steady rate.

(i) At 6 a.m. it is 100°, at 9 a.m. it is 180°; what is it at 8 a.m.?

(ii) At 6 a.m. it is $x°$, at 7 a.m. it is $y°$; what is it at 9 a.m.?

(iii) At 1 p.m. it is $p°$, at 4 p.m. it is $q°$; what is it at 2 p.m.?

4 Simplify (i) $\dfrac{2pq}{3pqr}-\dfrac{q}{6qr}$; (ii) $\dfrac{-9y^2}{(-3y)^2}$.

5 A grocer buys some eggs at 19p per dozen; he sells half of them at $2\frac{1}{2}$p each and the rest at 2p each, and thereby gains £1·60 on the whole transaction. How many eggs did he buy?

6 In △ABC, AB = AC; AB is produced to D so that BD = BC. Prove that ∠ACD = 3∠ADC.

Paper 26

1 Find six consecutive odd numbers whose sum is 120.

2 Find the total area of wall-surface of a room 4·8 m long, 4·2 m wide, 2·7 m high, allowing for two windows each 1·5 m by 0·9 m, a door 2·1 m by 1·2 m, and a fireplace 1·8 m by 1 m.

3 Simplify $a-\frac{a+b}{4}-\frac{a-b}{4}+\frac{a-7b}{12}$.

4 (i) A boy gets 3 marks for each sum right and loses 2 marks for each sum wrong. He obtains 37 marks for 24 sums. How many were right?

(ii) A boy gets r marks for each sum right and loses w marks for each sum wrong. He obtains n marks for k sums. Find in terms of r, w, n, k the number he did right.

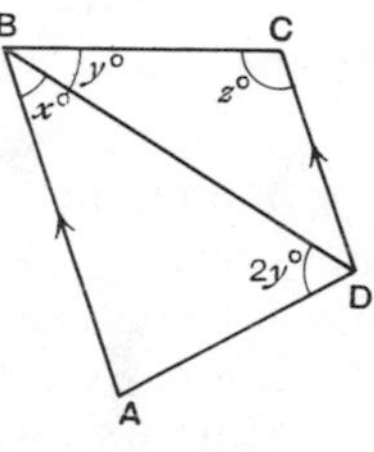

FIG. 235

5 Make r the subject of the formula $V = \frac{4}{3}\pi r^3$.

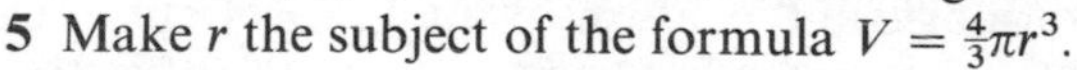

6 In Fig. 235, AB is parallel to DC. If $x = 1\frac{1}{3}y$ and $y = \frac{3}{8}z$, find ∠BCD, ∠ABC, ∠BAD.

Paper 27

1 Multiply 0·3625 by 0·464.

2 Beech trees planted for timber are allowed 40 m^2 per tree. The average yield per tree is 0·3 m^3 and is sold at £1·50 per load of 0·75 m^3. Find the value of timber per 10 000 m^2.

3 If $a = 3$, $b = -8$, $c = -4$, evaluate $(a-b)(b-c)+2c^2$.

4 For what value of t does $\frac{1}{5}t$ exceed $(\frac{1}{2}t-\frac{1}{3}t)$ by 1?

5 Find the sale price of an article costing £1·86, if the profit is one-third of the *sale price*.

6 Draw a straight line ABCD so that AB = 3 cm, BC = 2 cm, CD = 3 cm. Construct a point P such that PA = PD = 5 cm. *Measure* ∠PAD and use this measurement to calculate ∠APD. Join PB, PC and prove that PB = PC.

Paper 28

1 Bleriot's first Channel flight in 1909 took 37 min and the distance was 42 km. Find the speed in km per h, to 3 figures.

2 By selling my house for £5640 I lose at the rate of 6p in the £ on the price I paid for it. How much do I lose?

3 Simplify $\frac{3s^2}{t^2}\times(-3s)\times(-t)^3$.

4 If a clothes-line l metres long is attached to the tops of two posts d metres apart, the sag h metres is given approximately by the formula, $h = \sqrt{\{\frac{3}{8}d(l-d)\}}$. Find h in terms of d if $l = 1{\cdot}06d$.

5 A French taxi-driver charges 14·35 francs for a journey of 35 km. Taking £1 = 13·3 francs, find the cost in pence per km, to the nearest penny.

6 The sides AD, BC of the quadrilateral ABCD are parallel; ∠BAC = 50°, ∠ACD = 20°, CA = CB. Calculate ∠BCD and prove that CB = CD.

Paper 29

1 Find the cost of 209 tonnes at £3·40 per tonne.

2 Find the weight of an iron bar of length 6 m and cross-section 16 cm^2, if the iron weighs 7600 kg per m^3.

3 Simplify $\dfrac{6x-3y}{2}-\dfrac{5y-2z}{3}+\dfrac{6z-12x}{4}$.

4 Electricity is consumed at the rate of 1 watt if $\frac{1}{1000}$ of a unit is used per hour. Find to the nearest penny the cost of running an electric fan for 45 hours a week for 52 weeks if its rate of consumption is 150 watts and if electricity costs 1·75p per unit.

5 In △ABC, AB = AC; D is a point on AC such that AD = BD = BC. Calculate ∠BAC.

6 C degrees Celsius is the same temperature as F degrees Fahrenheit if $C = \frac{5}{9}(F-32)$. Find n if n degrees Celsius is the same temperature as $(-\frac{1}{3}n)$ degrees Fahrenheit.

Paper 30

1 Find the ratio $\frac{3}{4}:\frac{5}{6}$ in simplest form.

2 A rectangular swimming bath, 40 m long, 9·5 m wide, is being filled at the rate of 1·9 m^3 per minute. How long does it take for the water-level to rise 7·5 cm?

3 Evaluate $6x^2-2x+3$ when $x=-2$, $x=-1$, $x=0$, $x=1$, $x=1\frac{1}{2}$.

4 (i) Simplify $\dfrac{ac^2}{b}\times\dfrac{(-b)^2}{(-c)^3}$.

(ii) Divide $\frac{1}{2}n$ by $\dfrac{2}{n}$.

5 Fig. 236 shows the lengths in cm of the sides of an *isosceles* triangle. Find the numerical values of these lengths. [There is more than one set of answers.]

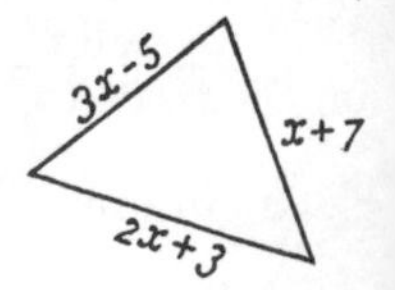

FIG. 236

6 The diagonals of the quadrilateral ABCD cut at K. If AB = BC = CD and if ∠ABC = ∠BCD, prove that (i) KB = KC, (ii) ∠AKD = ∠ABC.

Paper 31

1 2400 pocket-knives are sold wholesale for £192. What is the price per knife?

2 A plot of ground 80 m long, 32 m wide, has a fence round it, and a pathway 1·5 m wide is made round the edge of it inside the fence. Find the cost of the path at 3p per m^2.

3 If $x = 6$, $y = -3$, $z = -2$, $t = 0$, evaluate

$$\text{(i) } x \div (yz); \quad \text{(ii) } x + 4y + 3tz; \quad \text{(iii) } (x-y)(y-z).$$

4 For what value of W is $\frac{1}{2}(W + 1\frac{1}{2})$ equal to $\frac{1}{3}(W + 2\frac{1}{2})$?

5 Find v if v m per second equals $3(v+4)$ km per hour.

6 The bisector of $\angle$A of $\triangle$ABC cuts BC at D. Through C a line is drawn parallel to DA to meet BA produced at P. Prove AP = AC.

Paper 32

1 Use a subtraction method to find the cost of 136 m of material at 99p per metre.

2 A metal plate 20 cm long, 15 cm wide, weighs 4·68 kg. If 1 cm^3 of the metal weighs 7·8 g, find the thickness of the plate.

3 Simplify $(4c^2)^3 \times (2c^3)^2 \div (4c^3)^4$.

4 A rectangle is $(2n+1)$ cm long, $(3n-4)$ cm high. Find its perimeter in terms of k if $n = 2-k$.

5 If a certain number is added both to the numerator and to the denominator of $\frac{9}{17}$, the new fraction reduces to $\frac{5}{7}$. Find the number.

6 ABCD is a quadrilateral; AC is joined and CD is produced to E; AC = AB, $\angle$ABC = 55°, $\angle$ACD = 70°, $\angle$ADE = 140°. (i) Prove that DA = DC. (ii) Name with reasons two parallel lines in the figure.

Paper 33

1 7 m of cotton fabric cost £2·24. Find the cost of $11\frac{1}{2}$ m.

2 The air speed of a plane is 200 km per h. Find the difference in time taken to travel 210 km with and against a wind blowing at 25 km per h.

3 Simplify $\dfrac{y(y+z)}{z} - y\left(1 - \dfrac{1}{y}\right) + z\left(1 - \dfrac{1}{z}\right)$.

4 Find c if $x = -2$ is a root of

$$3x(x-c) - 9c = \tfrac{1}{2}x^2.$$

5 In Fig. 237, AB = AC and $\angle$BAC = 90°; BH, CK are perpendicular to AD. Prove that BH = AK. What can you say about the length of CK?

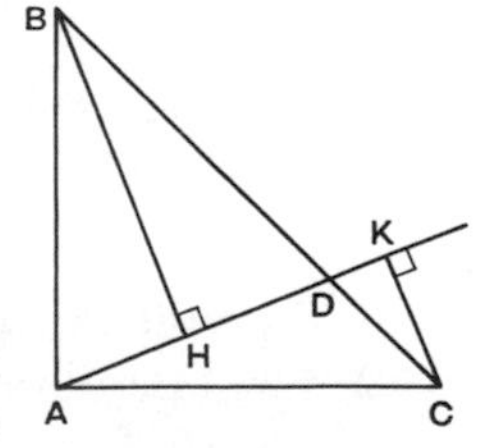

FIG. 237

6 After $\frac{7}{10}$ of a tank was filled, there was room for $(g+3)$ more litres. Find in terms of g how much the tank holds.

Paper 34

1 Decrease 3 hours in the ratio 5 : 6.

2 When driving at 72 km per h, a man is blinded for 4 seconds by a bright light. How far does he travel blind?

3 The rateable value of a house is $\frac{5}{6}$ of the annual rent. If the rent is £84 a year and if a rate of 56p in the £ is demanded, how much must be paid in rates?

4 If $x = 2, y = -1, z = 0$, evaluate (i) $\frac{x}{y} - \frac{y}{x}$; (ii) $2xyz$; (iii) $x^3 + y^3 + z^3$; (iv) $(x+y)^3$.

5 A man travels from A to B at 15 km per h and returns from B to A at 18 km per h. The total journey takes 1 h 28 min. How far is B from A?

6 *Using ruler and compasses only*, construct $\triangle$ABC so that AB = 4 cm, AC = 6 cm, $\angle$BAC = 45°. *Construct* the equilateral triangles ABP, ACQ so that P, C are on the same side of AB and so that Q, B are on the same side of AC. Prove CP = BQ.

Paper 35

1 The scale of a map is 10 cm to 1 km. Find its R.F.

2 A dealer makes a profit of 35p in the £ on the cost price of his goods. Find his profit on an article he sells for £4·32.

3 If $\frac{1}{3}x - \frac{1}{4}y = 2\frac{1}{2}$, express $5x + 6y$ in terms of x.

4 For what value of b is $x = -2$ a root of $b(x^3+6) = x(2b+5)$?

5 4 tickets at $(x-6)$ pence each cost the same as 3 tickets at $(x+6)$ pence each. Find the value of x and the cost of $(x+1)$ tickets at $\frac{1}{3}x$ pence each.

6 In Fig. 238, arrows indicate that lines are given parallel. Find c in terms of a, b.

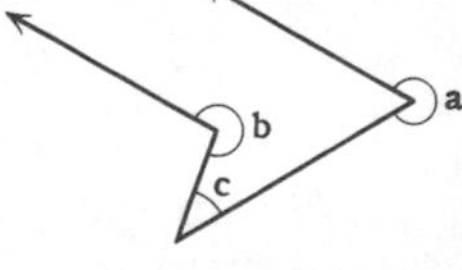

FIG. 238

PAPERS 36–50 (Ch. 1–11)

When using tables, work to 4 figures and give answers to 3 figures, unless otherwise stated.

Paper 36

1 Find the rate payable on £126 at 73p in the £.

2 What percentage of 80 m per second is 201·6 km per hour?

3 Simplify (i) $\frac{a}{2} + \frac{a}{3} + \frac{a}{6}$; (ii) $\frac{(-8x)^2}{-2x}$.

4 If a number of two digits is added to the sum of its digits, the result is 69; also the number is increased by 18 if its digits are reversed. Find the number.

5 In quadrilateral ABCD, $\angle A = \angle B = 90°$, $AB = 6$ cm, $AD = 5$ cm, $BC = 4$ cm. Calculate the acute angle at which AC cuts BD.

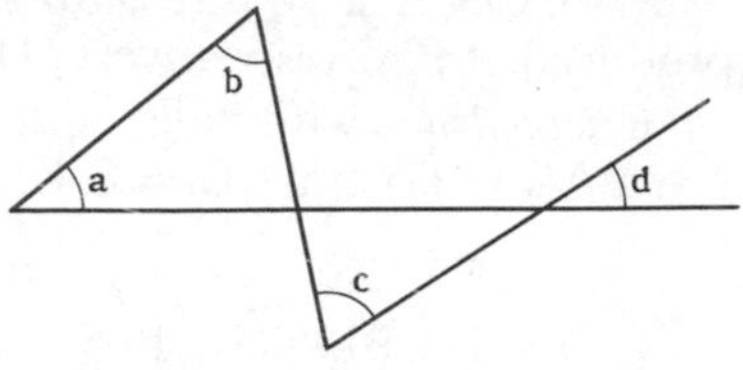

FIG. 239

6 In Fig. 239, find d in terms of a, b, c.

Paper 37

1 Find by factors the value of $5{\cdot}37^2 - 4{\cdot}63^2$.

2 If 56% of a sum of money is £2·10, find the value of 72% of the same sum.

3 Solve $2y - \frac{1}{3}z = 14$, $\frac{1}{4}y + 3z = 20$.

4 If $N + (35$ per cent of $N)$ equals 324, find N.

5 A conical funnel, vertical angle 75°, rests on the rim of a cylinder, height 9 cm, diameter 5 cm, internal measurements, see Fig. 240. Find the height of the apex of the funnel above the base of the cylinder.

6 ABC is an equilateral triangle; BC is produced to D so that BC = CD. Prove $\angle BAD = 90°$.

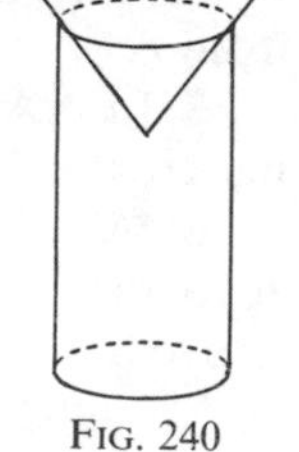
FIG. 240

Paper 38

1 Simplify $(3\frac{1}{4} - 1\frac{1}{2}) \div (2\frac{5}{6} - \frac{1}{2})$.

2 A hut costs £175 when new and is worth £105 after 2 years. By how much per cent has its value decreased?

3 Simplify (i) $a\left(1 + \dfrac{1}{a}\right) - b\left(1 + \dfrac{1}{b}\right)$; (ii) $(x+1)(x^2 - x + 1) - 1$.

4 If $x + y = 16$ and $x + 2y = 27$, find the value of $x + 3y$.

5 At a distance of 80 m from a tower, the angles of elevation of the top and bottom of a flagstaff on the tower are 53° 30′ and 51°. Find the length of the flagstaff.

6 The diagonals of the parallelogram ABCD cut at K. Any line through K cuts AD, BC at P, Q. Prove DP = BQ.

Paper 39

1 If 12% of a bill is deducted, £154 is left. How much is the bill?

2 A rug costs £2·50 and is sold for £3·80; find the gain per cent.

3 Simplify (i) $3(z - y - x) - 5(2x - 3y + z)$; (ii) $r^2s^3 \div \dfrac{r^3}{s^2}$.

4 Find two consecutive even numbers such that one-quarter of the smaller exceeds one-fifth of the larger by 4.

5 ABCD is a square side 6 cm; APQ is drawn to cut BC, DC produced at P, Q respectively; DQ = 7·5 cm. Find ∠CAQ and BP.

6 *Construct* with ruler and compasses the parallelogram ABCD, given AC = 6·8 cm, BD = 8·4 cm, BC = 4·0 cm. Measure AB.

Paper 40

1 Divide 0·01972 by 13·6.

2 Tea costing 50p per kg is mixed with tea costing 70p per kg in the ratio 3:5 by weight. The mixture is sold at 80p per kg. Find the gain per cent.

3 Simplify $\frac{2x-1}{4}-\frac{1+x}{6}$.

4 A car goes from A to B in 3 hours. If its average speed had been 18 km an hour slower, it would have taken 4 hours. How far is B from A.

5 In △ABC, ∠B = 32°, ∠C = 55°, BC = 8 cm; the perpendicular bisector of BC cuts AB, CA produced at H, K. Find HK.

6 In △ABC, ∠A is greater than ∠B; AN is the perpendicular from A to the line bisecting ∠ACB and when produced meets BC at K. Prove (i) ∠BAN = $\frac{1}{2}$(∠BAC − ∠ABC); (ii) AN = $\frac{1}{2}$AK.

Paper 41

1 When $\frac{3}{8}$ of a tank has been filled, there is still room for 45 litres. How much can the tank hold?

2 A man's income is £1800 a year. Find the error per cent in saying that it is £4·93 a day, taking 1 year = 365 days.

3 (i) Simplify $(3b^2)^3-(2b^3)^2$. (ii) Factorise $9(x+3)^2-25$.

4 The area of Fig. 241 is 72 cm² and the perimeter is 48 cm. Find the values of x and y.

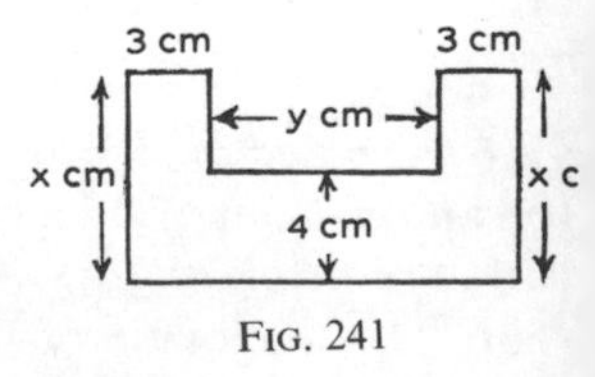

Fig. 241

5 ABCD is a straight line; AB = 3 cm, BC = 4 cm, CD = 2 cm; BP, CQ are drawn perpendicular to AD and on the same side of it; BP = 5 cm, ∠APQ = 140°. Calculate (i) ∠PAD, (ii) CQ, (iii) ∠QDA.

6 In △ABC, the bisector of ∠A meets BC at D and the bisector of ∠B meets AC at E. If ∠ADC = 79°, ∠BEC = 83°, find ∠BAC and ∠ABC.

Paper 42

1 A car travels 224 km in $3\frac{1}{2}$ h and a train travels 48 km in 50 mir Find the ratio of speed of car to that of train.

2 A border 50 cm wide all round the edge of the floor of a room 6·45 m by 6 m is stained. Find the cost at 12p per m^2.

3 If $a = 3$, $b = -2$, $c = 0$, $d = 1$, find the values of

$$\text{(i) } a^2+b^2+2cd; \qquad \text{(ii) } \frac{a-b}{c-d}; \qquad \text{(iii) } \frac{1}{b}-\frac{1}{3b}.$$

4 The lengths of the sides of a triangle are x cm, y cm, z cm; the perimeter of the triangle is 24 cm. If $2z-x = y = 3x-z$, find the values of x, y, z.

5 The angle of elevation of the top of a tower from a point on the ground 100 m from its foot is 18° 45′; what is it from a point on the ground 40 m nearer the tower?

6 In $\triangle ABC$, AB = AC; BC is produced to D so that BD = BA. If $\angle BAC = 2\angle CAD$, find $\angle ABC$.

Paper 43

1 Divide £8·22 between A, B, C so that for every 5p A gets, B gets 10p and C gets 15p.

2 A shop manager is paid a salary of £450 a year and receives a commission of $\frac{3}{4}$ per cent on the sales. What were the sales in a year in which he received £1170 in all?

3 If $\frac{1}{x}+\frac{1}{3} = \frac{1}{2}$ and $\frac{1}{y}+\frac{1}{4} = \frac{1}{5}$, find the value of xy.

4 I walk at 6 km per h and run at 9 km per h. I find that I can save 4 minutes by running instead of walking from my home to the station. How far off is the station?

5 In $\triangle ABC$, AB = AC, $\angle A = 49°$, BC = 10 cm; a line parallel to BC cuts AB, AC at H, K. If HK = 7 cm, calculate the distance between HK and BC.

6 ABCD is a parallelogram; BP, DQ are two parallel lines cutting AC at P, Q. Prove (i) $\triangle CBP \equiv \triangle ADQ$, (ii) BQ is parallel to PD.

Paper 44

1 The edges of two cubes are 4·5 cm, 6 cm. Find the ratio of (i) the areas of their surfaces, (ii) their volumes.

2 A boy obtains 37 marks out of 65. What percentage is this, to the nearest whole number?

3 Simplify $(-x)(y-z-x)+(-y)(z-x-y)-(-z)(y+z-x)$.

4 (i) Solve $d+\frac{1}{2}\pi d = 1$, if $\pi = 3\frac{1}{7}$.

(ii) Factorise $3(n+1)^2-48$.

5 ABCD is a quadrilateral; $\angle ABD = \angle BDC = 90°$, $\angle ADB = 48°$, $\angle CBD = 35°$, $BD = 6$ cm. Calculate the lengths of AB, CD and the acute angle at which AC cuts BD.

6 P is any point on the side AB of the square ABCD. The line from A perpendicular to DP cuts BC at Q. Prove DP = AQ.

Paper 45

1 In $\triangle ABC$, D is the foot of the perpendicular from A to BC; $BD = 10$ cm, $DC = 4$ cm, $\angle B = 27°$. Calculate AD and $\angle ACB$.

2 The duty on a ring is £3·30 and this is 24% of its value. If it is sold at 32% above this value, find the sale price.

3 Simplify (i) $(x-3y)(x-5y)-(x-4y)^2$; (ii) $\frac{1}{c}-\frac{c-1}{c^2}$.

4 Fig. 242 gives the lengths of the sides of a parallelogram in cm. Find the values of x, y and the perimeter.

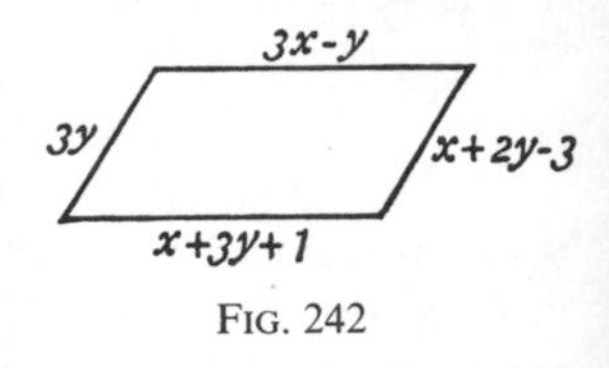

FIG. 242

5 The diagonals of a rectangle ABCD meet at X. Given $\angle AXB = 100°$, $AB = 4$ cm, draw the rectangle and measure AD. Also find the length of AD by calculation.

6 ABCD is a square; AXB is an equilateral triangle outside the square. Prove that $\angle ACX = \frac{1}{2}\angle ABX$.

Paper 46

1 Express $\frac{13}{32}$ as a decimal, correct to 3 decimal places.

2 In the course of a year, $\frac{1}{3}$ of my capital increases by 3%, $\frac{1}{5}$ of it decreases by 4%, and the rest increases by 6%. What is the percentage increase on the whole?

3 Solve $\frac{1}{2}(2-n) = \frac{1}{3}(2+n)-\frac{1}{4}(8+n)$.

4 In $\triangle ABC$, AB = AC and P is a point on AC such that BP = BC. If $\angle ABP = x°$ and $\angle ACB = 3x°$, find x.

5 From half-way up a tower the angle of depression of a mark on the ground is 47° 20′. What will it be from the top of the tower?

6 O is a point inside an equilateral triangle XYZ; OXP is an equilateral triangle such that O and P are on opposite sides of XY. Prove YP = OZ.

Paper 47

1 How many seconds does a torpedo travelling at 45 knots take to go 1000 m? [1 knot = 1850 m per hour.]

2 Find the cost of lining a box without a lid, 90 cm long, 60 cm wide, 20 cm deep, internal measurements, with material at £1·08 per m^2

3 Simplify (i) $\dfrac{4a^2-6ab}{2a}-\dfrac{6ab-4b^2}{2b}$; (ii) $\left(x+\dfrac{1}{2x}\right)^2-\left(x-\dfrac{1}{2x}\right)^2$.

4 Solve $10x+3y = 35$, $21x+2y = 52$.

5 In △ABC, ∠ACB = 47° and AD is the perpendicular from A to BC; P is a point on BC between D and B such that ∠CAP = 75°; AD = 4 cm, PB = 5 cm. Calculate CD, DP, ∠ABC.

6 In △ABC, ∠A = 115°, ∠C = 20°; AD is the perpendicular from A to BC. Prove that AD = DB.

Paper 48

1 (i) Divide 0·64 by 10·24.
(ii) Multiply 7·005 by 0·408.

2 A man saves 15% of his income. What is his income if he spends £1360 a year?

3 Find the value of z if $\dfrac{x}{3}-\dfrac{y}{4}=\dfrac{7}{z}$ when $x = -10$, $y = -11$.

4 A piece of wire 78 cm long can be bent into either of the shapes in Fig. 243, in which all corners are right-angled and dimensions are shown in centimetres. Find x, y and the area of each figure.

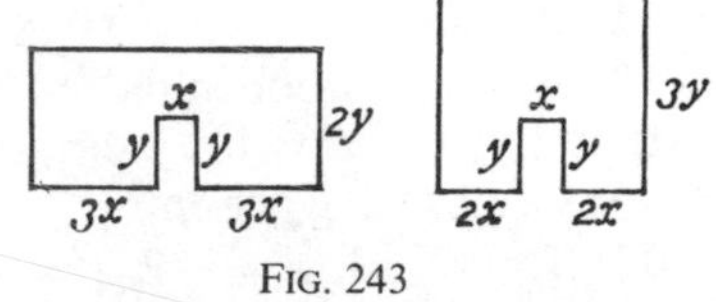

FIG. 243

5 ABC is a triangle on level ground in which ∠A = 90°, ∠B = 37°, AB = 70 m; AP is a vertical flagstaff at A; AP = 25 m. Find the angle of elevation of P from B and from C.

6 In △ABC, AB = AC; BC is produced to D so that CD = AB. Prove that ∠ABD = 2∠ADB.

Paper 49

1 Find correct to the nearest penny $4\frac{1}{2}$% of £4·86.

2 9720 litres of water are drawn from a tank, 5·4 m long, 4·5 m wide. Find how much the water-level falls.

3 Simplify $\dfrac{a^3}{b}\div\dfrac{b^2}{a^2}$. (ii) Expand $(4+3x)(2-7x-5x^2)$.

4 Find x and y if $2xy-3y = 10$ and $3xy+6 = 8y$.

5 Make a table showing the values of tan x° for $x = 0, 10, 20, 30, 40, 50, 60$, correct to 2 places of decimals, and use it to draw the graph of tan x°.

6 In △ABC, ∠C = 3∠B. From AB a part AD is cut off equal to AC. Prove that CD = DB.

Paper 50

1 A man walks from his house to a neighbouring town at 6 km per h and drives back at 40 km per h. The whole journey takes 1 h 32 min. Find the distance of the town from his house.

2 The length and breadth of a tank are each increased by 25% and the height is increased by 20%; find the percentage increase of volume.

3 Find the value of $\pi h(R^2-r^2)$ when $R = 2{\cdot}63$, $r = 1{\cdot}37$, $h = \frac{3}{4}$, $\pi = \frac{22}{7}$.

4 An article is bought for £$\left(\dfrac{3c}{4}\right)$ and sold for £$\left(\dfrac{4c}{5}\right)$; find the gain per cent.

5 In quad. ABCD, $\angle$A = $\angle$B = 90°; P is a point on AB such that AP = 3 cm, PB = 7 cm; also $\angle$APD = 50°, $\angle$CPB = 35°. Calculate AD, BC and the angle at which CD cuts BA when produced.

6 ABCD is a square; the bisector of $\angle$BCA cuts AB at P; PQ is the perpendicular from P to AC. Prove PB = PQ = AQ.

PAPERS 51–66 (Ch. 1–16)

Paper 51

1 Simplify $\frac{7}{30}$ of $(\frac{1}{2}+\frac{7}{15})\div(\frac{5}{6}-\frac{3}{5})$.

2 A hollow wooden closed box measures 1·1 m by 70 cm by 40 cm, externally; the wood is 1 cm thick. Find the weight of the box if the wood weighs 0·8 g per cm³.

3 If $x = -3$, $y = -2$, $z = -1$, evaluate $(y-z)^2-(x-y)(x-z)$.

4 A journey takes me $2t$ minutes if I walk at 6 km per h; find in terms of t how long it takes at 15 km per h.

5 B is 300 m N. 60° E. from A; C is 200 m east of B. How much is B (i) east, (ii) north of A? What is the bearing of C from A?

6 The sides AD, BC of the quadrilateral ABCD are produced to P, Q respectively. If $\angle$DCQ = $\angle$A, prove $\angle$PDC = $\angle$B.

Paper 52

1 Find the linear function in x which has the value 1 when $x = 1$, and the value -11 when $x = -3$.

2 The rain over an area of 120 m² is collected in a tank 3·6 m long, 2·25 m wide. What rise of level of the water in the tank is caused by 1 cm of rain? (Answer to nearest mm.)

3 Find the change of value of $2\frac{1}{4}-7x$ when x increases from -3 to $+2$, and solve the inequality $2\frac{1}{4}-7x > 5\frac{3}{4}$.

4 The value of 30 coins consisting of 10p, 5p and 1p coins is £1·42. There are two more 5p coins than 10p coins; find the number of 1p coins.

5 The legs of a pair of dividers are each 8 cm long. (i) If they are opened to an angle of 42° 30′, find the distance between their points. (ii) Through what angle have they been opened if their points are 14 cm apart?

6 In Fig. 244, ∠ABD = ∠C and APQ is the bisector of ∠BAC. Prove BP = BQ.

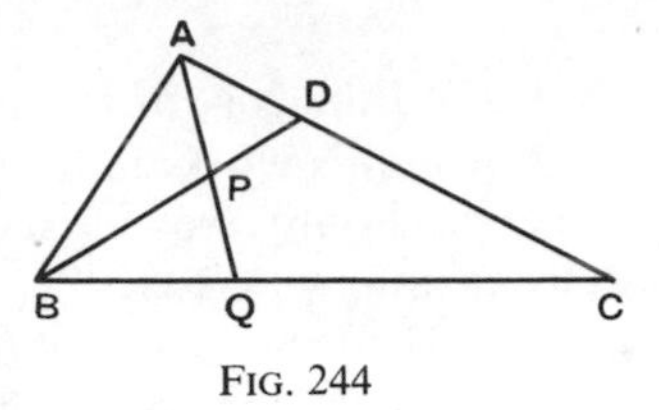

FIG. 244

Paper 53

1 Find to the nearest penny 23% of 88p.

2 A thin zinc plate is 20 cm long, 17 cm wide. From each corner a square of side 2·5 cm is cut away, and the remainder is then bent to form a lidless box. Find the volume of the box.

3 (i) Simplify $(\frac{1}{2}a+\frac{1}{2}b)^2-(\frac{1}{2}a-\frac{1}{2}b)^2$.
(ii) Factorise $(x+y)^2-(x+y)(2x-3y)$.

4 If $A=\frac{1}{2}n-1$ and $B=\frac{1}{3}n+1$, find the value of n for which $A=\frac{1}{4}B$.

5 In △ABC, AB = BC = 10 cm, AC = 7 cm; AP is the perpendicular from A to BC. Calculate ∠BAC and AP.

6 In △ABC, AB = AC; Q is any point on BC and H, K are points on AB, AC such that AHQK is a parallelogram. Prove that QH + QK = AB.

Paper 54

1 A barrel of beer when full weighs 142 kg and when half full weighs 77 kg; a litre of beer weighs 1 kg. How many half-litres can be drawn from a full barrel?

2 By selling an article for £14 a man gains 12%. What did it cost him? Find his gain % if he had sold it for £15.

3 Simplify $\frac{1}{2}\left(\frac{c}{3}-5\right)-\frac{2}{3}\left(\frac{c}{5}-2\frac{1}{2}\right)+\frac{1}{6}(c-1)$.
(ii) Factorise $1+x^2+xy+x^3y$.

4 In Fig. 245, the angles are shown in degrees; find x and y.

5 A man starts from O and walks 4 km in direction S. 32° W. to A and then 6 km in direction N. 38° W. to B. Find how far west and how far north B is from O.

6 Two equilateral triangles ABC, AYZ lie outside each other. If ∠CAY = 35°, find the acute angle at which BC and ZY intersect when produced.

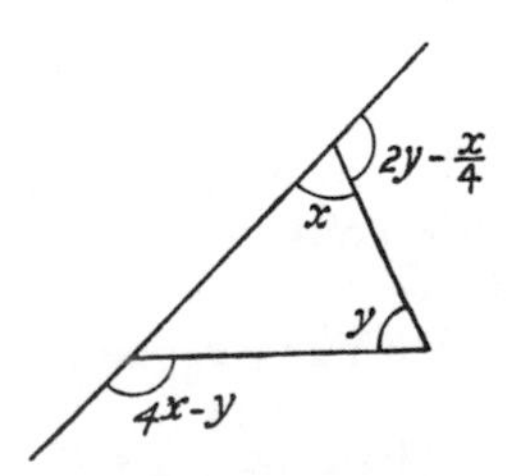

FIG. 245

Paper 55

1 Find the cost of 256 tonnes at £1·93 per tonne.

2 A man's working hours per day were increased by 25 %, and his wages per hour were increased by 20 %. By how much per cent were his daily earnings increased?

3 (i) Find the values of b, c if $y = bx^2+cx$ is satisfied by $x = y = 2$ and by $x = 3$, $y = 18$. (ii) Factorise $(3x+1)^2-(x-1)^2$.

4 A is 60 per cent of B; express $B-A$ as a percentage of $B+A$.

5 A regular pentagon is inscribed in a circle of radius 6 cm. Find (i) the length of its side, (ii) the distance of each side from the centre of the circle.

6 X, Y are the mid-points of the sides AB, AB of △ABC. BY is joined and produced to Q so that BY = YQ; CX is joined and produced to P so that CX = XP. Prove (i) AP = AQ, (ii) PAQ is a straight line.

Paper 56

1 Show in a diagram the region indicated by the inequalities $y > \frac{1}{2}x$, $y < 3x$, $x+y < 4$.

2 X motors at a steady speed from A to B, 78 km, leaving A at 9 a.m. and arriving at B at 11 a.m.; Y leaves B at 10 a.m. and bicycles to A along the same road at 16 km per h. Draw the travel graphs of X and Y and find at what time they pass one another.

3 (i) Simplify $(-3c)(c^2-4c)-(-3c)^2-(-2c)^3$.
(ii) Factorise $4t^2+12t-72$.

4 The nth number in the set of numbers 4, 11, 18, 25, 32, . . . is $7n-3$. Verify this if $n = 5$. Is 200 a number in this set? If so, where does it come?

5 In quad. ABCD, ∠A = ∠B = 90°; BC = 3 cm, CD = 5 cm, DA = 6·5 cm. Calculate ∠ADC and AB.

6 In Fig. 246, BAC is parallel to PDQ; AP, AQ are the bisectors of ∠BAD, ∠CAD. Prove PD = DQ.

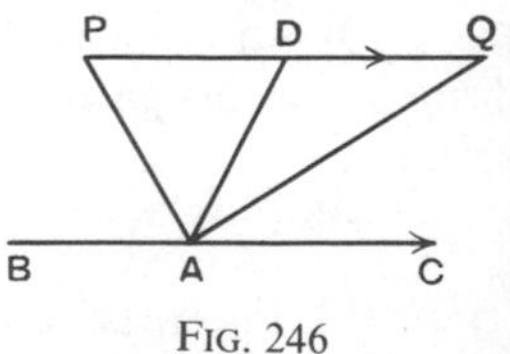

FIG. 246

Paper 57

1 Solve the inequality $2(x+1) > 3(x-1)$.

2 A rectangular tank is 3·5 m long, 1·8 m broad, 1·5 m deep, internally. Calculate in litres the amount of water it will hold. If 700 litres of water are drawn off, find in cm, to the nearest mm, how much the water-level sinks.

3 Solve $\frac{1}{2}\left(\frac{x}{7}+4\right)-\frac{2}{3}\left(\frac{x}{7}-3\right) = 3\frac{1}{2}$.

4 What is the gradient of the line joining the points (0, 3) and (3, 2)? Find in its simplest form the equation of this line.

5 ABC is an acute-angled triangle in which AB = 9 cm, $\angle C = 30°$; AD is the perpendicular from A to BC; AD = 8 cm. Find $\angle BAD$ and the lengths of AC, BC.

6 ABCD is a quadrilateral; the bisectors of the unequal obtuse angles ABC, BCD meet at P. Prove that the perpendiculars from P to AB and CD are equal.

Paper 58

1 A man has weeded $\frac{4}{9}$ of a field in 28 hours. How long should be taken to finish it at the same rate?

2 A grocer buys cheese at £30·50 for 100 kg. What is the least whole number of pence per kg at which he can sell it so as to make a profit of not less than 15 per cent?

3 Express $(2y-5)(3y+1)-(x-y)^2$ in terms of x, if $y = 2x+3$.

4 The value of $5x$ coins, each 10p, exceeds that of $(3x-4)$ coins, each 5p, by £4·40. Find x.

5 A man walks 200 m in the direction N. 75° 40′ E. and then 100 m in the direction N. 52° 10′ W. Find his bearing from his starting-point.

6 P is a point on the side AB of $\triangle ABC$ such that $AP = \frac{1}{4}AB$; PQ is drawn parallel to BC to meet AC at Q. By drawing sets of parallel lines, prove that $PQ = \frac{1}{4}BC$.

Paper 59

1 Solve the inequality $5-4x \leqslant 4(2x-1)$.

2 A metal sheet 17·5 cm long, 6·4 cm wide, weighs 280 g. It is cut down so as to be 16 cm long, 6 cm wide; what will it weigh? Find the percentage decrease in weight.

3 Factorise $72-12t-4t^2$.

4 How must £2·82 be shared between A, B, C so that B may get 42p more than A and so that C gets twice as much as A.

5 In $\triangle ABC$, $\angle B = 36°$, $\angle C = 69°$, BC = 9 cm; the perpendicular bisector of BC cuts BA, CA produced at P, Q. Calculate PQ.

6 The bisectors of $\angle B$, $\angle C$ of $\triangle ABC$ meet at I. If $\angle BIC = 135°$, prove that $\angle A$ is a right angle.

Paper 60

1 State the gradients of the lines $y = 3-5x$, $8y = 1+3x$, $4x-3y = 6$.

2 An open box 9 cm long, made of thin cardboard, has its two ends 3 cm square and is fitted with a cardboard lid which covers the top, the front, and the two square ends. Find the area of cardboard used for making the box.

3 (i) Complete $16a^2-4ab\ldots = (4a\ldots)^2$.

(ii) Solve $10r+3s = s-4r = 11$.

4 A leaves X at 9 a.m., walks for 30 min at 8 km per h, then stops for 30 min and then cycles on at 16 km per h; B leaves X at 10 a.m. and pursues A at 21 km per h. At what time will B overtake A? Solve by a graph *and* by calculation.

5 AB is a chord of a circle, centre O; radius 7 cm; AD is a diameter; $\angle AOB = 108°$. Find AB and BD.

6 In $\triangle ABC$, D is the mid-point of BC. Draw $\triangle ABC$ accurately, given AB = 5 cm, AC = 6 cm, AD = 4 cm. (See Fig. 247.) Measure BC.

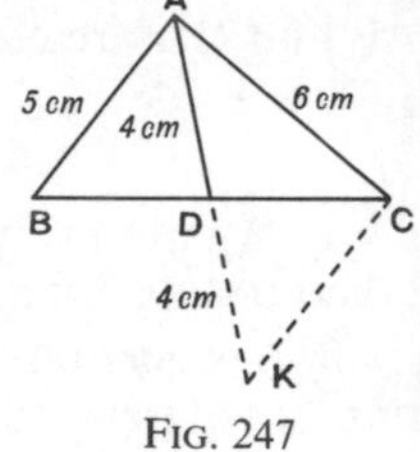

FIG. 247

Paper 61

1 Draw on the same axes the graphs of $2y = x+2$ and $4x+3y = 12$, and state the coordinates of their point of intersection.

2 A merchant mixes 3 grades of tea costing 57p, 66p and 75p per kg in the ratios 3:4:5 by weight respectively, and sells the mixture at 90p per kg. Find his gain %.

3 (i) If $a = -3$, $b = 5$, $c = 2$, evaluate $2abc^2$ and $(2a+b)^3$.

(ii) Factorise $3x^2+xy-2y^2$.

4 At an examination one-quarter of the candidates fail. At the next examination there are 84 more candidates and 8 more failures than at the first examination and one-fifth of the candidates fail. Find the number of candidates at the first examination.

5 A road 1000 m long is represented on a map, scale 1:20 000, by a line of length 4·86 cm. Find the average inclination of the road to the horizontal.

6 ABCDE is a regular pentagon; AC cuts BE at K. Find $\angle EKC$.

Paper 62

1 Find the equation of the line joining $(-1, 2)$ and $(3, 4)$.

2 The rent of my house is £64 a year and the rateable value is $\frac{7}{8}$ of the rent. During the year two rates are demanded, one at 41p in the £, the other at 34p in the £, on the rateable value. My income is £2250 a year. What percentage of it do I spend on rent and rates combined? (Answer to 1 decimal place.)

3 (i) Find the change in value of $(1-2x)(5x-2)$ when x increases from -1 to $+1$?

(ii) Find 4 factors of $(ax+by)^2-(ay+bx)^2$.

4 A man is now 3 times as old as his son. In 10 years' time the sum of their ages will be 76. How old was the man when his son was born?

5 In △ABC, AB = 4 cm, AC = 7 cm, ∠ABC = 90°; the circle, centre A, radius 6 cm, cuts BC at D and cuts CB produced at E. Calculate ∠EAD and ∠DAC.

6 In △ABC, AB > AC; D is the mid-point of BC and AE bisects ∠BAC; the perpendicular from C to AE meets AE, AB at H, K. Prove DH = $\frac{1}{2}$BK.

Paper 63

1 Find the value of $1\frac{1}{4}$% of £266·66, to nearest penny.

2 A metal box without a lid is 15 cm long, 11 cm wide, 5 cm high, externally; the metal is 5 mm thick. Find the weight of the box if 1 cm^3 of the metal weighs 7·2 g.

3 (i) From N subtract 60 per cent of 1·5 N.

(ii) Factorise $(p+2)^2-2(p+2)(p-1)-3(p-1)^2$.

4 If $y = mx+c$ where m, c are constants, and if $y = -3$ when $x = 2$, and $y = 5$ when $x = 4$, find m, c and the value of y when $x = 5$. State also the significance of the letters m and c.

5 In the quad. ABCD, AB = AD = 6 cm, CB = CD = 10 cm; P is a point inside ABCD such that PB = PD = 7 cm and ∠BPD = 69° 42′. Calculate BD, ∠ABP, ∠PDC.

6 In △ABC, ∠B = 90°, ∠C = 30°. Prove that AC = 2AB by producing AB to D so that AB = BD and joining CD.

Explain how to find the value of sin 30° *without* using tables.

Paper 64

1 36% of a certain number is 117; find the number.

2 Fig. 248 represent the top of a metal sheet 2 cm thick. The corners are right-angled and dimensions are shown in centimetres. Find the volume of the sheet and its weight if the metal weighs 7·6 g per cm^3.

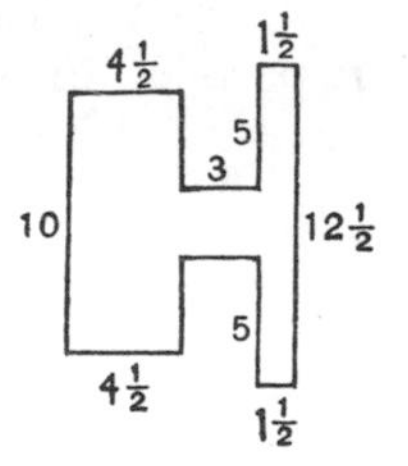

FIG. 248

3 (i) If $P = x^2-3xy+y^2$, $Q = 2x(x-y)$, $R = 3y(3y-x)$ and if $x = 2y$, express $P-Q-R$ in terms of x.

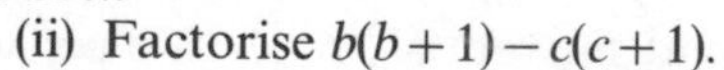

(ii) Factorise $b(b+1)-c(c+1)$.

4 The result of increasing N by r per cent is the same as that of increasing $\frac{3}{4}N$ by $3r$ per cent. Find the value of r.

5 In the quad. ABCD, ∠A = ∠B = 90°, BC = 12 cm, CD = 10 cm, ∠BCD = 23° 35′. Find AB and ∠ABD.

6 In △ABC, AB = AC and BC is less than AB. D is a point on BC produced such that BD = BA. Prove ∠ACD = 2∠ADB.

Paper 65

1 34·2 cm of wire weigh 5·7 g. Find the length of wire which weighs 0·95 kg.

2 By selling articles at 30p each, a man makes a profit of 20%. Find his gain or loss per cent if he sells them at 21p each.

3 (i) If $x = -2$, $y = -7$, evaluate $1-x+y$ and x^3y.
(ii) Find x and y if $3x-y = x+3y = 1$.

4 A man cycles 10 km at u km an hour and then motors 48 km at $3u$ km an hour; he takes 2 h 10 min altogether. Find u.

5 ABCD is a quadrilateral in which AD = 10 cm, BC = 7 cm; also $\angle$ABD = $\angle$BDC = 90°, $\angle$ADB = 66° 25′. Find BD, $\angle$DBC, $\angle$ACD.

6 ABCD is a quadrilateral in which AB = AD and $\angle$DBC is a right angle. Prove that the line through A parallel to BC bisects CD. [Let this line cut BD at N.]

Paper 66

1 On the same axes, draw the graphs of $y = 3x+1$ and $y = 8-x$. Read off the coordinates of the point where they intersect.

2 The base of a tank 1·8 m high is 1·35 m long, 1·2 m wide, internal measurements. Initially the tank is half full of water. Find the depth of water after 810 litres have been added.

3 Solve the equations $\frac{3}{4}x-1 = 2x-\frac{1}{3}y-9 = 2y-7$.

4 The cost of making x articles is £$(a+k\sqrt{x})$ where a, k, are constant. If the cost of making 25 articles is £20 and if the average cost per article, when 100 are made, is 30p, find the cost of making 16 articles.

5 In $\triangle$ABC, $\angle$A = 90°, $\angle$B = 41°, BC = 6 cm; BCPQ is a rectangle outside $\triangle$ABC; BQ = 5 cm. Find the lengths of the perpendiculars from P, Q (i) to AC, (ii) to AB.

6 ABCD is a quadrilateral in which the bisectors of $\angle$ABC, $\angle$ADC are parallel. Prove $\angle$A = $\angle$C.

PAPERS 67–81 (Ch. 1–21)

Paper 67

1 Find the value of $(0{\cdot}05)^3 \times (0{\cdot}16)^2 \div 0{\cdot}2$, *without* tables.

2 Interest is charged on a loan at the rate of 1p per 50p per week. Find the rate % p.a. [Take 1 year = 52 weeks.]

3 (i) What must be added to $r-2s+3t$ to make $t-r+s$?
(ii) Find the L.C.M. of x^2-2x-8, x^2+5x+6, $x^2-8x+16$.

4 The ratio of the numbers of sides of two convex polygons is 2 : 3, and the ratio of the sums of their angles is 3 : 5. Find the numbers of sides of each. [The angle-sum of an n-sided polygon is $(2n-4)$ rt. $\angle$s.]

5 In $\triangle$ABC, $\angle$B $= 41°$, $\angle$C $= 67°$, AB $= 8$ cm; AD is the perpendicular from A to BC. Calculate AD, BD, BC.

6 In $\triangle$ABC, $\angle$A $= 55°$, $\angle$C $= 35°$; K is a point on AC such that KB = KA. Prove KA = KC.

Paper 68

1 Evaluate by logarithms $\dfrac{0{\cdot}2079 \times 0{\cdot}0804}{0{\cdot}09072}$, to 3 figures.

2 Find to $\frac{1}{10}$ m the radius of a circular lake, area 25 000 m^2.

3 (i) If $x = 2t^2 - t - 4$ and $y = t^2 + 2t + 3$, find the values of x and y if $x = 2y$.

(ii) Factorise $6x^2 - 17xy - 14y^2$.

4 The simple interest on £256 for 15 months is £24. Find the rate % p.a.

5 In Fig. 249, PQR is a straight line and the lengths are in centimetres; calculate the lengths of BQ, CR, PR.

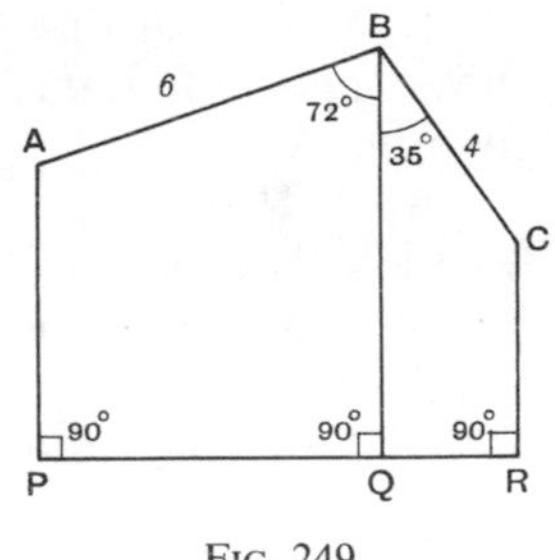

FIG. 249

6 P, Q are points on the sides BC, AC of an acute-angled triangle ABC such that P is equidistant from AB and AC, and Q is equidistant from A and P. Prove PQ is parallel to BA. (L)

Paper 69

1 Use tables to evaluate $\sqrt{(0{\cdot}2648)} - \sqrt{(0{\cdot}09174)}$.

2 In a certain year, there were 74 861 male students and 22 468 female students at the Universities in Great Britain. What percentage of the total, to 3 figures, were males?

3 (i) Solve $\frac{1}{4}x - \frac{1}{5}y = \frac{2}{3}y - \frac{2}{5}x = 2$.

(ii) Find the value of b if $x + 5$ is a factor of $x^2 + bx - 45$.

4 Find x if $\frac{1}{2}(x+4)$ exceeds $\frac{1}{3}(10-x)$ by 50 per cent of it.

5 ABCDEFG is a regular 7-sided polygon inscribed in a circle, radius 6 cm. Find the length of AB and of the perpendicular from E to AB.

6 ABCD is a square; AB and BC are produced to P and Q respectively so that BP = CQ. Prove (i) DP = AQ, (ii) DP is perpendicular to AQ.

Paper 70

1 When railway fares were increased in the ratio 7:4, the fare for a journey became 280p. What was the increase?

2 A cylindrical tankard, internal diameter 10·7 cm, contains a litre of liquid. Find the depth in centimetres, to 3 significant figures.

3 (i) Simplify $(x-1)(x-2)+(3-x)(4-x)-2(x-6)(1+x)$.
(ii) Find 4 factors of y^4-10y^2+9.

4 If a speed of v km an hour is half as fast as a speed of $(v-2)$ m per second, find v.

5 ABCD is a rectangle; AB = 7 cm, BC = 3 cm; P, Q are points on AB, DC such that AP = 5 cm, DQ = 4 cm. Find ∠DPC, ∠PQC.

6 In △ABC, AB = AC; D is a point on AC, or on CA produced, such that BD = BC. Prove ∠DBC = ∠BAC. [Draw two figures, one with ∠A < 60°, the other with ∠A > 60°.]

Paper 71

1 Find the simple interest on £480 for 2 years at $2\frac{1}{2}\%$ p.a.

2 A piece of thread is wrapped 10 times round a reel and when unwrapped is found to measure 121·5 cm. Find the diameter of the reel to 3 figures.

3 (i) If $P = 2x^2-xy-y^2$, $Q = (x+y)(x-y)$, $R = (2x-y)(x-2y)$, express $P-Q+R$ in terms of x, y. If $x+4y = 0$, express $P-Q+R$ in terms of y.
(ii) Factorise $bc+4-2b-2c$.

4 If $x+y = a$ and $x-y = b$, express x^2+y^2 and x^4-y^4 in terms of a, b.

5 ABCDE is a regular pentagon; AB = 6 cm. Find BD and the length of the perpendicular from A to CD.

6 *Construct with ruler and compasses* a quadrilateral ABCD in which ∠A = ∠B = 60°, ∠C = 90°, AB = 5 cm, BC = 4 cm. Construct a point P equidistant from AB, AD and such that PA = PC.

Paper 72

1 Evaluate by logarithms $\sqrt{\left\{\dfrac{9{\cdot}347\times 10{\cdot}73}{55{\cdot}09\times 0{\cdot}173}\right\}}$.

2 B exceeds A by 16%; C exceeds B by 25%. By how much per cent does C exceed A?

3 (i) Simplify $6a^3\times(-4b^2)^2\div\{(-2a)^4(-b^2)\}$.
(ii) Solve $\frac{1}{3}y = 1-z$, $\frac{1}{3}(2y-z) = \frac{1}{2}(y-3z)$.

4 Find to the nearest tenth of a unit the distance between the points $(-1, -2)$ and $(4, 1)$.

5 In △ABC, ∠B = 49°, ∠C = 63°, BC = 6 cm; the bisectors of ∠B, ∠C meet at I. Calculate the length of the perpendicular from I to BC.

6 In Fig. 250, AB = AC; KAD and BCD are straight lines. Prove that $r-s = 2p$.

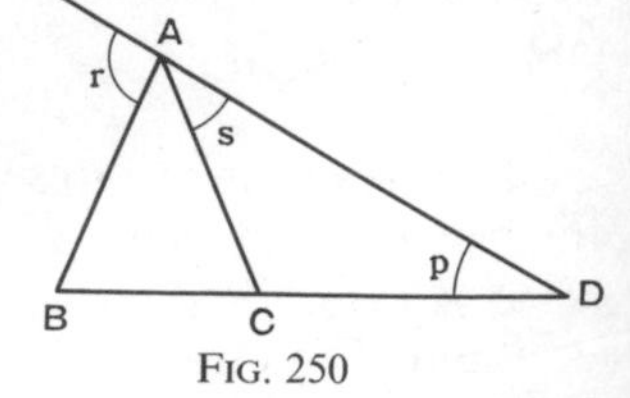

FIG. 250

Paper 73

1 Find, to 3 figures, $\sqrt{(0{\cdot}1)} \div \sqrt[3]{(0{\cdot}01)}$.

2 Find to 3 figures the number of litres required to fill a tank 8·55 m long, 4·71 m wide, 1·65 m deep.

3 (i) For what value of c is $2x^3 - 3x^2 - cx + 6$ zero when $x = -2$?
(ii) Factorise $1 - 4(a-b)^2$.

4 The angles of a quadrilateral taken in order are x, $x+20$, $x+30$, $x+50$ degrees. Find x, and prove that the quadrilateral is a trapezium.

5 A man starts from P and walks 3·5 km in direction N. 63° E. to Q; he then walks 4 km to a place R due east of Q. Find (i) the bearing of R from P, (ii) the distance of R from P.

6 AB, DC are the parallel sides of the trapezium ABCD; P, X, Q are the mid-points of DA, DB, CB. Prove that (i) PXQ is a straight line, (ii) $\text{PQ} = \frac{1}{2}(\text{AB} + \text{DC})$.

Paper 74

1 If $\frac{4}{3}\pi r^3 = 0{\cdot}1604$, find r to 3 figures.

2 At what rate % p.a. does the simple interest on £360 for 4 years amount to £50·40?

3 (i) Evaluate

$$a^2(b-c) + b^2(c-a) + c^2(a-b) + (b-c)(c-a)(a-b)$$

if $a = 2$, $b = -1$, $c = -2$.
(ii) Find c if $x - 3$ is a factor of $x^2 + cx - 12$. What is the other factor?

4 After $\frac{3}{5}$ of a tank has been filled there is room for N more litres. Find in terms of N the capacity of the tank.

5 In quad. ABCD, $\angle$A $= \angle$B $= 90°$, AB $= 8$ cm; P is a point on AB such that PA $= 5$ cm, PD $= 7$ cm, PC $= 4$ cm. Find $\angle$DPC, $\angle$ADC.

6 In $\triangle$ABC, $\angle$BAC is a right angle; AD is the perpendicular from A to BC. If P is a point on CB such that CP = CA, prove that AP bisects $\angle$BAD.

Paper 75

1 Divide £1 into 3 parts in the ratios 4:7:9.

2 Find to 3 figures the radius of a circle of area 6·25 cm².

3 (i) Solve $\frac{1}{2}(x+18) - \frac{1}{8}(3x-4) = 5(6-x)$.
(ii) Factorise $4a^2 - 4ab + b^2$ and $4a^2 - 4ab + b^2 - c^2$.

4 A walks 12 km in the same time that B walks 11 km, and B takes half an hour longer to walk 33 km than A. Find the speeds at which A and B walk.

5 A captive balloon is held by a rope 80 m long which is inclined at 62° to the horizontal. Find the height of the balloon. What angle does the rope make with the horizontal when the height of the balloon is 66 m?

6 PQRS is a parallelogram in which PQ is less than QR. The bisector PX of ∠P cuts QR at X; the bisector QY of ∠Q cuts PS at Y. Prove that PQXY is a rhombus.

Paper 76

1 On the same axes, draw the graphs of $3y = 2x$ and $3y = 4x+6$. Shade the region for which $4x+6 > 2x$, and check by solving this inequality.

2 Use logarithm tables to evaluate, to 3 figures,

$$\sqrt[3]{\frac{317{\cdot}6 \times 72}{0{\cdot}00515}}.$$

3 (i) Simplify $\dfrac{3x+y}{x^3y} - \dfrac{x-4y}{x^2y^2} - \dfrac{x^2+y^2}{x^3y^2}$.

(ii) Factorise

$$(x-2)(x-1)(x+2)-12(x-1).$$

4 If $5x+n = 3$ and $4y+n = 1$, find the value of n for which $2y = 3x$.

5 In Fig. 251, ABCD is a rectangle. Calculate the lengths of AB, AD, CP, CQ.

6 In △ABC, AB = AC; AB is produced to D. Prove that ∠ACD − ∠ADC = 2∠BCD.

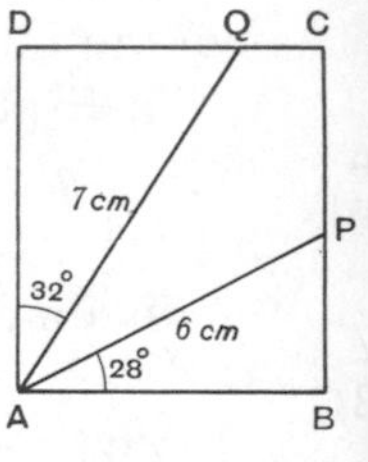

FIG. 251

Paper 77

1 Visitors to Great Britain increased from 733 260 in one year to 819 340 in the next year. Find the increase % to 3 figures.

2 Find to 3 figures the *total* area of the surface of a cylinder, height 19·4 cm, diameter 13·7 cm, closed at both ends.

3 (i) Solve $\frac{1}{3}(2a-b)-\frac{1}{5}(a-2b) = 3$, $3(a+2b+3) = 2(a+b)$.

(ii) Two numbers m and n are chosen so that their sum equals three times their difference. Find the value of $\dfrac{mn}{m^2+n^2}$.

4 A man leaves P at 1 p.m. and walks to Q at 6 km per h; another man leaves P at 2 p.m. and cycles at 16 km per h to Q, 32 km away, by the same road and returns at once to P at the same rate. Find graphically *and* by calculation the distances from P of the places where the men meet.

5 ABC is an equilateral triangle, side 10 cm; ABHK, ACPQ are squares outside $\triangle$ABC. Find the lengths of QK and PH, and the distance between the parallel lines QK, PH.

6 P is a point inside the quadrilateral ABCD such that PA, PB, PC bisect $\angle$A, $\angle$B, $\angle$C; PH, PK are the perpendiculars from P to DA, DC. Prove that (i) PH = PK, (ii) PD bisects $\angle$D, (iii) AB + CD = AD + BC. [Draw PX, PY perpendicular to BA, BC.]

Paper 78

1 Evaluate to 3 figures, $\dfrac{0{\cdot}2816 \times \sqrt{0{\cdot}7163}}{(0{\cdot}9293)^2}$.

2 A cylindrical tank holds 150 litres; its internal height is 64·6 cm. Find its internal diameter in cm, to 3 figures.

3 (i) Subtract $\dfrac{a}{10}+\dfrac{b}{4}-\dfrac{2c}{3}$ from $\dfrac{3a}{5}-\dfrac{3b}{4}+\dfrac{5c}{6}$.

(ii) Factorise $a^2(b+c)-(b+c)^3$.

4 Each angle of a regular polygon of x sides is $\frac{3}{4}$ of each angle of a regular polygon of y sides. Express y in terms of x, and find any values of x and y which will fit.

5 P, R, are points on the sides AB, AD respectively of the rectangle ABCD; APQR is a rectangle; AP = 8 cm, PB = 3 cm, AR = 5 cm, RD = 4 cm. Find $\angle$DPC and $\angle$RQC.

6 *Construct with ruler and compasses* $\triangle$ABC in which $\angle$A = 75°, $\angle$B = 45°, AB = 7 cm and construct in the simplest way a point P on BC such that AP + PC = BC.

Paper 79

1 The scale of a map is 4 cm to 1 km. Find in cm² the area on the map which represents an estate of 2 500 000 m².

2 Find in kg to 3 figures the weight of a cast-iron cylindrical pipe, internal diameter 10 cm, thickness 1 cm, length 1·2 m given that 1 m³ of iron weighs 7600 kg.

3 In Fig. 252, the straight line EF is the graph of $y = \dfrac{3x}{5}+3$, and the unit for each axis is 1 cm.

(i) What are the lengths of OF and OE?

(ii) If OM = 2 cm, find MP.

(iii) If ON = −7 cm, find NQ.

(iv) If MP = 6 cm, find OM.

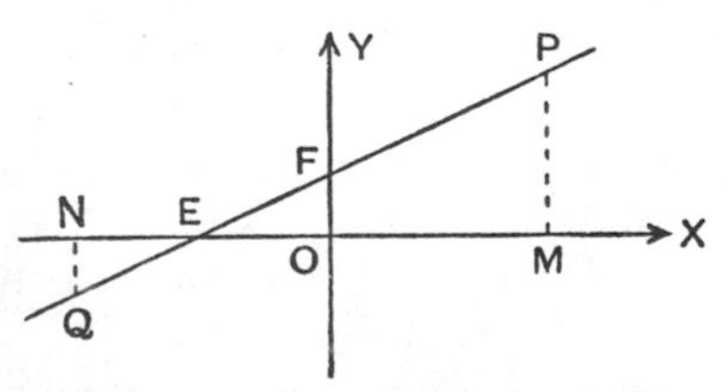

FIG. 252

4 (i) Solve $5P = 2Q$, $\frac{1}{4}P - \frac{1}{6}Q + 1 = 0$.
(ii) If $\sqrt[3]{0.01} \div \sqrt{0.001} = 10^n$, find n.

5 A is 3 km from O in the direction N. 46° W.; B is 7 km from O in the direction S. 52° W. How far north and east is A from B?

6 AB, BC, CD, DE are four consecutive sides of a regular 20-sided polygon. Find the acute angle at which AB and ED intersect when produced.

Paper 80

1 Evaluate by logarithms, $\dfrac{47.33 \times (8.971)^2}{698.4}$.

2 Indicate the region denoted by the inequalities $4x - y < 8$, $x - 2y > -2$, $3x + 2y > -6$, leaving the required region *unshaded*.

3 (i) Solve $\frac{2}{3}(\frac{1}{4} - x) = \frac{1}{2}(\frac{1}{3} - x) + \frac{1}{4}(\frac{1}{5} - x)$.
(ii) If $x = 3(y-1)$, express $x^2 - x - 12$ in terms of y in factors.

4 The angles of a triangle are $2x + y$, $x + 2y$, $5x - 2y$, degrees. If the triangle is isosceles, prove that it is equilateral and find x and y.

5 In △ABC, AB = 6 cm, ∠B = 90°; P is a point on BC such that ∠BAP = 25°, ∠PAC = 40°; the perpendicular at P to BC meets AC at N. Find BP, PC, PN.

6 ABCDE is a pentagon such that AB and DC when produced cut at right angles. If ∠A : ∠D : ∠E = 5 : 4 : 3, find ∠E, and prove that AB is parallel to ED.

Paper 81

1 Find to 3 figures the value of $\sqrt{\left\{\dfrac{3.2436 \times (5.321)^2}{19.836}\right\}}$.

2 A man borrows £1200 from a Building Society and repays £400 at the end of each year. He pays interest at 5% p.a. on the amount owing at the beginning of each year. How much does he owe at the beginning of the fourth year?

3 (i) Solve $\dfrac{25}{r} - \dfrac{8}{s} = 4$, $\quad \dfrac{5}{r} + \dfrac{12}{s} = 11$.
(ii) $p(q-t) = q^2(p-t)$, find t when $p = 2$, $q = -2$.

4 If $x^2 + ax + 15$ can be expressed in the form $(x+b)(x+c)$ where b, c are positive integers, find the possible values of a.

5 In △ABC, ∠A = 35°, ∠B = 90°; BD is the perpendicular from B to AC; DE is the perpendicular from D to AB; AD = 4 cm. Find AE, AB, BC.

6 In △ABC, ∠B = ∠C = 40°; the bisector of ∠ABC meets AC at D; Q is a point on BC such that BQ = BD. Prove that CQ = AD. [Take R on BC so that BR = BA; join DR.]

PROOFS OF THEOREMS

For convenience of reference, proofs of various standard theorems, not set out in full in the text, are given here.

(1) The opposite sides and angles of a parallelogram are equal.
(2) Each diagonal bisects the area of the parallelogram.

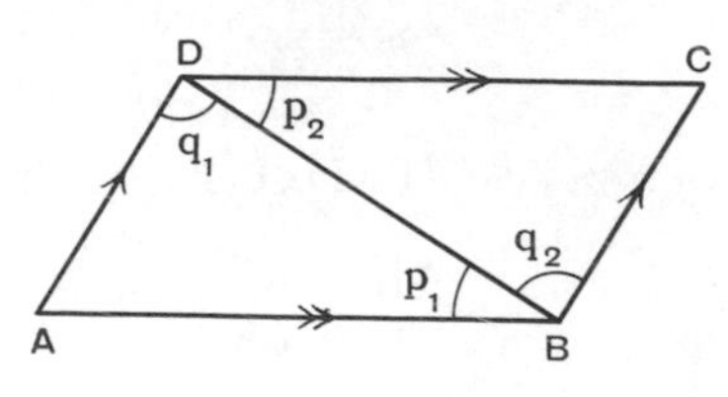

FIG. 253

Given a parallelogram ABCD and a diagonal BD.

To prove that

(1) AB = DC, AD = BC, ∠A = ∠C, ∠B = ∠D.

and

(2) Area of △ABD = Area of △CBD.

Proof (1) With the notation in the figure, in △s ABD, CDB,

$p_1 = p_2$ *alt.* ∠*s*, AB∥DC,
$q_1 = q_2$ *alt.* ∠*s*, AD∥BC,
BD = DB.

∴ △s ABD / CDB are congruent **ASA.**

∴ AD = CB, AB = CD,

ınd

∠A = ∠C.

(2) Since △ABD ≡ △CDB, the triangles are equal in area. Similarly, by joining AC it may be proved that

△ABC ≡ △CDA.

∴ ∠B = ∠D,

ınd AC bisects the area of the parallelogram.

Abbreviations for reference: (1) opp. sides ∥gram.
(2) opp. ∠s ∥gram.

The diagonals of a parallelogram bisect one another.

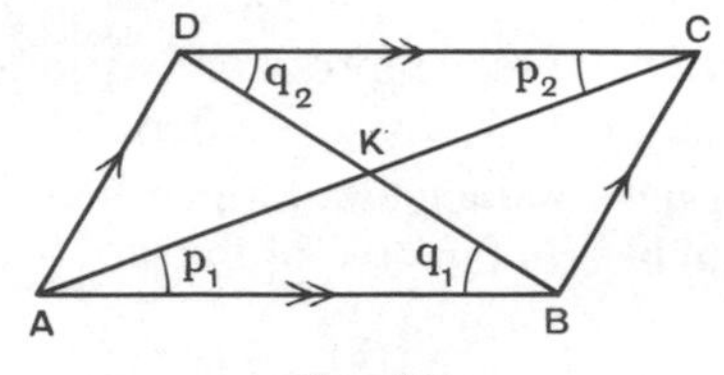

FIG. 254

Given a parallelogram ABCD whose diagonals AC, BD cut at K.

To prove that AK = KC
and BK = KD.

Proof With the notation in the figure, in $\triangle$s AKB, CKD,

$p_1 = p_2$ *alt.* $\angle$*s*, AB $\|$ DC,

$q_1 = q_2$ *alt.* $\angle$*s*, AB $\|$ DC,

AB = CD *opp. sides* $\|$*gram*.

$\therefore \triangle$s $\begin{matrix}\text{AKB}\\\text{CKD}\end{matrix}$ are congruent **ASA.**

$\therefore$ BK = DK

and AK = CK.

Abbreviation for reference: diags. $\|$gram.

Note The fact proved in this theorem is used to construct a parallelogram if the lengths of the diagonals and the length of one side are given. For example, if in Fig. 254, AB = 6 cm, AC = 10 cm BD = 8 cm, then AK = 5 cm, BK = 4 cm; the triangle KAB can then be constructed.

If one pair of opposite sides of a quadrilateral are equal and parallel, the other pair of opposite sides are also equal and parallel.

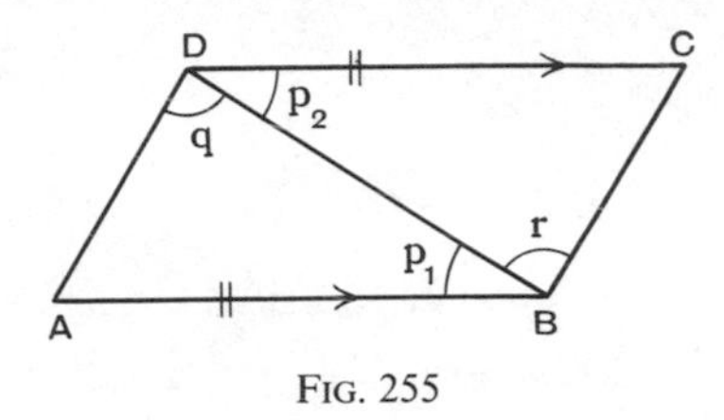

FIG. 255

Given a quadrilateral ABCD in which

$$AB = DC \quad \text{and} \quad AB \parallel DC.$$

To prove that

$$AD = BC \quad \text{and} \quad AD \parallel BC.$$

Construction Join BD.

Proof With the notation in the figure, in $\triangle$s ABD, CDB,

$AB = CD$ *given*,

$BD = DB$,

$p_1 = p_2$ *alt.* $\angle$*s*, $AB \parallel DC$.

$\therefore \triangle$s $\begin{matrix} ABD \\ CDB \end{matrix}$ are congruent **SAS.**

$\therefore AD = CB$,

and $q = r$,

but these are alt. $\angle$s,

$\therefore$ AD is parallel to BC.

Abbreviation for reference: 2 sides equal and $\parallel$

This theorem is also stated in the form:

A quadrilateral, which has one pair of equal and parallel sides, is a parallelogram.

The straight line drawn through the middle point of one side of a triangle parallel to another side bisects the third side.

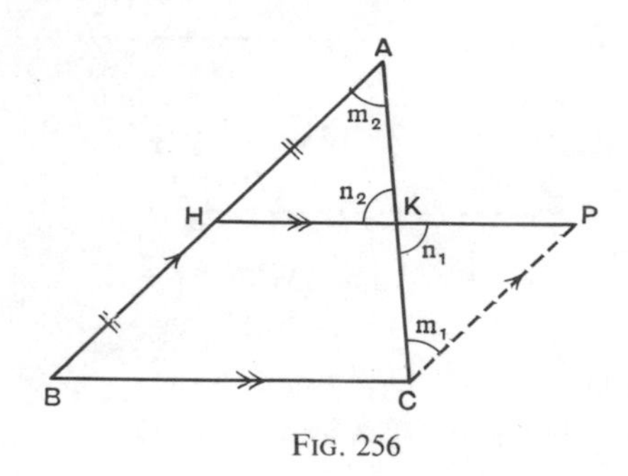

FIG. 256

Given the mid-point H of the side AB of $\triangle$ABC, and a line through H parallel to BC cutting AC at K.

To prove that AK = KC.

Construction Through C draw CP parallel to BA to meet HK produced at P.

Proof Since HP$\|$BC *given,*
and CP$\|$BH *constr.,*
BCPH is a parallelogram,

$\therefore$ CP = BH *opp. sides $\|$gram,*

but BH = HA *given,*

$\therefore$ CP = HA.

In $\triangle$s CPK, AHK, with the notation in the figure,

$m_1 = m_2$ *alt. $\angle$s,* CP$\|$BA,

$n_1 = n_2$ *vert. opp. $\angle$s,*

CP = AH *proved,*

$\therefore$ $\triangle$s CPK, AHK are congruent **AAS.**

$\therefore$ CK = AK.

Abbreviation for reference: mid-point theorem.

INDEX

ANSWERS

Page 1 EXERCISE 1

1 (i) -4, $+20$, -15, degrees; (ii) 5° below zero, 100° above zero.
2 39 m above, 396 m below sea-level.
3 Beyond target +, short of target −, direct hit O.K.
4 Upward velocity +, downward velocity −.
5 $+3$, -1, -4, $+4\frac{1}{2}$, $-1\frac{1}{2}$.
6 5 km west, 8 km south, 3 min slow, £42 in debt.
7 Ahead of Greenwich +, behind Greenwich −.

Page 3 EXERCISE 2

1 $(+2°)$ **2** $(-3°)$ **3** $(-7°)$ **4** $(-5°)$ **5** $(+7°)$ **6** $(-8°)$
7 $(+4°)$ **8** 160 m; $(+160)$ **9** £70; $(+70)$ **10** $(+16°)$
11 (-3), $(+4)$ **12** (-3); $(+3)$; (-4); (-3) **13** $(+4)$ **14** (-10)
15 (-2) **16** $(+1)$ **17** (-1) **18** 0 **19** (-5) **20** (-2)
21 (-4) **22** $(+4t)$ **23** $(-2c)$ **24** $(-6e)$ **25** 0 **26** $(+7s)$
27 $(-t)$ **28** $(+3n)$ **29** $(+3p)$ **30** 0 **31** $-d$ **32** 0
33 $-6f$ **34** Subtract $4a$ **35** Add $7b$ **36** Subtract $6c$ **37** Add $8d$

Page 6 EXERCISE 3

1 $+8000$; -3600; -6000; $+7000$ **2** ± 5, $\pm 3a$, $\pm \frac{1}{2}b$ **3** (-6)
4 $(+12)$ **5** (-10) **6** $(+5)$ **7** (-15) **8** $(+2)$ **9** $(+9)$ **10** 0
11 (-8) **12** (-3) **13** $(+12)$ **14** (-1) **15** (-2) **16** $(+10)$
17 (-4) **18** $(+1)$ **19** (-8) **20** 0 **21** $(+1)$ **22** $(+4)$
23 (-16) **24** (-1) **25** $(+16)$ **26** (-24) **27** 0 **28** (-4)
29 $(-\frac{1}{4})$ **30** (-2) **31** (-1) **32** 0 **33** (-4) **34** $(+\frac{1}{2})$
35 $(-\frac{1}{2})$ **36** 0 **37** $(+2)$ **38** (-16) **39** $(+1)$ **40** $(+64)$
41 41, 1 **42** 6, 0 **43** 5, 4 **44** 30 **45** $-\frac{1}{2}$, $-\frac{1}{5}$ **46** $2\frac{1}{2}$

Page 7 EXERCISE 4

1 $-2a$ **2** b **3** $-10c$ **4** $10d$ **5** $-2e^2$ **6** $-3f^2$
7 $9gh$ **8** km **9** 0 **10** $-3a$ **11** $4b$ **12** $-2c$
13 $2d^3$ **14** $3mn$ **15** $3s$ **16** $-5t^2$ **17** -1 **18** $-2a$
19 $3b$ **20** $-c$ **21** $-2d$ **22** $3e$ **23** f **24** 0
25 $7k-3h$ **26** $-m-5n$ **27** $s-3r$ **28** $4p-2q+2r$
29 $-s-4t$ **30** y^2-3y-2 **31** $2-5z+3z^2$ **32** $4b-4a$
33 $3c^2-c^3$ **34** $9a^2$ **35** $-8b^3$ **36** $-c^2d^2$ **37** $-27e^3f^6$
38 0 **39** 0 **40** $-8k$ **41** $-3n$ **42** $\frac{p}{q}$ **43** $-\frac{2r^2}{s}$
44 $-4t^2$ **45** $-\frac{1}{f}$ **46** $-8h^5$ **47** $\frac{1}{2}x$ **48** $-r$ **49** $\frac{1}{t}$
50 $-y^2$ **51** $a-3b$; $3b-a$ **52** $c-3d$ **53** q
54 $-\frac{pq}{rs}$ **55** $\frac{ab}{cd}$ **56** ± 1; $\pm \frac{1}{2}$ **57** ± 5; ± 1
58 -1; -1; 1

Page 9 EXERCISE 5

1 $2a+2b$ **2** $c+5d-2e$ **3** f^2-1 **4** $2h$
5 $2-2n$ **6** $3s-4r-2t$ **7** $-m^2-2m-3$ **8** a^2-2
9 $c-d$ **10** $-3m^2-1$ **11** $3d-3c$ **12** b^2+1
13 $-2f^2$ **14** $1-g$ **15** $2s-2r$ **16** $x-2y-2$
17 $4b+2c-3a$ **18** $2d-e-2c$ **19** $f-2g-2$ **20** h^3-2h^2+5h+6
21 m^2-m-1 **22** $-x^3+3x^2-2x-2$ **23** $bz-xz; 3rs-6s^2$
24 $2t^2-2t; 3h-h^2-1$ **25** $b-a; 2r-3r^2$ **26** $-p-q; 3q-4p$
27 $1-s^4; 2y^2+4xy-3x^2$ **28** $3a-2b; 2c-5d; 4e+9f; 2g+h$
29 $-2b; 6e-2d; g-f; m-2n$ **30** $-2r; -3s; -k-2x; 2z-y$
31 $3b-a-2$ **32** $8d-2c+4$ **33** $-4y-3$ **34** $y-z-2$

Page 11 EXERCISE 6

1 -8 **2** -20 **3** $-\frac{2}{5}$ **4** 3 **5** -2
6 -2 **7** -8 **8** 13 **9** $\frac{1}{3}$ **10** -4
11 $-1{\cdot}2$ **12** -3 **13** $-0{\cdot}35$ **14** -2 **15** $0{\cdot}35$
16 $4\frac{1}{3}$ **17** -17 **18** $1\frac{1}{2}$ **19** $1\frac{1}{3}$ **20** $-\frac{1}{5}$
21 -1 **22** -4 **23** $\frac{2}{3}$ **24** -4 **25** 3
26 -2 **27** $-3, -1, 1, 3, 5, 7$ **28** -7 **29** $-40°$
30 £20 overdrawn **31** 15 m

Page 13 EXERCISE 8

1 $b+2c+d, c$ **2** ∠ROT, ∠POS **3** $b = d, b+c = 90°$

Page 16 EXERCISE 9

1 65, 50° **2** 35, 75° **3** 47° **4** 35°; $90°-x°$
5 147°; 76° **6** 66° **7** 45° **8** 80°
9 $90°-\frac{1}{2}$C **10** $90°-$A **12** No, yes, no **13** 20
14 110° **15** 10° **17** 130°

Page 17 EXERCISE 10

4 $r = b+c-s$ **9** ∠B

Page 18 EXERCISE 11

1 $\frac{\text{ABC}}{\text{FED}}$; SAS **2** No **3** $\frac{\text{ABC}}{\text{EDF}}$; ASA **4** No **5** $\frac{\text{ABC}}{\text{EFD}}$; AAS
6 No **7** No **8** $\frac{\text{ABC}}{\text{EFD}}$; SSS **9** No **10** CA, ∠YXK
13 △BAP, △CAQ **19** △QXC; △XDQ, △XPB; ∠XDQ

Page 21 EXERCISE 12

1 12 km/h **2** 30p per hour **3** 12p per kg
4 15p per ticket **5** 2p per cloth **6** £90 a year
7 800 kg per m^3 **8** 40p in the £ **9** 30p in the £
10 13·6 g per cm^3 **11** 2:3 **12** 3:4 **13** 3:1 **14** 3:2
15 7:9 **16** 4:9 **17** 4:1 **18** 3:8 **19** 1:2
20 4:3 **21** 2:5 **22** 3:8 **23** 5:6 **24** 8:5
25 1:3 **26** 1:3 **27** 3:4 **28** 7:9 **29** 3:4

Page 22 EXERCISE 13

1 $\frac{8}{11}$ **2** $\frac{15}{8}$ **3** $\frac{2}{15}$ **4** $\frac{3}{40}$ **5** $\frac{7}{5}$
6 3:8 **7** 1:3 **8** 3:2; 2:5 **9** 25:21; 4:25 **10** 6:7; 5:6
11 3:4; 9:16 **12** 2:3 **13** 2:15 **14** 3:2 **15** 1:10
16 10:9 **17** 4:5

Page 23 EXERCISE 14

1 1:50 000 **2** 1:20 000 **3** 1:10 000 **4** 882·5 km
5 2·37 cm **6** 4·7 km **7** 1·75 cm **8** 7·2 km
9 1:200; 3 cm, 2 cm; 4 m^2 **10** 350 hectares
11 172 hectares **12** 102·5 cm^2 **13** 4·2 m **14** 8·3 m

Page 25 EXERCISE 15

1 252 **2** 63 **3** 90p **4** 105p
5 60p, $4\frac{1}{2}$p **6** 9·6 cm by 6·4 cm **7** 4:3 **8** 4:5
9 3:2 **10** 4:5 **11** $\frac{4}{3}$ **12** $\frac{4}{5}$
13 $\frac{6}{5}$ **14** $\frac{3}{5}$ **15** $\frac{5}{4}$ **16** 15:13 increase
17 90p **18** 12·25 km **19** $\frac{3}{2}$ **20** $\frac{7}{10}$
21 5:4 increase **22** 6:5 increase **23** 10:9 increase **24** 12 cm; 144:25

Page 26 EXERCISE 16

1 72p **2** 55p **3** £5·04 **4** 8 days **5** 4 h
6 50 min **7** £21 **8** 70p **9** 90 km/h **10** tv/x min
11 36 days **12** £13·50 **13** $y=\frac{1}{3}x$, $y=\frac{2}{3}x$, $y=x$, $y=2x$, $y=\frac{4}{5}x$; 20, 40, 60, 120, 48, km/h
14 1·6 cm **15** £79 **16** 405 revs **17** 8

Page 27 EXERCISE 17

1 £9 **2** 150 m **3** £2 **4** 12 days
5 $22\frac{1}{2}$ kg **6** 12 h **7** £33·75 **8** 40 days
9 $9\frac{1}{2}$ h per day **10** £4·50 **11** 180 men **12** £3·75
13 £$\frac{cxy}{bd}$ **14** $\frac{bdy}{cx}$ days **15** 24 days

Page 30 EXERCISE 18

1 £221·10 **2** £104·65 **3** £46·50 **4** £12 480
5 £215 298 **6** £1·07 **7** 92p in the £ **8** £6·40 increase
9 £542·80 **10** £151·40, £788·60 **11** £98·56, £1·90
12 £203 200, £14·08, 8·4p **13** 4·3p, £1·50, 6p
14 £24 000, $12\frac{1}{2}$p, £37

Page 33 EXERCISE 19

1 13 **2** 3; 58 cm **3** $1\frac{1}{3}$; 12 cm **4** 6 **5** 11
6 3 **7** 10 **8** 8 **9** 5 **10** $4\frac{1}{2}$
11 $1\frac{1}{2}$ km **12** 4, 41 cm; 5, 47 cm; 6, 53 cm **13** 24 **14** 104
15 9 dozen, 12 dozen **16** 14 **17** 20 **18** 6
19 300 **20** 18 km/h **21** 45p **22** 2.24 p.m. **23** 3 km

EXERCISE 19 (*continued*)

24 $1\frac{3}{5}$ km **25** $(2\frac{1}{2}-l)$ metres, $1\frac{1}{2}$ **26** 160 **27** 18 cm
28 66°C.; 16·8 min **29** 81p per kg **30** 20 kg
31 $1\frac{3}{5}$ km **32** 12

Page 36 EXERCISE 20

1 $7\frac{1}{2}$ **2** 1·3 **3** 20 **4** 39 **5** 27 **6** $1\frac{5}{11}$ **7** $26t$ **8** 24

Page 37 EXERCISE 21

1 $z = 180-x-y$ **2** $P = 500/v$ **3** $n = \frac{1}{2}r+2$
4 $T = 60-2R$ **5** $b = \dfrac{A}{l}$ **6** $p = 2(l+b); l = \frac{1}{2}p-b$
7 $2x+y = 180; x = 90-\frac{1}{2}y$ **8** $p = 2a+b; a = \frac{1}{2}(p-b)$
9 $a = 120-\frac{1}{2}b; b = 240-2a$ **10** $y = 2z-180; z = 90+\frac{1}{2}y$
11 $F = \frac{9}{5}C+32$ **12** $r = \dfrac{C}{2\pi}$ **13** $h = \dfrac{2A}{b}$
14 $t = \dfrac{s}{u+v}$ **15** $P = \dfrac{A}{1+R}$ **16** $3; l = \sqrt{A}, l = \sqrt[3]{V}$
17 $r = \sqrt{\left(\dfrac{A}{\pi}\right)}$ **18** $r = \sqrt{\left(\dfrac{V}{\pi h}\right)}$ **19** $d = \sqrt{\left(\dfrac{5H}{2n}\right)}$
20 $d = \sqrt[3]{\left(\dfrac{6V}{\pi}\right)}$ **21** $25; d^2$ **22** $p = \dfrac{2A}{b}+2b; A = \frac{1}{2}l(p-2l)$

Page 39 EXERCISE 22

1 20% **2** 70% **3** 30% **4** 6% **5** 28%
6 5% **7** $33\frac{1}{3}$% **8** $12\frac{1}{2}$% **9** $2\frac{1}{2}$% **10** 1·78%
11 95% **12** 30% **13** 15% **14** 84%; 840 per 1000
15 42%; 420 per 1000 **16** 35% **17** 40% **18** 44%

Page 42 EXERCISE 23

1 $\frac{1}{4}$, 0·25; $\frac{1}{2}$, 0·5; $\frac{3}{4}$, 0·75 **2** $\frac{1}{20}$, 0·05; $\frac{3}{20}$, 0·15; $\frac{17}{20}$, 0·85
3 $\frac{1}{50}$, 0·02; $\frac{1}{25}$, 0·04; 3 **4** $\frac{1}{40}$, 0·025; $\frac{9}{8}$, 1·125; $\frac{3}{8}$, 0·375
5 50, $33\frac{1}{3}$, 25, 20, 140, % **6** 5, 4, $3\frac{1}{3}$, $2\frac{1}{2}$, 144, %
7 35, 8, 140, $6\frac{1}{2}$, % **8** 150, 175, 400, $133\frac{1}{3}$, 230, %
9 75, 80, $83\frac{1}{3}$, $87\frac{1}{2}$, % **10** $53\frac{1}{3}$, $18\frac{3}{4}$, 85, 44, %
11 15p **12** 8p **13** £2·25 **14** 9p **15** £5
16 5 m **17** 68 cm **18** 550 g **19** 1350 m **20** 25%
21 $33\frac{1}{3}$% **22** 125% **23** 75% **24** $133\frac{1}{3}$% **25** 15%
26 48% **27** $2\frac{1}{2}$% **28** $133\frac{1}{3}$% **29** 324 **30** £5·25
31 75%; 750 per 1000 **32** $6\frac{1}{4}$% **33** 750, 150, 100
34 $\dfrac{x}{y}, \dfrac{100x}{y}$ per cent **35** £0·14 **36** £1·51 **37** £2·18
38 £0·58 **39** £5·74 **40** 42p **41** £1890
42 49p **43** £3750 **44** £455

Page 44 EXERCISE 24

1 £24, £36; £25, £35; £48, £12 **2** 18, 30 cm; 8, 40 cm; 27, 21 cm
3 £4, £8, £12 **4** 30p, 12p, 54p
5 $3\frac{1}{2}$, 2, $4\frac{1}{2}$ kg **6** 12, 6, 4, m
7 £0·70, £1·40, £1·75, £3·15 **8** (i) 4:3:6; (ii) 5:8:3
9 135, 60, 15p **10** $2\frac{1}{2}$, $6\frac{1}{4}$, $11\frac{1}{4}$, t
11 8, 12, 16, cm **12** £150, £200, £400 **13** £100, £150, £200
14 £245, £490, £735 **15** (i) 15:18:8; (ii) 15:3:1 **16** 21p, 14p
17 £36, £18, £12 **18** 60 kg **19** £480
20 £1·25 **21** £135, £45 **22** £22·50, £33·75, £15·75
23 £60, £64, £36 **24** £540, £675, £1200 **25** £2 **26** 9:8

Page 46 EXERCISE 25

1 75 **2** £1620 **3** 72 m **4** £3·50 **5** £1008
6 £1·56 **7** 21p **8** £1·25 **9** £840 **10** £1·60
11 59°F **12** 2.40 p.m. **13** £18·60 **14** £2460 **15** £4·70
16 40 kg **17** 2 cm **18** 11·5 cm **19** £260 **20** 21 m

Page 48 EXERCISE 26

1 2:5 **2** $2\frac{2}{3}$, 4 **3** $x = \frac{2}{3}y$; 5:1 **4** 25:4
5 25%, 20% **6** n:$(100-n)$ **7** £1·50, £8·25, £1·95
8 15, 27 **9** 360 **10** £6·16, £2·64
11 $u-v$, $u+v$ km/h; 7:2 **12** 24 min **13** 39:37
14 5:4 **15** 0·8:1 **16** $10.47\frac{1}{2}$ a.m.; 80 km/h

Page 51 EXERCISE 27

4 Untrue **5** True **6** Untrue **7** $90°-\frac{1}{2}t°$
8 $2x°$, $90°-x°$; 90° **9** $2y°$, $2z°$; ∠POQ = 2∠PRQ, reflex ∠POQ = 2∠PRQ

Page 52 EXERCISE 28

1 x, $180-2x$ or $90-\frac{1}{2}x$, $90-\frac{1}{2}x$, degrees **2** 72°, 72°, 36°
3 36°, 36°, 108° **4** 30° **5** $57\frac{1}{2}°$ **6** 35°, 125°, 20°
7 38° **8** 36° **14** $360-2y$ **15** $(45-\frac{1}{4}z)$ degrees

Page 56 ORAL EXAMPLES

1 8 rt. ∠s **2** 10 rt. ∠s **3** 16, 196, rt. ∠s **4** n; $2n$ rt. ∠s

Page 58 EXERCISE 30

1 62° **2** 80° **3** 18 **4** 54
5 36° **6** 56, 76, rt. ∠s **8** 24°, $\frac{4}{n}$ rt. ∠s **9** 9, 10
10 24, 18 **11** 17 **12** 6 rt. ∠s **14** 10 rt. ∠s

Page 59 EXERCISE 31

9 4 rt. ∠s

Page 60 EXERCISE 32

1 3·00, 3·75 cm; 3·75, 0·75; 3·00, 0·75 **2** 3·2, 4, cm; 0·8, 0·8
3 4·2, 3·5, 2·8, cm; 83°, 56°, 41° **4** 7·5, 6, cm; 83°, 56°, 41°; 3:2
5 83°, 56°, 41°

Page 61 EXERCISE 33

1 △QPM, 4·8, 7·2, cm: △HKR, 13·5, 7·5, cm **2** QBP; 9, 18, cm
3 GKC; 1·8, 7·6, cm **4** 9, $2\frac{1}{2}$, cm **5** 10, $7\frac{1}{2}$, $7\frac{1}{2}$

Page 66 EXERCISE 34

1 5·6 cm, 4·2 cm, 0·7 **2** 3·6 cm, 3·0 cm, 0·6 **3** 25·5 m, 1·5
4 0·3640, 0·8391, 1, 1·1918, 1·7321, 3·7321
5 14° 2′, 41° 59′, 58°, 66° 30′
6 1·8807, 0·6346, 1·2045, 1·9970, 2·0057, 0·3178, 0·4805 or 0·4806, 1·2024, 0·2997, 1·3654 or 1·3655, 5·1608, 5·6235, 3·5938, 4·1126, 4·5801, 7·332, 6·730
7 23°, 71°, 15° 24′, 45° 48′, 60° 18′, 71° 36′, 63° 30′, 88°, 14° 14′, 13° 40′, 26° 35′, 51° 46′, 59° 39′, 34° 32′, 45° 4′, 48° 13′, 58° 33′, 76° 2′, 77° 44′, 79° 9′, 84° 32′.
8 38° 40′, 56° 19′, 62° 59′, 36° 52′ **9** 3·70, 6·88, 6·54, 20·5
10 5·21 m **11** N. 14° 2′ E.; N. 26° 34′ E. **12** 66·2 m
13 327 m **14** 21° 48′ **15** 30° 58′ **16** 36° 1′

Page 69 EXERCISE 35

1 2·26 cm, 11·1 cm, 1·80 cm **2** 2·02 cm **3** 48° 49′
4 2·36 cm **5** 63 26′, 26° 34′; 30° 58′, 59° 2′; 90°, 90°
6 77° 19′ **7** 47° 31′ **8** 5·69; 196, 83·9; 6·15
9 12° 55′ **10** 124° 31′, 86° 49′, 88° 12′, 60° 28′ **11** 16° 42′, 6·10 cm
12 17·6 m **13** 6·86 cm **14** 14·4 cm **15** 5° 5′
16 5·1 cm **17** 17° 23′, 14° 31′; no **18** 4·68, 15·1, cm
19 61° 44′ **20** 4·37, 3·97, km **21** 4·5 m **22** 25° 38′
23 82·0 m **24** 66° 48′, 5·02(2) cm **25** 8·83, 6·43, cm, 56° 46′
26 53° 58′ **27** 3·92 cm **28** 3·79 (4) cm **29** 37·6 m

Page 76 EXERCISE 38

1 58° **2** 23° **3** 62° **4** 110° **5** 55°, 35°
6 32° **7** 125° **8** 18°, 27° **9** 30°, 30° **11** $22\frac{1}{2}$°
13 126°, 30°, 24° **14** 35°, 115°, 30° **15** 150° **16** 72°
17 $67\frac{1}{2}$° **18** $22\frac{1}{2}$°, 135°, $22\frac{1}{2}$°

Page 78—Example 1. 6·78 cm **Example 2.** 7·36 cm

Page 79 EXERCISE 40

1 4·24 cm **2** 4·47 cm **3** 8·66 cm **4** 8·32 cm
5 4·74 cm **6** 3·54 cm **7** 6·69 cm **8** 5·41 cm
9 5·84, 3·98 cm **10** 6·32(5) cm **11** 8·66 cm **12** 10·46 cm
13 7·13, 3·63, cm. **14** 6·78 cm **15** 6·09 cm **16** 8·64 cm
18 4·26 cm **19** $117\frac{1}{4}$° **20** 3·08 cm **22** 3·55 cm

Page 81 EXERCISE 41

1 $P = 110+30n$; $P = 600-40n$; after 7 years, £320
2 $n = 10+9t$; $n = 25+6t$; 5 years, £55
3 $T = \frac{1}{2}(P-200)$; $T = \frac{2}{5}(P-160)$; £360, £80
4 $l = 20+\frac{1}{50}W$, $l = 14+\frac{1}{20}W$, 200 g, 24 cm

Page 83 EXERCISE 42

1 $y = 3x-11$ **2** $y = 6-2\frac{1}{2}x$ **3** $y = \frac{1}{12}x-4$ **4** $x = 3-5y$ **5** $x = -2\frac{1}{4}y$
6 $x = \frac{3}{5}(y-1)$ **7** y **8** x **9** y

Page 85 EXERCISE 43

1 7, 6 **2** 9, 3 **3** 2, 5 **4** 4, 5 **5** 11, 3
6 $2\frac{1}{2}$, 4 **7** 24, 4 **8** 8, 3 **9** 5, 1 **10** 16, 3
11 5, 3 **12** 3, -1 **13** 3, 4 **14** 2, 7 **15** 12, 4
16 2, 3 **17** 2, 5 **18** 21, 9 **19** 3, 4 **20** 4, 1
21 7, 0 **22** -1, 2 **23** 2, 12 **24** $-\frac{10}{13}$, $-\frac{1}{13}$ **25** 3, -12
26 $\frac{5}{7}$, $\frac{1}{7}$ **27** 3, 4 **28** 19, -17 **29** $-\frac{7}{8}$, $-\frac{1}{4}$ **30** 7, 2
31 2, 3 **32** -2, 3 **33** -3, 4 **34** 2, 3 **35** 16, $-12\frac{1}{2}$
36 $\frac{2}{3}$, $-5\frac{2}{3}$ **37** 20, -10 **38** 3, 2 **39** $1\frac{1}{3}$, $\frac{1}{3}$ **40** 5, 2
41 3, -1 **42** $\frac{1}{5}$, $\frac{1}{3}$ **43** $\frac{1}{5}$, -1 **44** $1\frac{1}{2}$, $1\frac{2}{3}$ **45** $\frac{1}{3}$, $2\frac{1}{3}$
46 -1, 3 **47** 6, 2; 38, 114 **48** 20, ($a = 6$, $b = -1$)
49 2, ($m = -7$, $a = 23$) **50** 9, ($a = 2$, $b = -3$)

Page 86 EXERCISE 44

1 25, 14 **2** 75, 69p **3** 25, 30p **4** £30, £9 **5** 9p
6 3, 5; 21 cm **7** 14, 7; 875 cm^2 **8** 17, 14 **9** 27, 63
10 3 t, $\frac{3}{4}$ t **11** 44 years, 12 years **12** 72°, 36°, 72°
13 100° **14** A £11, B £13 **15** A £3·50, B £4 **16** 84
17 54 **18** 653 **19** 14, 6 **20** 12 km **21** $\frac{11}{17}$
22 $\frac{23}{32}$ **23** 5, 30 **24** 30 **25** $5\frac{1}{2}$, $3\frac{1}{2}$
26 6 cm by 6 cm by 9 cm **27** $2\frac{1}{2}$ cm, 3 cm **28** 19, 22 **29** 4·8, 6·4 km/h
30 1·8, 32; 212 **31** 600, 0·9; 1410 kg **32** 400 m

Page 90 EXERCISE 45

1 12, 2 **2** 16, 9 **3** 18, 10 **4** 6, -1 **5** $3\frac{1}{2}$, $1\frac{1}{3}$ **6** $\frac{1}{4}$, $\frac{1}{5}$
7 15, 3 **8** -4, $3\frac{1}{2}$ **9** 10, 4 **10** $\frac{1}{2}$, 1 **11** 7, -4 **12** 11, 5
13 $\frac{1}{2}$, $\frac{1}{4}$ **14** 4·3, 1·5 **15** 3, 7 **16** 4, 5 **17** 5, 2 **18** 7, 13
19 $\frac{1}{4}$, $\frac{1}{5}$ **20** 7, 5 **21** 5, 8 **22** 5, 1 **23** 10, 12 **24** -2, 1
25 $1\frac{1}{4}$, $1\frac{1}{5}$

Page 91 REVISION EXERCISE R 1

1 $1-9x-3x^2$ **2** $12d-3b-6c$ **3** $1\frac{1}{12}$ **4** $\frac{1}{3}$ **5** $2x-y$
6 $\frac{1}{3}(a+2b)$ **7** 3 **8** $1\frac{67}{168}$ **9** 1·472 **10** 0·019
11 12·5 **12** 0·475, 0·4375, 0·0156 **13** 1 **14** 4
15 4 **16** 18 **17** $x^2+2xy+2xz$
18 $2+2x+5x^2-2x^3$ **19** $2x^4-x^3+3x^2-2x$ **20** $-2n$
21 -2 **22** -1 **23** 10·4 **24** 18 **25** 6
26 $\frac{1}{2}$ **27** 3 **28** 12 or 2 **29** 1 : 20 000 **30** $17\frac{1}{2}$ min

REVISION EXERCISE R1 (*continued*)

31 8:9 **32** £1·32 **33** 60p **34** £2·42 **35** £2·10, £1·40, £0·70
36 $x = \frac{1}{6}(y-3)$ **37** 2:3 **38** 9:11 **39** $6\frac{2}{3}$% **40** $1\frac{1}{4}$, $\frac{3}{4}$
41 $y = x+5$ **42** -15; -4, -6 **43** -640 **44** 11
45 8 **46** £$(a-c+10)$ **47** $\frac{1}{2}$ km **48** 8
49 0·9601, 2·3946, 4·0560, 4·2747 **50** 23° 44′, 55° 15′, 63° 52′, 78° 16′
51 3·45(1) cm **52** 4·98(9) cm **53** 55°
54 26° 34′, 10° 18′; 42° 1′, 47° 59′, 35°; 29° 3′ **55** 24° 14′ **56** 182 m
57 3·66(5) cm **58** 74° 29′ **59** 19° **60** 63°, 27 **61** 15°, 75°
62 5·38 cm **66** 32°, 48° **71** -2, 5 **72** -2, 3 **73** 5, 2
74 2 **76** 11, 15 **77** 10p, 30p **78** $x = 45 - \frac{1}{4}y$; 20, 100
79 12, 3 **80** 2, -3

Page 95

EXERCISE 46

1 1·17 **2** 1·83 **3** 1·7 **4** 1·2 **5** 2·39 **6** 0·91
7 0·63 **8** 0·39 **9** 0·7 **10** 0·6 **11** 324 **12** 320
13 72 **14** 78 **15** 93:100; 7:93; $\frac{100}{7}$ **16** 113:100; 81:100
17 $\frac{21}{20}$; $\frac{20}{21}$ **18** $\frac{22}{25}$; $\frac{25}{22}$ **19** $\frac{3}{50}$; $\frac{3}{53}$ **20** $\frac{2}{23}$; $\frac{2}{25}$ **21** $\frac{100}{100+x}$ **22** $\frac{100}{100-y}$

Page 97

EXERCISE 47

1 108 **2** 45 **3** 135 **4** 936 **5** 28 **6** 224
7 £20 **8** 120 **9** 135 **10** £160 **11** £240 **12** £1748
13 16% **14** £1650 **15** 45%; 480 **16** 160 **17** £2025
18 £30 **19** £360 **20** $6\frac{1}{4}$% **21** 1·0% **22** £1·35 **23** £6·75
24 12 g **25** £30·47 **26** 44% **27** 90% **28** 13%

Page 99

EXERCISE 48

1 $\frac{11}{10}$ **2** $\frac{3}{2}$ **3** $\frac{9}{10}$ **4** $\frac{4}{5}$ **5** $\frac{13}{10}$ **6** $\frac{21}{20}$
7 $\frac{3}{4}$ **8** $\frac{23}{25}$ **9** 22p **10** 18p **11** £8 **12** £75
13 $\frac{5}{6}$ **14** $\frac{5}{9}$ **15** $\frac{10}{9}$ **16** $\frac{5}{3}$ **17** $\frac{20}{21}$ **18** $\frac{25}{27}$
19 $\frac{4}{3}$ **20** $\frac{25}{22}$ **21** £10 **22** £10 **23** £10 **24** £20
25 £5 **26** £40 **27** £30 **28** £20 **29** 12% g **30** 20% l
31 25% l **32** 20% g **33** 25% g **34** 25% l **35** 20% l **36** $33\frac{1}{3}$% g
37 16% g **38** 20% l **39** $11\frac{1}{9}$% g **40** 40% l

Page 100

EXERCISE 49

1 £15·75 **2** 8p **3** £69 **4** 33p **5** £56·43
6 130p **7** 30% l **8** $11\frac{1}{9}$% l **9** $27\frac{7}{9}$% g **10** $16\frac{2}{3}$% g
11 $14\frac{2}{7}$% l **12** 15% g **13** £10 **14** 25p **15** 225p
16 £30 **17** £8 **18** 90p **19** 25% **20** $16\frac{2}{3}$%
21 63p **22** £135 **23** £12 **24** £5250 **25** 40%
26 44% **27** 40% **28** £25 **29** £10 **30** £135
31 25% **32** $66\frac{2}{3}$% **33** 20% **34** 78p **35** 3p each

Page 101

EXERCISE 50

1 $37\frac{1}{2}$% **2** 5% **3** 2·4% **4** £26·50 **5** $\frac{5}{9}$
6 1·3%, (1·256) **7** 45 **8** $37\frac{1}{2}$p per kg **9** 40%

EXERCISE 50 (*continued*)

10 $25\frac{1}{3}\%$ **11** £22 **12** 20% **13** £800, £1200, £1600
14 50% **15** $16\frac{2}{3}\%$ **16** $7\frac{1}{2}\%$ **17** 25 km/h **18** 25%
19 6 kg **20** $1\frac{1}{2}$p

Page 103 EXERCISE 51

1 $2c+d$ **2** $4a-5$ **3** $y+1$ **4** $3t+2$
5 $p-1$ **6** $b-1$ **7** $2x-3y$ **8** $1+n$
9 $3(2c-3)$ **10** $4(n+1)$ **11** $x(y+z)$ **12** None
13 $6a(a-b)$ **14** $10r(2r+1)$ **15** $2(2+s^2)$ **16** $bc(b-2c)$
17 $5(r+2s)$ **18** $c(b-d)$ **19** $2y(3y-1)$ **20** $xz(x^2+z^2)$
21 $a(b+1)$ **22** $d(3d^2-1)$ **23** None **24** $p^2(1-r)$
25 $3(3pq-7rs)$ **26** $s^2(r^2+1)$ **27** $4b(ab-1)$ **28** $y(1-cz)$
29 $a(p+q+r)$ **30** $x(x^2-3x-3)$ **31** None **32** $y(y+z+1)$
33 $cd(1+c+d)$ **34** $r(3r-s-t)$ **35** $(r+s)(a+b)$ **36** None
37 $(x+y)(x-y)$ **38** $(a+b)(a+b+c)$ **39** None **40** $(r-t)(r-t+1)$
41 $(c-d)(b+x)$ **42** $(p+q)(a+b)$

Page 103 EXERCISE 52

3 (i) $(x+2)(x+3) = x^2+5x+6$;
(ii) $(a+b)(x+y+z) = ax+ay+az+bx+by+bz$

Page 106 EXERCISE 53

1 $by+bz+cy+cz$ **2** $ax+ay-bx-by$ **3** $cy-cz+dy-dz$
4 $ax-az-dx+dz$ **5** $a^2-ac+ab-bc$ **6** $x^2-ax-bx+ab$
7 a^2+5a+6 **8** b^2+3b-4 **9** $c^2-8c+15$
10 $9-6m+m^2$ **11** $16-n^2$ **12** $5-4p-p^2$
13 $a^2-5ab+6b^2$ **14** $c^2-3cd-4d^2$ **15** $x^2+2xy-3y^2$
16 $9x^2-1$ **17** $25y^2-4$ **18** $8z^2-10z+3$
19 $8r^2+6rs-9s^2$ **20** $6x^2-11xy-10y^2$ **21** $a^2+2ab+b^2$
22 $a^2-2ab+b^2$ **23** a^2-b^2 **24** $9c^2-12cd+4d^2$
25 $16x^2+8xy+y^2$ **26** $9y^2-24yz+16z^2$ **27** $25y^2+20yt+4t^2$
28 $9z^2-30zp+25p^2$

Page 107 EXERCISE 54

1 $ax+ay+bx+by$ **2** $cy+cz-dy-dz$ **3** $qa-qb-ra+rb$
4 $cx-cy+dx-dy$ **5** $a^2-ac+ab-bc$ **6** $cd-cz-dz+z^2$
7 $mn+2m-n-2$ **8** $y^2+11y+28$ **9** $z^2+11z+10$
10 $a^2+3a-10$ **11** $b^2-4b-21$ **12** $c^2-8c+15$
13 $5-4n-n^2$ **14** $9-6p+p^2$ **15** $16-q^2$
16 $12-t-t^2$ **17** z^2-49 **18** $10-7s+s^2$
19 $21+4t-t^2$ **20** $40-6r-r^2$ **21** $12b^2+23b+10$
22 $12y^2+y-1$ **23** $25z^2-1$ **24** $12c^2-29c+14$
25 $x^2+2xy+y^2$ **26** $x^2-2xy+y^2$ **27** x^2-y^2
28 $4c^2+20cd+25d^2$ **29** $49f^2-42fg+9g^2$ **30** $25y^2-49z^2$
31 x^3+4x^2+5x+6 **32** a^3+b^3 **33** $6c^3-13c^2+3c+2$
34 $2-13x-13x^2-3x^3$ **35** x^3-8y^3 **36** $6r^3+7r^2s-7rs^2-6s^3$
37 2; $2pq-2q^2$ **38** $4rs$; $4x^2-y^2$ **39** $-28t$ **40** -9; 1
41 7; -11 **42** 0; -37 **43** -19; 12 **44** -16; 1
45 -57; 0 **46** 1 **47** $-\frac{1}{2}$ **48** $-\frac{1}{2}$

EXERCISE 54 (*continued*)

49 n^3-n **50** $8m^3-8m$ **51** $12p^2+12p+11$
52 4 cm^2; $(4r+12)$ cm, $(4r+12)$ cm **53** $(6l^2+12l+6)\text{ cm}^2$
54 $(14a^2-14ab)\text{ m}^2$

Page 108

EXERCISE 55

1 3; -3 **2** 9; 9 **3** 27; -27 **4** -1 **5** $-\frac{1}{3}$
6 -2 **7** $\dfrac{2(f+g)}{3(g-f)}$ **8** $\dfrac{p-q}{p+q}$ **9** 1 **10** $-\dfrac{xy}{ab}$
11 $(p+q)^2$ **12** $-(c-d)^2$ **13** $y-x$ **14** 1 **15** $-\dfrac{x}{z}$
16 1

Page 110

EXERCISE 56

1 $a^2+10a+25$ **2** b^2-6b+9 **3** $9-6b+b^2$
4 $a^2+10a+25$ **5** $16c^2+8cd+d^2$ **6** $16e^2-8ef+f^2$
7 $4p^2+12pq+9q^2$ **8** $25s^2-70rs+49r^2$ **9** $x^2-2xy+y^2$
10 $y^2-2xy+x^2$ **11** $25r^2+30rs+9s^2$ **12** $16y^2-40yz+25z^2$
13 $x^2+x+\frac{1}{4}$ **14** $y^2-2+\dfrac{1}{y^2}$ **15** $9a^2b^2-24abc+16c^2$
16 $a^2+2+\dfrac{1}{a^2}$ **17** $x-1$, $1-x$ **18** $a+3$; $-a-3$
19 No **20** $c-2$, $2-c$ **21** No
22 $4-k$, $k-4$ **23** No **24** $1-p$, $p-1$
25 $2-3x$, $3x-2$ **26** $3s+10$, $-3s-10$ **27** No
28 $t+\frac{1}{2}$, $-t-\frac{1}{2}$ **29** 25; $\pm(x+5)$ **30** 16; $\pm(y-4)$
31 $9b^2$; $\pm(a+3b)$ **32** $\frac{9}{4}$; $\pm(c+\frac{3}{2})$ **33** $49z^2$; $\pm(y-7z)$
34 $\frac{49}{4}$; $\pm(d-\frac{7}{2})$ **35** $\frac{1}{4}$; $\pm(a+\frac{1}{2})$ **36** $\frac{1}{4}d^2$; $\pm(c-\frac{1}{2}d)$

Page 111

EXERCISE 57

1 $(y+z)(y-z)$ **2** $(3+x)(3-x)$ **3** $(c+4d)(c-4d)$
4 $(2n+5)(2n-5)$ **5** $(1+2t^2)(1-2t^2)$ **6** $(4p^3+7q^4)(4p^3-7q^4)$
7 $(x+2y+z)(x+2y-z)$ **8** $(b-c+3)(b-c-3)$ **9** $(p+q-r)(p-q+r)$
10 $(a+5)(a+1)$ **11** $(b+1)(b-9)$ **12** $(3c-4d)(c+4d)$
13 $a(b+c)(b-c)$ **14** $3(d+2)(d-2)$ **15** $5(1+3e)(1-3e)$
16 $(2a+3b+3c)(2a-3b-3c)$ **17** $(5c-4d)(4d-3c)$
18 $(3e+2)(e-2)$ **19** 49 **20** 1840
21 8 **22** $5\frac{1}{2}$ **23** $(b+c)$; $(1-3b)$

Page 113

EXERCISE 59

1 3·6, 3, cm; 4·8, 4, cm **2** 2·5, 3, 3·5, cm **3** 1·8, 1·6, cm
4 $\text{AQ} = \frac{3}{7}\text{AC}$, $\text{PQ} = \frac{3}{7}\text{BC}$ **5** 9, 8
6 $7\frac{1}{2}$ **7** 18 **8** 16, 7·5 cm **9** 6 cm
10 5·5 cm **11** 1·5 cm **12** 5 cm **13** $8\frac{1}{2}$ m

Page 122

EXAMPLE FOR CLASS DISCUSSION

MP = 3·42 cm, OM = 9·40 cm; NQ = 6·43 cm, ON = 7·7 cm
KR = 9·40 cm, OK = 3·42 cm

Page 123 EXERCISE 62

1 (i) 0·423, 0·906; (ii) 0·574, 0·819; (iii) 0·906, 0·423; 1, 0, 0, 1.
2 0·2924, 0·6820, 0·8988, 0·9994, 0·3987, 0·4003, 0·3990, 0·4000, 0·6205, 0·9002, 0·7551, 0·1536, 0·1181.
3 0·9703, 0·8829, 0·5592, 0·0175, 0·3971, 0·3955, 0·3963, 0·3958, 0·4591, 0·7540, 0·8768, 0·9840, 0·6213.
4 $\sin\theta$, $\cos\phi = 0·6$, $\cos\theta$, $\sin\phi = 0·8$; $\sin\theta$, $\cos\phi = \frac{5}{13}$, $\cos\theta$, $\sin\phi = \frac{12}{13}$; $\sin\alpha$, $\cos\theta = 0·8$, $\cos\alpha$, $\sin\theta = 0·6$; $\sin\beta$, $\cos\phi = 0·6$, $\cos\beta$, $\sin\phi = 0·8$; $\sin\theta$, $\cos\phi = 0·28$, $\cos\theta$, $\sin\phi = 0·96$.

Page 124 EXERCISE 63

1 $z = 4·54, y = 8·91$; $a = 2·65, b = 1·41$; $p = 3·11, q = 2·52$; $e = 84·8, f = 53·0$.
2 30·9 m **3** 1120, 1660, m **4** 88·3 m **5** 6·41 cm
6 23·9 cm **7** 4·695 cm **8** 6·71 cm **9** 0·997 cm
10 4·88, 10·96, cm **11** $\cos\theta = 0·6$, $\cos\phi = 0·28$, $\cos\alpha = \frac{1}{2}$, $\tan\beta = 1\frac{1}{3}$
12 5·88 cm **13** 0·809, 0·809

Page 126 EXERCISE 64

1 $23\frac{1}{2}°$, $48\frac{1}{2}°$, 72°, 36° **2** $44\frac{1}{2}°$, 67°, 51°, $24\frac{1}{2}°$
3 23°, 74°, 50° 18′, 26° 42′, 26° 48′, 26° 44′, 26° 46′, 13° 34′, 18° 39′, 74° 45′, 36° 46′, 41° 39′.
4 56°, 38°, 29° 24′, 79° 36′, 79° 42′, 79° 40′, 79° 37′, 40° 40′, 30° 20′, 10° 28′, 75° 27′, 50° 52′.
5 31° 46′, 71° 56′, 32° 15′, 58° 39′, 0° 39′, 71° 37′
6 36° 52′, 53° 8′; 22° 37′, 67° 23′; α, ϕ = 53° 8′, β, θ = 36° 52′; 16° 16′, 73° 44′
7 14° 29′ **8** 5° 44′ **9** 45° 35′ **10** 24° 2′
11 30° **12** 89° 36′ **13** 11° 29′ **14** 2·40, 1·01 km
15 52·9, 140, m **16** 51° 24′; 51° 50′

Page 128 EXERCISE 65

1 55·9 m **2** 21·2 cm **3** 49° 14′ **4** 3·2 cm
5 90° **6** N. 53° 34′ E. **7** 1560 m **8** 128 m
9 2·41, 3·19, 4·37 cm **10** 14° 56′ **11** 46·9 m
12 11·9, 1·83, cm **13** 5·02 cm **14** 51° 19′, 6·24(5) cm, 37° 20′
15 60·7(5) cm **16** 3·79 cm
17 (i) 2·82, 2·65, 5·47, cm; (ii) 6·71 cm; (iii) 35° 26′, 88° 34′
18 14° 8′; 6·99, 0·994, 2·34, cm **19** 46° 22′, 88° 51′
20 70 m; 2159 m; 1° 51′ **21** 3·51 m, 64° 41′, 6·24 m

Page 131 EXERCISE 66

1 2; 3; no **2** 4; -1; $b+1$ **3** No; -1; no **4** No; -1; no
5 2; no; z **6** No; -1; $x+y$ **7** c; -1; $-a-b$ **8** No; $c-d$; -1
9 $x+y$ **10** No **11** No **12** $a-b$
3 No **14** $x+1$ **15** No **16** $(c+d)(a-b)$
7 No **18** No **19** $(c+d)(a-b)$ **20** $(a-b)(c-d)$
1 $(y+1)(2x+z)$ **22** No **23** $(x-a)(1-b)$ **24** $(x+y)(a+1)$
5 $(p+q)(x-1)$ **26** No **27** $(y-z)(c-1)$

Page 132 EXERCISE 67

1 $(x-y)(a+b)$ 2 $(a+b)(a+c)$ 3 $(c+d)(a-b)$ 4 None
5 $(x+y)(x+3)$ 6 $(a+c)(a-5)$ 7 None 8 $(c^2+d^2)(a^2+b^2)$
9 $(r+s)(p-q)$ 10 $(m-n)(c-k)$ 11 $(p-q)(d-4)$ 12 $(x-c)(x-d)$
13 $5(x-y)(c-d)$ 14 $(2a-x)(3b+y)$ 15 $(2x-y)(2x-3z)$
16 $(a+3x)(b-4y)$ 17 None 18 $(b+c)(2a+1)$ 19 $(3d+n)(2c-3m)$
20 $(2c-3q)(5p-2a)$ 21 $(a+b+c)(x+y)$
22 None 23 $(r-s)(a-b)$ 24 $(b+c)(a^2-bc)$ 25 $(x+1)(a-3)$
26 $(x+y)(y-1)$ 27 $(x+1)(x^2+1)$ 28 $(a-b)(c-1)$ 29 $(a+q)(p-1)$
30 $(y^2+1)(x-1)$ 31 $(b-c)(a+d)$ 32 $(x-a)(x-2b)$ 33 $(a-d)(c+b)$
34 $(2x-y)(p-q)$ 35 $(c-a)(a-d)$ 36 None 37 $(x-1)(x-y)$
38 $(1-x)(4-c)$ 39 None 40 $(a-x)(r-z)$ 41 $(1+c^2)(1+cd)$
42 $(a-x)(2a^3-1)$ 43 $(x+y)(x-y+a)$ 44 $(z-1)(z+1+c)$
45 $(a-b)(a+b-c)$ 46 $(1+m)(1-m-t)$
47 $(r-2s)(t-r-2s)$ 48 $(3c+d)(y-3c+d)$

Page 134 EXERCISE 68

1 11 2 -11 3 -5
4 5 5 1 6 -27
7 -5 8 2 9 -15
10 5, 3 11 15, 1 12 9, 5
13 16, 3 14 6, 2 15 18, 6
16 15, 4 17 9, 6 18 -8, -3
19 9, -2 20 15, -3 21 6, -16
22 $(x+1)(3x+2)$ 23 $(y+3)(2y+1)$ 24 $(3z+2)(z+2)$
25 $(a+2)(a+4)$ 26 $(b+1)(b+9)$ 27 $(c-3)(c-4)$
28 $(d-7)(d+2)$ 29 $(x+10y)(x-3y)$ 30 $(r+6s)(r-14s)$
31 $(2z-1)(z+3)$ 32 $(2p-3)(p-4)$ 33 $(t+4)(3t+1)$
34 $(2a+1)(a+5)$ 35 $(3b-4)(b-2)$ 36 $(4c-1)(c+3)$
37 $(d+2)(d+5)$ 38 $(t-2)(t+8)$ 39 $(n+9)(n-2)$
40 $(x-7)(x+4)$ 41 $(y+1)(y+14)$ 42 $(z+10)(z-7)$
43 $(5-a)(3+a)$ 44 $(4+b)(1-b)$ 45 $(12-c)(1+c)$
46 $(4x+y)(x+3y)$ 47 $(4a-b)(3a-2b)$ 48 $2(y-3z)(y-8z)$
49 $2(c-5d)(c-6d)$ 50 $3(3r+2s)(r-5s)$ 51 $(p-5q)(p-8q)$
52 $(4m+n)(2m-3n)$ 53 $(2+3k)(1-2k)$ 54 $(6+x)(3-x)$
55 $(3-2y)(2+3y)$ 56 $3(a+21)(a-1)$ 57 $2(b+4)(b-12)$
58 $3(n+3)(4n-1)$ 59 $3(3t-2)(3t+4)$ 60 $3(5-y)(4+y)$
61 $(2z+3)(2z+5)$ 62 $2(4-5x)(3+2x)$ 63 $(2a-3b)(3a-2b)$

Page 135 EXERCISE 69

1 $(a+2)(a+3)$ 2 $(b+3)^2$ 3 $(c-3)(c-4)$
4 $(d+2)(d-5)$ 5 $(k+4)(k-2)$ 6 $(n-3)(n+2)$
7 $(p+6)(p-5)$ 8 $(r+6)^2$ 9 $(t-10)(t+5)$
10 $(x+5)(x+9)$ 11 $(y+12)(y-5)$ 12 $(z-14)(z+5)$
13 $(m+8)^2$ 14 $(n+8)(n-8)$ 15 $(r-8)^2$
16 $(a+2)(a+6)$ 17 $(b-4)^2$ 18 $(c+5)(c-1)$
19 $(d+3)(d+6)$ 20 $(k-7)(k+4)$ 21 $(n-5)(n-6)$
22 $(p-7)^2$ 23 $(q+9)(q-3)$ 24 $(r-6)(r-12)$

EXERCISE 69 (*continued*)

25 $(t+6)(t-6)$ **26** $(x+10)(x-9)$ **27** $(y-9)(y+7)$
28 $(z-15)(z+8)$ **29** $(m-10)^2$ **30** $(s+10)(s-10)$
31 $(x-4y)(x-8y)$ **32** $(a+5b)(a-2b)$ **33** $(c-8d)(c+4d)$
34 $(1+5a)(1-2a)$ **35** $(1-6b)(1+4b)$ **36** $(1-5c)(1-7c)$
37 $(1-7d)^2$ **38** $(1+7e)(1-7e)$ **39** $(1+7g)^2$
40 $(1-5k)(1+4k)$ **41** $(8+n)(3-n)$ **42** $(4-p)(7-p)$
43 $(4+a)(3-a)$ **44** $(7+c)(5-c)$ **45** $2(x+5y)(x-2y)$
46 $3(1+3z)(1-3z)$ **47** $(ab-2c)(ab-5c)$ **48** $(1-6xy)(1+3xy)$
49 $(6r-s)(4r+s)$ **50** $(2p+3)(p-5)$ **51** $(5q-1)(q-3)$
52 $(4r-3)(r+2)$ **53** $(5x-4)(2x-1)$ **54** $(4y-5)(3y+1)$
55 $(5z-3)(2z-3)$ **56** $2(a+3)(a-2)$ **57** $5(b-6)(b+5)$
58 $4(c+5)(c-5)$ **59** $(2d-1)(d-2)$ **60** $(3e-1)(e+2)$
61 $(2b-3)^2$ **62** $(9c+10)(c-1)$ **63** $(3m+8n)(3m-8n)$
64 $(5x+4y)^2$ **65** $(7c-6d)(6c+5d)$ **66** $(10y-3)(y-4)$
67 $x+5$; 7 **68** $x-4$; -1 **69** -11
70 12 **71** -14 **72** -1

Page 137 — EXERCISE 70

1 2·5 m^2 **2** 0·9 m^2 **3** 105 m^2 **4** 24·32 cm^2 **5** 1·6 m^2
6 6·25 m^2 **7** 12 m **8** 8 m **9** 9·33 m **10** 5·4 cm **11** 10 cm^2
12 55·2 cm^2 **13** 223 cm^2 **14** 29·2, 30·3 cm^2
15 63; 153 cm^2 **16** 47·5 m^2 **17** 48·6 m^2 **18** 376 cm^2
19 79·5 cm^2 **20** 11·75 m^2 **21** 132 m^2 **22** 5·52 m^2

Page 139 — EXERCISE 71

1 148 m^2; 108 m^2 **2** 1836 cm^2 **3** 390 cm^2 **4** 150 cm^2
5 5·04 m^2 **6** 12·5 cm^2; 185 cm^2 **7** £9 **8** £1·80
9 19·4 m^2

Page 140 — EXERCISE 72

1 46·75 m^3 **2** 540 cm^3 **3** 61·6 m^3 **4** 243 000 cm^3
5 77175 **6** 12 800 **7** 480 l **8** 2·16 l **9** 420
10 36 kg **11** 38·4 kg **12** 300 g **13** 5·04 kg **14** 3·2 cm
15 40 cm **16** 42·9 cm **17** 14·4 cm **18** 210 **19** 60 cm
20 100 000 **21** 1·5 kg **22** 2·75 **23** 1152 kg **24** 16 cm

Page 142 — EXERCISE 73

1 1728 cm^3 **2** 1675 cm^3 **3** 600 cm^3 **4** 503 cm^3 **5** 2·12 kg
6 20·5 t **7** 507

Page 143 — EXERCISE 74

1 1600 cm^3 **2** 3·52 m **3** 0·375 m^2 **4** 0·145 cm^2 **5** 6·5 cm
6 40 cm^2 **7** 330 cm^3, 2·51 kg **8** 11·7 kg **9** 41·7 cm
10 1·8 cm **11** 7250 cm^3, 56·55 kg **12** 283 kg **13** 143 cm^2
14 3456 **15** 71·4 cm^2 **16** 1500 **17** 0·8 mm^2 **18** $1\frac{1}{3}$ kg

Page 145 — REVISION EXERCISE R 2

1 14 kg **2** 1 km **3** £1800 **4** 3·6 cm **5** £0·72 **6** 7·56 m^2
7 $9\frac{1}{2}$, $-1\frac{1}{2}$ **8** 25, 5 **9** 4, 7 **10** 4, $-\frac{1}{2}$ **11** $4\frac{3}{4}$, $-2\frac{1}{4}$

REVISION EXERCISE R2 (*continued*)

12 $-3, 8$ **13** 0·02, 0·08

14 A walks at $7\frac{1}{2}$ km per h; B goes out at 50 km per h, halts for 1 hour, returns at 16·7 km per h; about 2.14 p.m., 11·8 km from Oxford; about 3.39 p.m., 22·4 km from Oxford.

15 15·4, 30·2, t; 2·17, 3·12, cm **16** 60p **17** £75·65

18 63:72:80 **19** 0·68 g per cm^3

20 0·5397, 0·8418; 0·9413, 0·3376; 0·9899, 0·1415

21 70°, 20°; 26° 38′, 63° 22′; 8° 52′, 81° 8′ **22** 72°; 56°

23 N. 54° 28′ W. **24** 0·877, 2·869, km **25** 0·901 cm

26 67° 23′, 112° 37′ **27** 10° 48′ **28** 3·89(7) cm

29 31° 48′ (49′), 5·58 cm **30** 106° 16′ **31** 7° 42′

32 62 m **33** 40p, 12p **34** A 11p, B 13p **35** 11, $\frac{1}{5}$; 27 kg

36 $\frac{11}{31}$ **37** 35 **38** 1·8, 2·4, 3·6, 4·2, cm

39 75% **40** 16p **41** $33\frac{1}{3}$% **42** 4%

43 £204·80 **44** £62·50 **45** $(x-2)(x-3)$

46 $(x+3)(x+7)(3x+1)(3x+5)$ **47** $\pm(a-2)(a-3)(a+5)$

48 $5x^2-x+3$ **49** No **50** 3 **51** -15

52 233 cm^2 **53** 300 cm^2 **54** 12 cm **58** 9, 8; $17\frac{1}{2}$, $22\frac{1}{2}$

59 4·5, 6·9 m

Page 149 **Oral Work on Fig. 180**

(i) $(-4, -3)$, $(6, -4)$, $(8, 0)$, $(0, 1)$, $(-2, 0)$

Page 151 **Oral Work on Fig. 181**

(i) 14·4, 5·8, 19·4, 6·8 (ii) 3·16, 4·80, 2·24, 3·74

Page 156 EXERCISE 75

1 33° 41′ **2** (11·91, 5·54), (11·91, −3·54)

3 (7, 3·5), (7, −1·5), (−1, −1·5), (−1, 3·5); $x = 7$, $x = -1$, $y = 3{\cdot}5$, $y = -1{\cdot}5$; (3, 0·5), (3, −4·5), (−5, −4·5), (−5, 0·5); $x = 3$, $x = -5$, $y = 0{\cdot}5$, $y = -4{\cdot}5$

4 1·2, 5·4

5 Straight lines through origin; slope k; upward slope if k is positive; downward slope if k is negative.

6 (0, 2), (0, c) **7** -3 **8** Increase by 10

9 Decrease by $1\frac{1}{2}$ **10** 2·4, 0·1, $(2\frac{3}{7}, \frac{1}{7})$ **11** $y = \frac{1}{2}(6-3x)$; 0·5, 2·25

12 3, −1, −4, $\frac{1}{3}$, $-\frac{1}{4}$; $y = 3x$, $y = -x$, $y = -4x$, $3y = x$, $4y = -x$

13 $y = 5x-9$ **14** $y = 3x-7$ **15** 10·7 km

Page 158 EXERCISE 76

9 $y > 1$, $y < x$ **10** $-3 < x < 1$, $1 < y < 3$ **11** $y < x+1$, $x+y < 3$

Page 160 EXERCISE 77

1 $x > 1\frac{1}{2}$ **2** $x < -1\frac{1}{2}$ **3** $x > -\frac{2}{3}$ **4** $x < -1$ **5** $x > -3$

6 $x < 2$ **7** $x < 1$ **8** $x > 18$ **9** $x > -5$ **10** $x < 3$

11 $x > -\frac{2}{3}$ **12** $x > -4$ **13** No x **14** $x < 1\frac{1}{3}$ **15** $x < 6\frac{1}{3}$

16 $x > \frac{1}{2}$ **17** $x < 5$ **18** $x < \frac{9}{11}$ **19** $x < 4$ **20** $x > 8$

21 $0 < x < 3$ **22** $x > 4$ **23** All x **24** $x < 5\frac{4}{5}$

Page 162 EXERCISE 78

1 2209 **2** 22·09 **3** 5184 **4** 518 400 **5** 68·89
6 84 100 **7** 9 610 000 **8** 5·29 **9** 0·0576 **10** 435 600
11 17·06 **12** 5256 **13** 383 200 **14** 3·098 **15** 8 237 000
16 998·6 **17** 10·05 **18** 0·1197 **19** 0·8010 **20** 483 000
21 0·1069 **22** 25 400 000 **23** 0·000 102 0 **24** 0·4942
25 0·008 836 **26** 0·004 007 **27** 5013 **28** 10 110 000 **29** 14·6 (4)
30 145*4* **31** 150 *1*00 **32** 14 960 000 **33** 41·6 (4) **34** 0·413 (1)
35 0·004 18 (9) **36** 420*4* **37** 81·5 (6) **38** 823*8* **39** 82 0*4*0 000
40 0·00812 (0) **41** 640*6* **42** 28 *1*80 000 **43** 160 *6*00
44 0·496 (6) **45** 3·23 (3) **46** 0·0274 (3) **47** 221·(4)
48 0·000 427 (3) **49** 9·99 (9) **50** 10·00 (5) **51** 0·0965 (9)
52 10 *1*60 000 **53** 0·0403 (6) **54** 904·(8) **55** 10 1*4*0 **56** 1 01*8* 000
57 72 0*3*0 000 **58** 847 *8*00 **59** 0·981 (5) **60** 810*2* **61** 20·9 (3)
62 267·(9) **63** 51 9*1*0

Page 164 EXERCISE 79

1 1·612 **2** 5·099 **3** 16·37 **4** 0·5177 **5** 8·544
6 27·02 **7** 85·67 **8** 0·2709 **9** 24·82 **10** 7·727
11 0·9607 **12** 0·2939 **13** 7·120 **14** 55·14 **15** 0·07273
16 0·02468 **17** 4·17 (5) **18** 1·32 (0) **19** 18·8 (8) **20** 0·597 (0)
21 26·6 (0) **22** 7·80 (1) **23** 303·(4) **24** 0·969 (9) **25** 0·621 (5)
26 2·00 (1) **27** 75·6 (2) **28** 0·225 (3) **29** 277·(6) **30** 239·(6)
31 0·455 (4) **32** 28·5 (6)

Page 166 EXERCISE 80

1 89·4 m **2** 117 m **3** 224 m **4** 69·6 m **5** 9·37
6 7·75 **7** 5·73 **8** 28·9 **9** 10·1 cm **10** 66·5 m
11 156 m **12** 4·00 m **13** 5·20 cm **14** 7·98 cm **15** 6·24 (5) cm
16 7·21 **17** 6·48 m **18** 7 cm **19** 263 cm **20** 8·01 (5) cm

Page 168 EXERCISE 81

1 0·3571 **2** 0·035 71 **3** 3·571 **4** 0·000 357 1 **5** 0·3448
6 0·003 448 **7** 0·012 99 **8** 0·001 414 **9** 0·022 94 **10** 1·898
11 12·25 **12** 0·000 164 2 **13** 2·445 **14** 0·099 01 **15** 9·434
16 49·02 **17** 0·378 (1) **18** 0·377 (7) **19** 2·03(5) **20** 0·018 2 (7)
21 4·91 (2) **22** 0·000 049 8 (0) **23** 49·5 (7) **24** 0·094 0 (6)
25 9·03 (2) **26** 0·447 **27** 0·168 **28** 0·0274 **29** 0·244
30 0·271 **31** 0·756 **32** 6·03 (5) **33** 15·0 **34** 0·212
35 0·155 **36** 0·843 **37** 0·712 **38** 3·56 **39** 83·4
40 0·710 **41** 0·412 **42** 1% **43** 5·38 cm **44** 718·(5) m
45 13·9 m **46** 4·62 cm; 128 cm^2

Page 171 EXERCISE 82

1 Circle **2** Arc of circle **3** Straight line **4** Arc of circle
5 Straight line; circle **6 and 7** Arc of circle
8 Parts of three lines **9 and 10** Part of surface of sphere
11 Circle **12** Circle, sphere, radius 3 cm

EXERCISE 82 (*continued*)

13 Two lines parallel to AB, surface of cylinder, axis AB.
14 Sphere, centre A, radius 5 cm
15 Circle, radius 2 cm
16 Circle, centre Q, radius 4 cm
17 Circle, centre C, radius 5 cm
18 Circle, centre 2 m below O, radius 1 m
19 Quadrant of circle, radius 2 m

Page 174 EXERCISE 83

5 3·66 cm

Page 175 EXERCISE 84

1 1·1 cm **2** 3·57 cm **3** 5·04 cm **4** 3·76 cm **5** 2·16 cm **8** 9·50 cm

Page 176 EXERCISE 85

1 4, 3, cm **2** 9 cm **3** (ii) 12 cm, $(12-r)$ cm; (iii) 9·38 cm ($9\frac{3}{8}$)

Page 176 EXERCISE 86

1 13 cm **2** 4·47 cm **3** 11·5 cm **4** Circle, radius 6 cm
5 8 cm **6** 8·58, 0·583, cm **7** 13·0 cm **8** 11·3 cm
9 5·83 cm **10** 8·94 cm **11** 9 cm **12** $8\frac{1}{3}$ cm, 11·1 cm
13 7·04 cm **14** 3·46 cm **15** 5·38 cm **16** 4·8 cm; 3·6, 6·4, cm

Page 177 EXERCISE 87

3 $l = 2\sqrt{(R^2-p^2)}$

Page 182 EXERCISE 90

9 3·73 cm

Page 184 EXERCISE 91

1 a^5 **2** b^3 **3** c^6 **4** d^6 **5** a^6 **6** b^3
7 c^5 **8** d^4 **9** e^8 **10** f^8 **11** g^8 **12** h^{15}
13 k^4 **14** l^8 **15** m^6 **16** n^{10} **17** $a^m \times a^n = a^{m+n}$
18 $a^m \div a^n = a^{m-n}$ **19** $(a^m)^n = a^{mn}$ **20** x^{12}; y^{12}

Page 186 EXERCISE 92

1 $x^4, x^5, x^{4\frac{1}{2}}$ **2** $a^4, a^{1\frac{1}{3}}, a^{\frac{1}{3}}$ **3** b^2; $\sqrt[3]{x^2}$ **4** b^4; $\sqrt[3]{x^4}$ **5** $\sqrt[4]{a}$
6 $\sqrt[5]{b^2}$ **7** $\sqrt{c^3}$ **8** $\sqrt[3]{d^5}$ **9** 1 **10** 4 **11** 3
12 4 **13** 27 **14** 9 **15** 32 **16** 8 **17** 27
18 100 **19** 32 **20** $x^{-2}, \frac{1}{x^2}; x^{-4}, \frac{1}{x^4}; x^{-5}, \frac{1}{x^5}; x^\circ, 1; x^\circ, 1$
21 $a^2; \frac{1}{a^3}$ **22** $b^3; \frac{1}{b}$ **23** c^2; 1 **24** $\frac{1}{9}$ **25** $\frac{1}{4}$ **26** 1
27 $\frac{1}{8}$ **28** $\frac{1}{3}$ **29** $\frac{1}{4}$ **30** $\frac{1}{2}$ **31** $\frac{1}{3}$ **32** $\frac{1}{4}$
33 9 **34** 2 **35** 1 **36** $\sqrt{10}$ **37** $\sqrt[4]{10}$ **38** $\sqrt{10}$
39 $\sqrt[4]{1000}$ **40** $\sqrt{1000}$ **41** -2 **42** 0·0721, 0·81

Page 188 ORAL EXAMPLES

1 1·6 **2** 2·0 **3** 5·0 **4** 2·6 (3) **5** 6·9 (2) **6** $10^{0\cdot30}$
7 $10^{0\cdot48}$ **8** $10^{0\cdot7}$ **9** $10^{0\cdot78}$ **10** $10^{0\cdot86}$ **11** $10^{0\cdot58}$
12 0·45, 0·50, 0·95, 8·9 (5)

Page 189 EXERCISE 93

The indices are as follows:

1 0·5051	**2** 0·8062	**3** 0·9823	**4** 0·4771	**5** 0·9542
6 0·3979	**7** 0·6990	**8** 0·6128	**9** 0·9138	**10** 0·8865
11 0·6294	**12** 0·8420	**13** 0·4829	**14** 0·7482	**15** 0·0453
16 0·0294	**17** 0·0792	**18** 0·9571	**19** 0·6474	**20** 0·7803
21 0·408 (7)	**22** 0·409 (6)	**23** 0·872 (3)	**24** 0·873 (2)	**25** 0·667 (0)
26 0·668 (1)	**27** 0·670 (0)	**28** 0·663 (3)	**29** 0·802 (6)	**30** 0·803 (4)
31 0·948 (2)	**32** 0·987 (1)	**33** 0·849 (2)	**34** 0·845 (3)	**35** 0·706 (0)
36 0·699 (4)	**37** 0·4843	**38** 0·477 (8)	**39** 0·035 (7)	**40** 0·031 (1)
41 0·001 (2)	**42** 0·020 (4)	**43** 0·002 (5)	**44** 0·300 (0)	**45** 0·999 (5)

Page 190 EXERCISE 94

1 4·2	**2** 5·9	**3** 6	**4** 3·49	**5** 3·56
6 5·08	**7** 8·14	**8** 9·36	**9** 5·03	**10** 3·75
11 6·04	**12** 8·02	**13** 2·52 (3)	**14** 3·35 (6)	**15** 5·72 (9)
16 7·31 (5)	**17** 3·04 (4)	**18** 5·05 (8)	**19** 6·02 (7)	**20** 8·03 (9)
21 3·35 (4)	**22** 4·57 (4)	**23** 1·88 (5)	**24** 1·58 (2)	**25** 1·58 (9)
26 4·07 (8)	**27** 5·17 (3)	**28** 5·78 (2)	**29** 7·00 (6)	**30** 8·00 (5)
31 9·33 (4)	**32** 1·07 (1)	**33** 1·07 (5)	**34** 1·00 (6)	**35** 1·00 (7)
36 9·00 (8)				

Page 191 EXERCISE 95

1 7·64 (5)	**2** 8·68 (2)	**3** 7·70 (9)	**4** 8·38 (9)	**5** 7·42 (0)
6 8·51 (5)	**7** 8·56 (1)	**8** 8·52 (0)	**9** 7·58 (5)	**10** 2·63 (5)
11 1·83 (6)	**12** 3·47 (5)	**13** 3·42 (1)	**14** 1·92 (0)	**15** 1·23 (1)
16 1·93 (2)	**17** 6·44 (0)	**18** 1·13 (5)	**19** 9·29 (4)	**20** 5·73 (6)

Page 192 EXERCISE 96

1 1	**2** 2	**3** 3	**4** 1	**5** 0	**6** 4
7 5	**8** 4	**9** 7	**10** 2	**11** 4	**12** 3
13 5	**14** 4	**15** 6	**16** 1	**17** 1	**18** 4
19 5					

Page 192 EXERCISE 97

1 1·4150	**2** 2·5798	**3** 1·7952	**4** 3·8739	**5** 1·620 (3)
6 3·506 (8)	**7** 2·911 (0)	**8** 1·001 (7)	**9** 2·301 (6)	**10** 4·7300
1 5·631 (7)	**12** 6·302 (5)	**13** 3	**14** 5	**15** 7
6 0	**17** 4·804 (4)	**18** 2·570 (1)	**19** 5·903 (5)	**20** 2·712 (9)
1 3·8476	**22** 8·9845	**23** 38·40	**24** 534·0	**25** 7700
6 300·0	**27** 85 200	**28** 904 000	**29** 70 300 000	**30** 8 550 000
1 335·(5)	**32** 46·63	**33** 3004	**34** 20·0 (4)	**35** 40 080
6 100·6	**37** 503 500	**38** 6123	**39** 100·0	**40** 100 000
1 12·5 (9)	**42** 25 120			

age 194 EXERCISE 98

3 586	**2** 11 080	**3** 6 556	**4** 9·21 (2)	**5** 5·40 (5)
479·(2)	**7** 10 040	**8** 26 560	**9** 380·(6)	**10** 75·0 (5)
233·(8)	**12** 126·(8)	**13** 150 000	**14** 16·3 (2)	**15** 15 420 000

EXERCISE 98 (*continued*)

16 35·1 (6) **17** 21 500 **18** 80·9 (7) **19** 473·(5) **20** 387·(9)
21 13 160 **22** 2·70 (9); 8·56 (6) **23** 8·51 (8); 3·95 (4); 1·83 (5)
24 21·3 (0) **25** 7·97 (5) **26** 9·65 (0) **27** 2·18 (4) **28** 3·76 (4)
29 2·10 (0) **30** 4·64 (2) **31** 5·62 (4)

Page 194

EXERCISE 99

1 3 424 **2** 13 860 **3** 329 800 **4** 2 791 **5** 25·9 (7)
6 43·0 (3) **7** 11·2 (0) **8** 12·3 (9) **9** 2·86 (8) **10** 108·(0)
11 7·15 (1) **12** 457·(5) **13** 6·93 (5) **14** 40·4 (8) **15** 195·(9)
16 30·2 (3) **17** 3·03 (5) **18** 1·12 (6) **19** 7·00 (7) **20** 2·03 (1)
21 289·(2) **22** 1·65 (1) **23** 2·22 (5) **24** 3629 **25** 4·98 (6)
26 3·94 (9) **27** 1·97 (8) **28** 8·05 (5)

Page 195

EXERCISE 100

1 2300 cm^3 **2** 3210 **3** 1·36 kg **4** 7·42 (5) g **5** 14·2%
6 3·86% **7** 54·4; 1·78 **8** 1·37; 6·95 **9** 1690; 27·6 **10** 8·63
11 118 m **12** 2·88, 104 **13** 22·0 cm^2 **14** 6·71

Page 197

EXERCISE 101

1 $\bar{1}$·5340 **2** $\bar{2}$·6839 **3** $\bar{3}$·8808 **4** $\bar{1}$·9552 **5** 0·3096
6 $\bar{3}$·8751 **7** 1·0004 **8** $\bar{4}$·4771 **9** $\bar{3}$·3979 **10** $\bar{1}$·0043
11 $\bar{3}$ **12** $\bar{2}$·8476 **13** $\bar{2}$·6535 **14** $\bar{1}$·6034 **15** $\bar{3}$·8536
16 $\bar{2}$·0418 **17** $\bar{1}$·9031 **18** $\bar{3}$·8751 **19** $\bar{3}$·5228 **20** $\bar{2}$·4472
21 $\bar{5}$·6053 **22** $\bar{6}$·7856 **23** $\bar{8}$·6314 **24** 0·391 **25** 0·0732
26 0·000 81 **27** 0·3 **28** 0·0406 **29** 0·005 09 **30** 0·704
31 0·000 093 4 **32** 0·333 4
33 0·000 196 7 **34** 0·014 62 **35** 0·001 354 **36** 0·020 03
37 0·110 3 **38** 0·001 013 **39** 0·010 07 **40** 0·1 **41** 0·01
42 0·000 275 (4) **43** 0·001 107

Page 198

EXERCISE 102

1 $\bar{3}$·7 **2** $\bar{2}$·8 **3** $\bar{4}$·4 **4** $\bar{1}$·5 **5** 3·5 **6** 0·6
7 $\bar{2}$·3 **8** $\bar{5}$·5 **9** 7·4 **10** $\bar{3}$·3 **11** $\bar{3}$·7 **12** $\bar{5}$·8
13 4·7 **14** $\bar{3}$·5 **15** 1·68 **16** $\bar{3}$·2 **17** 2·6 **18** 3·3
19 $\bar{4}$·6 **20** $\bar{5}$·2 **21** $\bar{9}$·6 **22** $\bar{3}$·4 **23** $\bar{3}$·2 **24** $\bar{2}$·7
25 $\bar{2}$·5 **26** $\bar{1}$·8 **27** $\bar{3}$·9 **28** $\bar{1}$·9 **29** $\bar{2}$·7 **30** $\bar{1}$·68
31 $\bar{1}$·4 **32** $\bar{1}$·5 **33** $\bar{1}$·8 **34** $\bar{8}$·6 **35** 4·2 **36** $\bar{2}$·2
37 1·2 **38** 1·4 **39** $\bar{1}$·4 **40** 0·45

Page 199

EXERCISE 103

1 0·758 (1) **2** 0·604 (8) **3** 40·9 (6) **4** 0·079 7 (3)
5 0·000 655 (1) **6** 0·018 4 (4) **7** 22·2 (1) **8** 0·412 (6)
9 0·063 9 (1) **10** 0·147 (7) **11** 0·076 8 (5) **12** 1·36 (0)
13 15·7 (5) **14** 38 560 **15** 12 920 **16** 0·006 25 (1) **17** 3·19 (3)
18 26·9 (1) **19** 0·080 6 (5) **20** 0·007 15 (8) **21** 0·051 4 (6)
22 0·000 000 001 29 (6) **23** 0·293 (8) **24** 0·718 (8) **25** 0·850 (4)
26 0·364 (7) **27** 0·925 (0) **28** 0·371 (1) **29** 0·607 (1) **30** 0·049 1 (7)
31 1·29 (6) **32** 0·003 63 (8) **33** 0·122 (3) **34** 0·007 03 (2) **35** 20 300

EXERCISE 103 (*continued*)

36 112·(8) **37** 1·98 (7) **38** 0·000 301 (1) **39** 0·107 (5)
40 225·(1) **41** 1215

Page 200 EXERCISE 104

1 1·86 m **2** 429 kg **3** 48·9 % **4** 12·7 **5** 3·39 **6** 0·869
7 1·93 **8** 2·48 g **9** 18·0 s **10** $4{\cdot}49 \times 10^{-10}$ **11** 1·21 (5)
12 0·506 **13** 1·18 **14** 1·95 (5)

Page 202 EXERCISE 105

1 $\bar{1}$·5919 **2** $\bar{1}$·5972 **3** $\bar{1}$·5978 **4** $\bar{1}$·5987 **5** $\bar{1}$·9562 **6** $\bar{1}$·9564
7 $\bar{1}$·9892 **8** $\bar{1}$·9893 **9** $\bar{1}$·8817 **10** $\bar{1}$·8810 **11** $\bar{1}$·8815 **12** $\bar{1}$·7820
13 $\bar{1}$·5711 **14** $\bar{1}$·9451 **15** $\bar{1}$·4692 **16** $\bar{1}$·2773 **17** $\bar{1}$·0046 **18** $\bar{2}$·9970
19 $\bar{2}$·9286 **20** $\bar{2}$·8613 **21** $\bar{2}$·8156 **22** $\bar{2}$·7400 **23** $\bar{2}$·9907 **24** $\bar{2}$·9503
25 $\bar{1}$·7865 **26** 0·1814 **27** 0·5035 **28** 1·0244 **29** $\bar{1}$·0068 **30** $\bar{1}$·0093
31 $\bar{1}$·0336 **32** $\bar{1}$·0807 **33** 1·2855 **34** 1·4221 **35** 1·3080 **36** 1·4323
37 37° 36′ **38** 37° 39′ **39** 59° 6′ **40** 59° 16′, (17′)
41 71° 24′, (23′, 25′) **42** 71° 26′ (27′) **43** 6° 18′ **44** 6° 21′
45 74° 12′ **46** 74° 10′ **47** 51° 42′ **48** 51° 38′ **49** 58° 14′ **50** 40° 39′
51 31° 27′ **52** 26° 44′ **53** 63° 58′ **54** 48° 2′ **55** 84° 12′
56 6° 42′, (41′ to 46′) **57** 83° 10′ **58** 83° 38′ **59** 86° 38′ **60** 87° 26′
61 31° 42′ **62** 47° 48′ **63** 84° 12′ **64** 84° 24′ **65** 31° 15′ **66** 55° 9′
67 75° 16′ **68** 78° 59′ **69** 84° 8′ **70** 85° 21′ **71** 45° 59′ **72** 49° 48′

Page 203 EXERCISE 106

1 36° 5′ **2** 67° 25′ **3** 197 m **4** 7·49 cm **5** 3°
6 81·(1) m; N. 6° 22′ W. **7** 5·85, 3·70, km; N. 32° 19′ E.
8 22·6, 30·5(5), km; S. 53° 28′ E., 38·0 km **9** 2·10, 3·66, 4·47, cm, 25° 10′
10 5·66, 4·42, 7·18 (5), cm, 18° 29′ **11** 961·(6) m, 28·5 m, 1° 42′
12 9·69 m, 18° 23′

Page 207 EXERCISE 108

1 31·4 cm **2** 7·54 cm **3** 518 m **4** 18·8 (5) cm **5** 314 m
6 1080 m **7** 2·45 m **8** 39·8 m **9** 1·18 cm **10** 280 m
11 79·5 m **12** 50·3, 16·7 (5) cm **13** 18 cm **14** 114 m
15 455 **16** 3·14 (6) **17** 19·8 km/h **18** 31·4 m; yes **19** 25·1 cm
20 1·75 cm **21** 101 **22** 1100 cm/s **23** 4·17 cm, ($4\frac{1}{6}$)
24 19·4 cm **25** 11·7 **26** 398

Page 208 EXERCISE 109

1 28·28 cm^2, 3·14 **2** 1257, 12·57 cm^2, 3·14 **3** 7852, 314 cm^2, 3·14

Page 211 EXERCISE 110

1 113 cm^2 **2** 9·62 cm^2 **3** 5670 m^2 **4** 50·3 cm^2 **5** 86·6 cm^2
6 49 100 m^2 **7** 2·65 cm **8** 17·8 m **9** 49·6 (5) m **10** 3·5 cm
11 2·07 m **12** 27·8 m **13** 8·55 cm^2 **14** 693 cm^2 **15** 1·93 cm^2
16 5030, (5027) **17** 12 100 m^2, 15 400 m^2 **18** 22·0 cm^2 **19** 3·43 cm^2
20 107 m^2 **21** 7·2 kg **22** 20·1 cm^2

Page 213 EXERCISE 111

1 3·5 cm; 55 cm^2; 9·62 cm^2 **1** 22, 6, cm; 132 cm^2 **3** 126 cm^2
4 291·(5) cm^2 **5** 1885 cm^2 **6** 56·5 (5) cm^2 **7** 88·0 cm^2
8 660 cm^2 **9** 11·0 m^2 **10** 228 cm^2 **11** 66·0 cm^2 **12** 217 cm^2
13 264 cm^2 **14** 99·0 m^2 **15** 490 cm^2 **16** 11·4 m^2 **17** 6·68 (5) cm
18 4·22 cm **19** 2·65 cm **20** 16·5 cm

Page 215 EXERCISE 112

1 302 cm^3 **2** 13 300 cm^3 **3** 25 400 cm^3 **4** 170 cm^3
5 18·9 cm **6** 6·22 cm **7** 4·33 cm **8** 6·96 cm **9** 176 kg
10 1·32 kg **11** 1700 **12** 78·6 cm **13** 70·7 cm^3 **14** 322 cm^3
15 84 **16** 18 cm **17** 97·5 g **18** 678·(5) cm^3
19 51·9 cm per min **20** 66 800

Page 216 EXERCISE 113

1 5·72 m^2 **2** 3·93 cm **3** 79 600 m^2 **4** 455
5 3·86 cm^2 **6** 1·05 m; 0·433 m^2 **7** 4 kg
8 31·5 cm^3 **9** 15·8 km/h **10** 77·6 g **11** 8·98 m
12 40·6 m^2 **13** 374 cm^2 **14** 484 cm^2 **15** 5·65 m^2
16 1220 m^3 **17** 5000 cm^3, 1610 cm^3 **18** 11·5 (5) cm
19 129° **20** 3·06 m

Page 218 EXERCISE 114

1 £3, £103 **2** £12, £112 **3** £7, £107 **4** £4, £104
5 £8, £208 **6** £21, £321 **7** £24, £424 **8** £96, £896
9 £20, £220 **10** £70, £470 **11** £5200; £5408

Page 221 EXERCISE 115

1 £63, £413 **2** £52·50, £472·50 **3** £18·40, £202·40
4 £177·45, £1022·45 **5** £102, £867 **6** £4·50, £67
7 £45, £420 **8** £8·80, £200·80 **9** £20·22, £188·72
10 £3·84, £51·84 **11** £20·40, £292·40 **12** £27·72, £555·72
13 £24·96, £440·96 **14** £38·71, £670·71 **15** £30·00
16 £4·76 **17** £45·35 **18** £103·23
19 £5·89 **20** £10·10

Page 222 EXERCISE 116

1 £138; 5% **2** £56; $3\frac{1}{2}$% **3** £27; $2\frac{1}{2}$ yr **4** £1158; $3\frac{3}{4}$ yr
5 £720; £768 **6** £560; £602 **7** £40·50; £400·50
8 £0·19, £10·19 **9** £1805·70; $4\frac{1}{2}$% **10** £2·05; $2\frac{1}{2}$%
11 £2422·50; $6\frac{1}{2}$ yr **12** £6825; £7166·25 **13** £73·12, £560·62
14 £73·50; $2\frac{1}{2}$ yr **15** £846·51; 6% **16** £44·80; 4% **17** $7\frac{1}{2}$%
18 2 yr **19** £168 **20** 48% **21** £4·05
22 £88; 4·4% **23** £1·92 **24** 120 **25** £550

Page 223

Q.R. 1: **1** $1\frac{3}{8}$ **2** 0·01 **3** 0·025 **4** 54 **5** 3:4 **6** 18p **7** -8 **8** 1 **9** $\frac{2}{5}$ **10** $4n$ mm **11** 35 **12** 171p

Q.R. 2: **1** 4·5 **2** 101·09 **3** 0·01 **4** b^8 **5** 12 cm **6** 3:20 **7** 14 g **8** 40p **9** $3+k$ **10** 3759 **11** 5, -10 **12** 90°

Page 224

Q.R. 3: **1** $\frac{2}{3}$ **2** 1 **3** 0·0009 **4** 11p **5** 250 g **6** 10 cm, 14 cm **7** 81 **8** -24 **9** 0·05 **10** $6x^5y^5$ **11** (s/v) hours **12** 55

Q.R. 4: **1** $\frac{2}{5}$ **2** 0·08508 **3** 100 **4** 54p **5** 3:4 **6** 1:200 000 **7** 60% **8** $3\frac{3}{4}$ **9** $2a^2b^3$ **10** 66 **11** 12 **12** £$3x$

Q.R. 5: **1** 12·6 **2** $\frac{3}{5}$ **3** 0·001728 **4** 0·002 **5** £60 **6** $\frac{1}{4}(x+3y)$ **7** 4:9 **8** £9 **9** 30 **10** 14 **11** 40 **12** -18

Q.R. 6: **1** $\frac{4}{5}$ **2** 0·0016 **3** 130 **4** £35·64 **5** 20p, 30p, 70p **6** 20 m^2 **7** 12·6 km **8** -18 **9** $2/b$ **10** $24n$ **11** ± 9 **12** 30°

Page 225

Q.R. 7: **1** 0·794 **2** 1 **3** 0·1111 **4** £77 **5** 24p **6** 40% **7** $85P$ pence **8** 3·9 kg **9** 75, 22 **10** $9a^2-30ab+25b^2$ **11** $(t-9)(t+8)$ **12** 40

Q.R. 8: **1** $1\frac{1}{5}$ **2** 0·01548 **3** 7 cm **4** 300 **5** $8\frac{1}{2}$ **6** 10:1 **7** 35% **8** $x^2+7x+12\frac{1}{4} = (x+3\frac{1}{2})^2$ **9** $4(b+3c)(b-3c)$ **10** 9, 3 **11** 25% **12** 125

Q.R. 9: **1** 1 **2** 0·06 **3** 100 **4** $\frac{3}{40}$ **5** 8 **6** 45% **7** £385 **8** £2·50 **9** $125nt$ **10** $(x-2)(2x-3)$ **11** $49p^2-56pq+16q^2$ **12** 5 cm

Page 226

Q.R. 10: **1** £1·53 **2** $\frac{5}{8}$ **3** 1·01 **4** 60p **5** 40 **6** £28, £12, £8 **7** $2b^2-2ab$ **8** 8% **9** $\frac{7}{12}$ **10** $\frac{1}{5}$ **11** $y(y-3)(y+2)$ **12** 8, 4, cm

Q.R. 11: **1** 0·5 **2** $2\frac{1}{4}$ **3** 149·85 m **4** 14 kg **5** $3\frac{1}{3}$% **6** 63° 22′ **7** $\sqrt{(2s/g)}$ **8** 7:4:9 **9** $9\frac{1}{11}$% **10** $12x$ **11** -2

Q.R. 12: **1** 700 **2** $\frac{1}{3}$ **3** $\frac{1}{12}$ **4** £17·01 **5** 400 m **6** $1\frac{1}{2}$ N. **7** y^2 **8** 3 **9** 35° 32′, 0·8138 **10** $(1+5t)(1-3t)$ **11** $(2-p)(3-q)$

Q.R. 13: **1** 0·005 **2** 30p **3** $\frac{7}{8}$ **4** $26\frac{2}{3}$% **5** 1·924 **6** 9°·36′ **7** $5(t+3)(t-3)$ **8** $2C^7$ **9** $nr/(n-2)$ **10** 3 **11** 1·6 m

Page 227

Q.R. 14: **1** $\frac{1}{12}$ **2** 3000 cm^2 **3** 0·00308 **4** 0·006
5 0·5385 **6** -2 **7** 17·5 **8** 36° 52′, 67° 23′
9 $(x+y)(6x-5y)$ **10** 0 **11** 0·02600 **12** 6·5, 10·5, cm

Q.R. 15: **1** 1 **2** £342·20 **3** 0·00705 **4** 7372 **5** 66·83
6 120 **7** $(2a-1)(3b+2)$ **8** £100·80 **10** 2·29 cm
11 -1

Q.R. 16: **1** £16·92 **2** 0·1406 **3** $1\frac{4}{9}$ **4** $4x^2(x+2)(x-2)$
5 50° 46′ **6** 25% **7** 20% **8** 15 400 m^2 **9** 5
10 8 **11** $\bar{2}$·56, $\bar{1}$·88

Page 228

Q.R. 17: **1** 40p, 60p **2** 37p **3** 0·16 **4** 4·64
5 51° 50′ **6** $3\frac{1}{4}$ **7** 10:3 **8** $7x-3$ **9** -10 **10** £98
11 $2V/S$ **12** Mid-point of AB

Q.R. 18: **1** 806 **2** 0·4748 **3** $\frac{8}{125}$ **4** £1·44
5 89 m 60 cm **6** $x < 3$ **7** 64° 37′, 6·32 (5) cm **8** 0·464
11 $2(t+3)(3t-2)$ **12** $2\frac{1}{2}$%

Page 228

Test 1: **1** $3^2.11$ **2** $2\frac{49}{60}$ **3** 0·008 483 **4** 1
5 9 **6** $(204-5n)$ degrees

Test 2: **1** $\frac{1}{8}$ **2** 0·0043 **3** $9c^6$ **4** 0·00 1728
5 $1\frac{1}{4}$ **6** $(7t+10)$ years

Page 229

Test 3: **1** $2\frac{1}{2}$ **2** 78·306 **3** $9x^2-10xy+6y^2$ **4** 72
5 $\dfrac{3a}{b}$ **6** 77

Test 4: **1** $2.3.5^2.7^2$ **2** $\frac{1}{72}$ **3** 0·8 **4** $\dfrac{3a^2b^3}{c^4}$ **5** $\dfrac{x+5y}{12}$
6 $1\frac{2}{3}$

Test 5: **1** $\frac{3}{5}$ **2** 0·024 **3** 79·2 km/h **4** $2a+3c$
5 17 **6** $(150-n)$ degrees

Test 6: **1** 16·2 cm **2** 4·6 m^2 **3** £174·72 **4** $24b^9$ **5** $r+2t$
6 $20/x$

Test 7: **1** 126 **2** 0·203 **3** 426 cm^2 **4** 0 **5** 20
6 $6a-4b$

Test 8: **1** $\frac{29}{50}$ **2** 0·000 000 04 **3** 47·6 **4** $\dfrac{1}{a}$
5 23 **6** $\frac{1}{4}x^4y^2$

Page 230

Test 9: **1** 4·8 **2** $24a^4b^3$ **3** 0·545 **4** 41·75 m^2
5 $\frac{2}{3}$ **6** 1·3

Test 10: **1** 175 **2** 2·85 cm **3** £89·76 **4** $\frac{5}{11}$ **5** 37
6 $(70-1\frac{1}{2}x)$ degrees

Test 11: **1** 5 **2** $4r+2$ **3** £199·29 **4** 1704 cm^3
5 7; 5 **6** $3s/(2n)$

Test 12: **1** 0·003 136 35 **2** £19·25 **3** 0·61 g **4** $\frac{13x-10}{20}$
5 $2x+y-6z$ **6** -36

Page 231

Test 13: **1** 22·4% **2** 16 **3** $\frac{1}{a^2b^2}$ **4** 587 500 m^2
5 8, -8 **6** 6

Test 14: **1** £4560 **2** 0·0538 **3** 100 **4** $\frac{p^2+q^2}{pq}$ **5** 2
6 $-\frac{2}{5}$

Test 15: **1** $\frac{2}{15}$ **2** $\frac{1}{4}(7s-r)$ **3** 0·485 **4** £7·74 **5** $-\frac{1}{5}$
6 $n = 1+\frac{l-a}{d}$

Test 16: **1** 0·001 111 **2** £7·70 **3** 1·09
4 £50, £31·25, £18·75 **5** $-\frac{3}{4}$ **6** $-2\frac{1}{2}$

Test 17: **1** 560 **2** $\frac{1}{2}(h-k)$ **3** £596·30 **4** 84
5 $5x^3-10x^2+x-1$ **6** 0·04

Page 232

Test 18: **1** 26p **2** 831 **3** £1·75 **4** 16; 3 **5** $1\frac{1}{2}x^2-2y^2$
6 1:4

Test 19: **1** $2\frac{5}{8}$ **2** 0·000 001 44 **3** 191
4 $\frac{12rs-2r^2-3s^2}{6rs}$ **5** $4\frac{1}{2}$ **6** $A = \frac{pb}{n}-b^2$

Test 20: **1** 0·98 **2** 567 m 30 cm **3** 3:8 **4** $3y^2-4x^2$
5 -7 **6** $66\frac{2}{3}$%

Test 21: **1** $67\frac{1}{2}$p **2** 2 **3** £3·49 **4** £1·96 **5** 70
6 $(3p-1)(p+2)$ **7** 59° 2′

Test 22: **1** 0·253 504 **2** $7\frac{1}{2}$% **3** £11 880 **4** $y = \frac{3x}{2}-3$
5 4, -3 **6** $(b-c)(a-b)$ **7** 55° 32′

Page 233

Test 23: **1** 6304 **2** £251·10 **3** 1620 t **4** $3q-p-7r$
5 $\frac{2c}{b}$ **6** $6c^3-5c^2-21c-10$ **7** 59° 13′

Test 24: **1** $72\frac{1}{2}$% **2** $(1-4t)(1+3t)$ **3** 525 l **4** £2·10
5 $12x$ **6** 11 **7** 9·71 cm

Test 25: **1** 0·064 468 **2** 684 **3** £73·80 **4** $(x-y+z)(x+y-z)$
5 $\dfrac{15b-14c-3a}{30}$ **6** 15:8 **7** 117° 59′

Test 26: **1** 33·9 **2** $6(1+3z)(1-3z)$ **3** 6·2% **4** £6·48
5 3 **6** $3rs^3-2r^4$ **7** 71° 4′

Page 234

Test 27: **1** 0·37 **2** $(14-3c)/10$ **3** 120 **4** 0·1674
5 £7·20 **6** $27x-13x^2$ **7** 4, 9

Test 28: **1** £608·16 **2** a^4; $5b/(12c)$ **3** 0·116
4 £15·12 **5** $16r^3$ **6** $(x+2)(2x^2+x-7)$
7 462 m

Test 29: **1** 75% **2** £4·20, £2·80, £2·10
3 £8 **4** $3(x-6y)^2$ **5** $2-19y+2y^2-9y^3$
6 8, −6 **7** 195 m

Test 30: **1** 0·18 **2** 10·74 m
3 84p **4** −80 **5** $(5x-2y)(x-6y)$
6 $3n^6$; $6x-6y$ **7** 52° 43′ or 127° 17′

Page 235

Test 31: **1** 4th, 2nd, 3rd, 1st **2** 0·293 (5) **3** $\dfrac{4-5a}{12}$
4 204 cm³ **5** $\dfrac{3N}{5}$ **6** $(x+3y)(x-5y)$ **7** 35° 52′, $\bar{1}$·9087

Test 32: **1** $17\frac{1}{2}$ **2** £15·39
3 0·3081 **4** $(1+a)(1-a)(1+b)(1-b)$
5 $3x^3-16x^2-7x-5$ **6** $1-x^4$; $n-23$ **7** 58° 42′

Test 33: **1** £18 **2** 17·8 cm **3** 0·579
4 $\dfrac{b+1}{6}$ **5** $-2c^2-2d^2$ **6** $2\frac{1}{2}$, 2 **7** 50° 37′

Test 34: **1** 75% **2** 3·33 **3** $(1+3c)(1-2c)$
4 5 cm **5** 0·620 (5) **6** $\dfrac{25b^2}{4}, -\dfrac{5b}{2}$ **7** 74° 7′

Page 236

Test 35: **1** 0·16 **2** 0·009 98 **3** 46 293
4 $22ab-7a^2-16b^2$ **5** −4, −3 **6** −8 **7** 12·7 (5) cm

Test 36: **1** 0·638 **2** 28; 5 cm **3** 1·26 cm **4** −14
5 $2(a+1)(2a-9)$ **6** $2\frac{1}{3}$ or −1 **7** 39° 34′ or 140° 26′

Test 37: **1** £35·88 **2** £2·87 **3** 0·853 (5)
4 $\dfrac{1}{x}$ **5** $\frac{4}{9}$ **6** $6\frac{1}{4}$ **7** 7·86

Page 237

Test 38: 1 8 t   2 18·8 t   3 667
4 $(5+y)(1-2y)$ 5 2·04   6 7, −3   7 25·3

Test 39: 1 $31\frac{1}{2}$p   2 0·348   3 251 m^2
4 $(5a+b)(3a-5b)$ 5 $\bar{1}$·9448   6 ± 3 7 44° 14′

Test 40: 1 82·4%   2 0·0190   3 1500 l
4 4·26   5 $\frac{1}{c^6}$ 6 $(p+q)(m-1)$ 7 44° 9′

Page 238

Paper 1: 1 $2.7^3.11$; 154; 484   2 26   3 42 cm^2
4 $l+12m-n$ 5 7

Paper 2: 1 £1·38   2 66 l   3 0
4 $36a^4b^6, 9a^4b^4$ 5 30°, 10°   6 114°

Paper 3: 1 176057   2 7·56 m^2, 1·25 cm   3 $24a^3b^2c^3$
4 18   5 $(180-3p)$ min; 12   6 70

Page 239

Paper 4: 1 0·638   2 14·6 t   3 6   4 $a-1$ 5 9

Paper 5: 1 0·0237   2 1147·5 cm^2 3 $\frac{a^2}{b^4}$ 4 $h > 3$
5 $\frac{2}{5}, \frac{1}{3}$ 6 108

Paper 6: 1 $1\frac{5}{9}$ 2 £2·33   3 $1-5c-3c^2$ 4 4
5 30; 100, 0; 75   6 5·22, 1·72, cm

Page 240

Paper 7: 1 1020   2 $9b^7$ 3 1180 m
4 $14xy-2x+7y$ 5 6; 45 cm

Paper 8: 1 £10·32   2 50 min   3 $\frac{p^2}{9}$
4 $c = \frac{1}{10}(150-5d-e)$ 5 216 m

Paper 9: 1 96·51   2 1280 cm^3; 2810 cm^2
3 $\frac{2z}{x}$ 4 0   5 105
6 60°

Page 241

Paper 10: 1 $2^3.3^3.5.7$ 2 600°   3 $\frac{3a}{2}$ 4 8
5 6 km/h

Paper 11: 1 £6378·01   2 25 000   3 1   4 55; 18 000
5 $\frac{2k}{v}$ g   6 10·73 cm

Paper 12: 1 $1\frac{1}{2}$ 2 7200 l   3 $\frac{1}{4}(5b+c+6)$ 4 3
5 $\frac{13c}{3}$ cm; c^2 cm^2 6 $e = b+c+d$

Page 242

Paper 13: **1** 0·005 06 **2** 182 g **3** $\frac{25A}{9}$
4 10 **5** 11.22½ a.m.; 11·25 km **6** 8 cm

Paper 14: **1** $1\frac{1}{2}$ **2** 150 m **3** $2x-12y$ **4** $40n$ **5** 35p

Paper 15: **1** $\frac{3}{5}$ **2** 80 cm **3** $6n+8$; $4\frac{1}{2}$ **4** $2t^4$
5 35 sea miles

Page 243

Paper 16: **1** 0·020 706 **2** 15 cm^2; 78·75 g **3** $4n$
4 2, 3, 6 **5** 42 **6** 100°

Paper 17: **1** 0·000 15 **2** $4\frac{2}{5}$ **3** $\frac{3}{4}xy^5$
4 2·4 **5** £13·50, £7·50, £9

Paper 18: **1** £26·10 **2** 26 min **3** $\frac{1}{2}(a+6b)$ **4** 0·15 **5** 4

Page 244

Paper 19: **1** 0·626 **2** 90 000 **3** $\frac{4x^5z}{3y^2}$ **4** 46
5 17·5 km **6** 28

Paper 20: **1** 0·012 75 **2** 30p **3** $2a-4b$ **4** $-\frac{4}{5}$; 400
5 56 km/h **6** 72°

Page 245

Paper 21: **1** (i) 0·000 108 (ii) 0·000 304 **2** £1700
3 225 **4** 8 km **5** $2rt/(t-3)$

Paper 22: **1** $\frac{3}{4}$ **2** 630 g **3** $6x^3-24x^2+16x+2$
4 £2·55 **5** n years

Paper 23: **1** 40 **2** 540 000 m^2 **3** $7c-2-3c^2$
4 -6 **5** 43·3 % **6** 36°

Page 246

Paper 24: **1** 849 cm^3 **2** 784 **3** £1584 **4** $36xy^2$
5 18·3 km

Paper 25: **1** £200 **2** 63
3 $153\frac{1}{3}°$; $(3y-2x)$ degrees; $\frac{1}{3}(2p+q)$ degrees **4** $1/(2r)$; -1
5 20 dozen

Page 247

Paper 26: **1** 15, 17, 19, 21, 23, 25 **2** 41·58 m^2
3 $\frac{1}{2}(a-b)$ **4** 17; $\frac{n+kw}{r+w}$ **5** $r=\sqrt[3]{\frac{3V}{4\pi}}$ **6** 96°, 84°, 60°

Paper 27: **1** 0·1682 **2** £150 **3** -12 **4** 30
5 £2·79 **6** 37°, 106° (36° 52′, 106° 16′)

Paper 28: **1** 68·1 km/h **2** £360 **3** $9s^3t$ **4** $h = 0{\cdot}15d$ **5** 3p **6** 100°

Page 248

Paper 29: **1** £710·60 **2** 73·0 kg **3** $\frac{1}{6}(13z-19y)$ **4** £6·14 **5** 36° **6** -15

Paper 30: **1** 9:10 **2** 15 min **3** 31, 11, 3, 7, $13\frac{1}{2}$ **4** $-ab/c$; $\frac{1}{4}n^2$
5 11, 11, 7 if $x = 4$; 13, 13, 15 if $x = 6$; 19, 19, 15 if $x = 8$.

Page 249

Paper 31: **1** 8p **2** £9·81 **3** 1; -6; -9 **4** $\frac{1}{2}$ **5** 20

Paper 32: **1** £134·64 **2** 2 cm **3** 1 **4** $(14-10k)$ cm **5** 11 **6** AB, DC

Paper 33: **1** £3·68 **2** 16 min **3** $\dfrac{y^2+z^2}{z}$ **4** $3\frac{1}{3}$ **5** CK = AH **6** $(3\frac{1}{3}g+10)$ litres

Page 250

Paper 34: **1** 2h 30 min **2** 80 m **3** £39·20 **4** $-1\frac{1}{2}$; 0; 7:1 **5** 12 km

Paper 35: **1** 1:10 000 **2** £1·12 **3** $13x-60$ **4** -5 **5** 42; £6·02 **6** $c = a-b$

Paper 36: **1** £91·98 **2** 70% **3** a; $-32x$ **4** 57 **5** 73° 29′ **6** $d = a+b-c$

Page 251

Paper 37: **1** 7·4 **2** £2·70 **3** 8, 6 **4** 240 **5** 5·74 cm

Paper 38: **1** $\frac{3}{4}$ **2** 40% **3** $a-b$; x^3 **4** 38 **5** 9·32 m

Paper 39: **1** £175 **2** 52% **3** $12y-13x-2z$; $\dfrac{s^5}{r}$ **4** 88, 90 **5** 6° 20′, 4·8 cm **6** 6·52 cm

Page 252

Paper 40 **1** 0·001 45 **2** 28% **3** $(4x-5)/12$ **4** 216 m **5** 3·21 cm

Paper 41: **1** 721 **2** 0·03% **3** $23b^6$; $(3x+4)(3x+14)$ **4** 8, 6 **5** 59° 2′, 6·38 cm, 72° 36′ **6** 58°, 50°

Paper 42: **1** 10:9 **2** £1·37 **3** 13; -5; $-\frac{1}{3}$ **4** 6, 10, 8 **5** 29° 30′ **6** 72°

Page 253

Paper 43: **1** £1·37, £2·74, £4·11 **2** £96 000 **3** -120 **4** 1·2 km **4** 3·29 cm

Paper 44: **1** (i) 9:16; (ii) 27:64 **2** 57% **3** $x^2+y^2+z^2$ **4** $\frac{7}{18}$; $3(n+5)(n-3)$ **5** 6·66 cm, 4·20 cm, 61° 5′

Page 254

Paper 45: **1** 5·09(5) cm, 51° 52′ **2** £18·15 **3** $-y^2$; $1/c^2$ **4** $5\frac{1}{2}$, $2\frac{1}{2}$; 43 cm

Paper 46: **1** 0·406 **2** 3% **3** 4 **4** $22\frac{1}{2}$ **5** 65° 15′

Paper 47: **1** 43·2 s **2** £1·23 **3** $-a-b$; 2 **4** 2, 5 **5** 3·73 cm, 2·50 cm, 28° 4′

Page 255

Paper 48: **1** (i) 0·0625; (ii) 2·858 04 **2** £1600 a year **3** -12 **4** 3, 6; 234, 252, cm² **5** 19° 39′, 25° 21′

Paper 49: **1** £0·22 **2** 40 cm **3** $\frac{a^5}{b^3}$; $8-22x-41x^2-15x^3$ **4** $2\frac{1}{3}$, 6

Page 256

Paper 50: **1** 8 km **2** $87\frac{1}{2}$% **3** 11·88 **4** $6\frac{2}{3}$% **5** 3·57 (5), 4·90, cm; 7° 33′

Paper 51: **1** $\frac{29}{30}$ **2** 23·14 kg **3** -1 **4** $\frac{4t}{5}$ min **5** 260, 150, m; N. 71° 56′ E.

Paper 52: **1** $3x-2$ **2** 14·8 cm **3** 35 decrease; $x < -\frac{1}{2}$ **4** 12 **5** 5·80 cm; 122° 6′

Page 257

Paper 53: **1** 20p **2** 450 cm³ **3** ab; $(y+x)(4y-x)$ **4** 3 **5** 69° 31′; 6·56 cm

Paper 54: **1** 260 **2** £12·50; 20% **3** $\frac{1}{5}(c-5)$; $(1+x^2)(1+xy)$ **4** 48, 72 **5** 5·81, 1·34, km **6** 85°

Page 258

Paper 55: **1** £494·08 **2** 50% **3** 5, -9; $8x(x+1)$ **4** 25% **5** 7·05, 4·85, cm

Paper 56: **2** 10.43 a.m **3** $5c^3+3c^2$; $4(t+6)(t-3)$ **4** 29th number **5** 45° 35′, 3·57 cm

Paper 57: **1** $x < 5$ **2** 9450 litres; 11·1 cm **3** 21 **4** $-\frac{1}{3}$; $3y+x = 9$ **5** 27° 16′; 16, 18·0, cm